ANCIENT PIÑON-JUNIPER WOODLANDS

ANCIENT PIÑON-JUNIPER WOODLANDS

A NATURAL HISTORY OF MESA VERDE COUNTRY

Edited by
M. LISA FLOYD

Technical Editors
DAVID D. HANNA
WILLIAM H. ROMME
MARILYN COLYER

UNIVERSITY PRESS
OF COLORADO

© 2003 by the University Press of Colorado

Chapter 8 was written and prepared by U.S. government employees on official time and therefore is in the public domain and not subject to copyright.

Published by the University Press of Colorado
5589 Arapahoe Avenue, Suite 206C
Boulder, Colorado 80303

All rights reserved
Printed in the United States of America

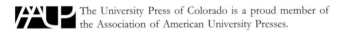 The University Press of Colorado is a proud member of the Association of American University Presses.

The University Press of Colorado is a cooperative publishing enterprise supported, in part, by Adams State College, Colorado State University, Fort Lewis College, Mesa State College, Metropolitan State College of Denver, University of Colorado, University of Northern Colorado, and Western State College of Colorado.

The paper used in this publication meets the minimum requirements of the American National Standard for Information Sciences—Permanence of Paper for Printed Library Materials. ANSI Z39.48-1992

Library of Congress Cataloging-in-Publication Data

Ancient piñon-juniper woodlands : a natural history of Mesa Verde country / edited by M. Lisa Floyd.
    p. cm.
Includes bibliographical references.
  ISBN 0-87081-740-X (hardcover : alk. paper) — ISBN 0-87081-749-3 (pbk. : alk. paper)
  1. Natural history—Colorado—Mesa Verde National Park. 2. Old growth forest ecology—Colorado—Mesa Verde National Park. 3. Pinus edulis—Colorado—Mesa Verde National Park. 4. Juniper—Colorado—Mesa Verde National Park. I. Floyd, M. Lisa.
  QH105.C6 A67 2003
  508.788'27—dc21

200301095

Design by Daniel Pratt

12 11 10 09 08 07 06 05 04 03   10 9 8 7 6 5 4 3 2 1

CONTENTS

		Preface	ix
		Acknowledgments	xi
1		Introduction and Overview —*M. Lisa Floyd, David D. Hanna, and William H. Romme*	1
	PART I	WHAT ARE THE BIOLOGICAL AND ECOLOGICAL CHARACTERISTICS OF OLD-GROWTH PIÑON-JUNIPER WOODLANDS IN MESA VERDE COUNTRY?	9
2		Gnarly Old Trees: Canopy Characteristics of Old-Growth Piñon-Juniper Woodlands —*M. Lisa Floyd, Marilyn Colyer, David D. Hanna, and William H. Romme*	11
3		Beneath the Trees: Shrubs, Herbs, and Some Surprising Rarities —*M. Lisa Floyd and Marilyn Colyer*	31
4		At the Ground Level: Fungi and Mosses —*Jayne Belnap and J. Page Lindsey*	61
5		Magnificent Microbes: Biological Soil Crusts in Piñon-Juniper Communities —*Jayne Belnap*	75

6	Mesa Verde Country's Woodland Avian Community —*George L. San Miguel and Marilyn Colyer*		89
7	Mammals of the Old-Growth Piñon-Juniper —*Preston Somers, David D. Hanna, Marilyn Colyer, and M. Lisa Floyd*		111
8	Bats of the Piñon-Juniper Woodlands of Southwestern Colorado —*Alice L. Chung-MacCoubrey and Michael A. Bogan*		131
9	Reptiles and Amphibians of the Piñon-Juniper Woodlands —*David D. Hanna and Timothy B. Graham*		151
10	Insects Associated with the Piñon-Juniper Woodlands of Mesa Verde Country —*David A. Leatherman and Boris C. Kondratieff*		167
PART II	**WHAT CHARACTERIZES THE GEOLOGY, WEATHER, AND SOILS OF MESA VERDE COUNTRY?**		181
11	Bedrock Geology —*Mary Griffitts*		183
12	Landforms and Surficial Deposits —*Mary L. Gillam*		197
13	Soils of Mesa Verde Country —*Doug Ramsey*		213
14	Water Resources in Mesa Verde Country —*Marilyn Colyer*		223
15	The Climate History of Southwestern Colorado —*Nolan J. Doesken and Thomas B. McKee*		247
PART III	**WHAT PROCESSES OF CHANGE ARE OCCURRING IN MESA VERDE COUNTRY?**		259
16	Fire History —*M. Lisa Floyd, William H. Romme, and David D. Hanna*		261
17	Effects of Fire on Insect Communities in Piñon-Juniper Woodlands in Mesa Verde Country —*Deborah M. Kendall*		279
18	A Personal Perspective on the Ethnobotany of Old-Growth Piñon-Juniper Woodlands —*William J. Litzinger*		287

19	Some Ethnobotanical Uses of Plants from Piñon-Juniper Woodlands —*Marilyn Colyer*	295
20	The Ancestral Puebloans and Their Piñon-Juniper Woodlands —*Patricia Robins Flint-Lacey*	309
21	"Only Man Is Vile" —*Duane A. Smith*	321
PART IV	LOOKING INTO THE FUTURE: WHAT ARE THE HOPES AND CONCERNS FOR THE ECOLOGICAL INTEGRITY OF MESA VERDE COUNTRY?	337
22	Threats to the Piñon-Juniper Woodlands —*William H. Romme, Sylvia Oliva, and M. Lisa Floyd*	339
23	Epilogue: Management Considerations for Conserving Old-Growth Piñon-Juniper Woodlands —*George L. San Miguel*	361
	Contributors	375
	Index	381

PREFACE

THE PIÑON-JUNIPER WOODLAND IS ONE OF THE MOST WIDESPREAD HABITAT TYPES in the West, yet much of it is biologically depauperate. In rare pockets of the Mesa Verde Country, we can see what it once was and perhaps could be again. In the following chapters we develop the story of piñon and juniper woodlands for the 13,000 square kilometers surrounding Mesa Verde National Park. These trees have served as sources of food, medicine, shelter, and fuel for indigenous populations for centuries. They provide homes, nest sites, food, and protection for scores of bird, mammal, amphibian, and reptile species. Furthermore, the woodlands stabilize watersheds and prevent erosion. Nonetheless, the piñon-juniper communities are often considered invasive, undesirable, and "unpopular." In the twentieth century, piñon and juniper trees became the targets of scorn, the objects of an intensive federally sanctioned eradication program favoring grasslands over woodlands and food for cattle over stable woodland watersheds. As a result, woodland understories, once dominated by a dense cover of native grasses and forbs, have given way to bare, hardpan soil surfaces. Lack of success in these land management activities and a maturing of attitudes toward ecosystem-based management in the past decade have renewed an appreciation for the conifers of the mid-elevations in the southwestern United States. Despite extensive eradication, grazing pressures, and encroachment of human developments, we still find among the 6.8 million hectares of Colorado Plateau's piñon-juniper woodlands a few mature stands that have reached their full growth and developmental potential. Mesa Verde Country harbors some of these ancient stands.

In this volume scientists and historians who share a love for the Four Corners region of the Colorado Plateau celebrate the characteristics of the old-growth piñon-juniper woodlands. Many of us have roamed this area for decades, accumulating observations, naturalist notes, and reams of data describing the ecosystem in myriad ways. It is time to condense and integrate our observations. We focus on Mesa Verde National Park, the most protected portion of Mesa Verde Country and the source for the bulk of our data, but we include observations from Dolores Canyon, the Durango area, and other nearby sites as well. The book is divided into four parts. In the first part contributors discuss the biological characteristics of piñon-juniper woodlands. The second part describes the physical factors needed to support the ancient woodlands. The third part addresses the effects of fire and human interaction with the ancient woodland, based on archaeological, botanical, and historical information. In Part 4 we examine current threats to the ancient woodlands and their implications for future management of this important natural resource.

All authors donated their time for the preparation of this work. Any royalties generated will be used to establish a fund for research into the natural resources of Mesa Verde.

—M. Lisa Floyd

ACKNOWLEDGMENTS

Throughout the two years from conception to completion of this project, we relied on numerous naturalists, ecologists, park employees, and friends for their observations and perceptions of the ancient piñon-juniper woodlands. We would like to thank those who are not shown as authors but nonetheless were contributors. Dr. Albert Spencer (professor emeritus, Fort Lewis College) provided a wealth of observations and data accumulated over many years of study in Mesa Verde Country. He was an important player in conceiving the idea for this work. Mesa Verde National Park personnel—Allan Loy, Howard Dimont, Steve Budd-Jack, Guy Keene (and the Helitack crew), Tim Oliverias, Jane Anderson, Linda Towle, Linda Martin, Bob Heyder, Larry Weise—have been continually supportive of our work. We thank Mark Santee, Bureau of Reclamation, for helicopter support through the years. Many students from Prescott College and Fort Lewis College contributed over the years to fieldwork and data analysis, including Ross Martin, Matt Levy, Kirsten Alicki, Anne DaVega, Tiffany Rice, Lissa Bartlett, Dan Tinker, Laura Bohland, Chris Bergh, Lyn Chenier, Carianne Funicelli, Stephen VanVleet, and Douglas Estes. Charlotte Thompson, Bill Koons, Dustin Hanna, Lani Hanna, Caleb Chambers, Emily Larson, and Nicolas Spencer also assisted with fieldwork. We wholeheartedly thank Sharon Brussel for reading and editing the entire manuscript. Dr. John Kricher and Gary Salamacha read several chapters. Dr. Gary Gianniny and Terri Cook reviewed the geology chapters. We thank all the illustrators—Noni Floyd, Susie Harvin, Walt Anderson,

Acknowledgments

Amy Wendland, Agnes Suazo, Mary Vozar, Sarah Luecke, Elizabeth Griffitts, Pat Oppelt, and Marilyn Colyer—for adding beauty to our words. We are grateful to Dr. Bob Schiller and Dr. Bob Moon, National Park Service, for their support of many of the studies presented here.

SPECIFIC ACKNOWLEDGMENTS BY CHAPTER

Chapter 3

Leslie Stewart, Bureau of Land Management, assisted with the plant listings.

Chapter 8

A variety of individuals named above have assisted in fieldwork at Mesa Verde National Park. We appreciate the field assistance of Suzy Collins, Bob Finley, Clyde Jones, Rick Manning, Cindy Ramotnik, Donna Shay, and Ernie Valdez.

Chapter 17

Identifications by the following taxonomists were most appreciated: Dr. Howard Evans (Colorado State University), Dr. Boris Kondratieff (Colorado State University), and Dr. Robert Davidson (Carnegie Museum of Natural History).

ANCIENT PIÑON-JUNIPER WOODLANDS

INTRODUCTION AND OVERVIEW

M. LISA FLOYD, DAVID D. HANNA, AND WILLIAM H. ROMME

MESA VERDE COUNTRY IS HOME TO AN ANCIENT WOODLAND of broken, twisted junipers and stout old piñon pines (Figures 1.1 and 1.2). The old-growth woodland, now rare in much of the southwestern United States, is relatively undisturbed in the deep canyons and high ridgetops of Mesa Verde. As each tree ages over the centuries, it balances the need for new leaves, cones, or pollen with the wood fiber required to resist the stress and strain of wind and snow loads. Yet a critical component of the woodland structure results when years of accumulated lignin finally give way to these physical burdens. The old trees and their detached limbs become home for scores of birds, mammals, reptiles, and even some amphibians. Germinating on the woodland floor in the shade of downed limbs sprout the newest generation of bitterbrush and sagebrush, lupines and buckwheats, muttongrass and ricegrass. Within the geographic landscape of Mesa Verde Country—given enough time free from widespread fire and intense human activity—the woodland trees grow more ragged, more twisted, and more dispersed, and the woodland's biological inhabitants become more and more diverse.

It is unusual that Mesa Verde has remained undisturbed long enough for the ancient woodland to develop. Although the area supported a large Ancestral Puebloan population a thousand years ago, which probably altered the vegetation substantially, human influences were greatly diminished following abandonment of the region around A.D. 1300. Utes occupied the region for several centuries thereafter, but their ecological impact appears to have been localized and relatively

Figure 1.1. A stout Utah juniper from Mesa Verde. Illustration by Susie Harvin.

unimportant in the rugged, inaccessible terrain of Mesa Verde. Even with the livestock and lumbering that Euro-American settlers brought to southwestern Colorado in the late 1800s, and despite recent droughts and extensive wildfires, a

Figure 1.2. This piñon pine is nearly 1 m in diameter and 10 m tall. Illustration by Noni Floyd.

substantial portion of today's dense, ancient woodland has remained relatively unaffected by human-induced changes. Thus the southern portions of Mesa Verde and pockets of protected habitats in the region today support ancient piñon-juniper woodlands unknown in most other areas on the Colorado Plateau.

The following chapters summarize our current scientific understanding of the biology, ecology, geology, and climatology of this interesting portion of southwestern Colorado; but the book is also a celebration of the beautiful and remarkable natural history of the Mesa Verde region. The chapters have been contributed by a diverse set of authors, many of whom have spent nearly their entire lives in view of Mesa Verde. Their love for the land comes through as strongly as their technical expertise. As editors, we have smoothed the text, removed redundancies, and rectified technical inconsistencies among chapters. All of the chapters have been reviewed by experts to guard against errors of fact or interpretation. We have not attempted to change the basic flavor of each chapter, however. As you read the book, you will see that some chapters have a rigorous analytic underpinning and present concepts and interpretations of broad application. Others are reflections on natural history, honed by decades of close and patient observation and rich in local detail. Some chapters focus almost exclusively on Mesa Verde itself, while others range widely over the surrounding area depicted in Figure 1.3.

We wrote this book to elucidate and to celebrate the fascinating but often underappreciated natural history of Mesa Verde and the surrounding countryside. Almost everyone in Colorado and surrounding states is aware of Mesa Verde's famous archaeological sites; but few realize the exceptional nature of the ancient woodlands that, along with striking topography, form a backdrop for the abandoned cities of an ancient culture. Few recognize the myriad ways in which seemingly insignificant things—broken branches on centuries-old trees or microscopic organisms in the dry soil—are essential components of an ecological system that has supported the human and nonhuman inhabitants of the area from Ancestral Puebloan times to the present. For reasons explained in the following discussion, piñon-juniper woodlands are especially well developed in Mesa Verde, so this unique region illuminates the ecological importance of piñon-juniper woodlands and reveals some of their fascinating secrets of natural history.

REGIONAL GEOGRAPHY

We have carved out a naturally defined landscape as the focus of this book, and it is important to understand the boundaries of this region because they define a portion of piñon-juniper woodland unlike others in the Southwest. Figure 1.3 depicts this 13,000-square-km area along the eastern boundary of the Colorado Plateau geologic province, where the plateau meets the southern Rocky Mountains. It lies within the Four Corners region of the United States, an area described by the intersection of the political boundaries of Colorado, New Mexico, Arizona, and Utah. The greatest part of Mesa Verde Country lies in Colorado, with small incursions into New Mexico, Arizona, and Utah. The region is defined by a natural landform drained by major and minor rivers that originate in the San Juan Range of the southern Rocky Mountains. It is bounded by the Pine River to the east, the San Juan River to the south, Montezuma Creek to the west, and the salt

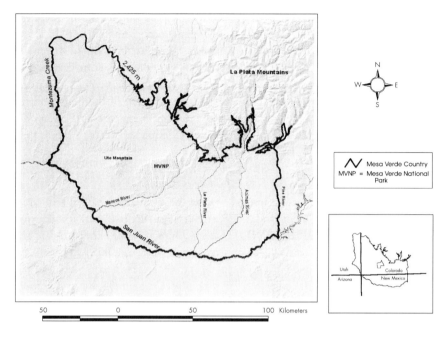

Figure 1.3. The region enclosed by the black line bounded to the east by the Pine River, the south by the San Juan River, the west by Montezuma Creek, and the north by the 2,425 m contour is what we call Mesa Verde Country. Note Mesa Verde National Park in the center of this geographic region.

basin structures of Paradox Basin, Disappointment Valley, and the 2,425 m contour to the north. Our region includes the north side of the San Juan Basin and the southern edge of the Paradox Basin; it extends up onto the slopes of the San Juan Mountains and abuts the Uncompaghre Plateau.

Extensive topographic relief characterizes this area. The landscape's surface is dissected into mesas, cuestas (similar to mesas but with one side decidedly higher than the other), buttes, canyons, and valleys. The bulk of the region sits atop layer after layer of sandstone, shale, coal, and other sedimentary rocks, in some places more than 9,000 m in aggregate thickness. Over time these strata have been lifted, tilted, folded, faulted, and eroded by water and wind to form a maze of canyons and a puzzle of disjointed surfaces (Chapters 11 and 12). The centerpiece is Mesa Verde (meaning "green table" in Spanish), a prominent cuesta that is visible throughout most of the area. Mesa Verde National Park encompasses the northeastern portion of the cuesta; most of the remainder lies within the Ute Mountain Ute Reservation. The country surrounding Mesa Verde is more or less similar in terms of geology, climate, and vegetation, so we refer to the whole area as "Mesa Verde Country." Here the high peaks of the southern Rocky Mountains, which lie to the east and north, promote an uplift of airmasses (an orographic effect) that shower plateaus and mesas with up to 46 cm of annual precipitation—a lot of moisture for the generally arid Four Corners region. The relative abundance of water (Chapter 14); unique geology (Chapters 11 and 12); diverse soils that vary with slope angle, the dip of the beds, and the nature of the parent strata (Chapter 13); and a long growing season (Chapter 15) provide the physical foundation for Mesa Verde Country's distinctive piñon-juniper woodland.

REGIONAL VEGETATION PATTERNS

To get oriented, let's look in a general way at the vegetation patterns in Mesa Verde Country. Between 1,666 and 2,275 m elevations in Arizona, New Mexico, Colorado, and Utah, woodlands of *Pinus edulis* (Colorado piñon pine) and *Juniperus osteosperma* (Utah juniper) form the dominant vegetative community. At the lowest and driest elevations, where annual precipitation may be less than 30 cm and few shrubs or forbs persist, the woodland is represented by only scattered junipers. Junipers become more abundant as mean annual precipitation increases with elevation. Piñons appear and become more common than junipers as the average annual precipitation approaches 35 cm and increases to 45 cm. Piñon-juniper stands in this upper portion of the zone grow taller and closer together than lower stands. The piñon-juniper stands continue to occupy sites on steep slopes with south exposures to about 2,600 m, especially those protected against fire by cliffs or barren areas (Chapter 16). The Utah juniper of lower elevations is replaced by *Juniperus scopulorum* (Rocky Mountain juniper) as moisture increases and mean annual temperature declines. Piñon pine and Rocky Mountain juniper continue as subordinate elements of the Gambel oak–ponderosa pine community up to about 2,900 m.

As a rule, the distribution of the *Pinus edulis–Juniperus osteosperma* (piñon-juniper) woodland is restricted in elevation and latitude to landscapes that receive between 30 cm and 50 cm of precipitation. The upper elevation boundary, however, is not a physiological limit for either species; the moister climate at higher elevations permits such mesophytic species as *Pinus ponderosa* (ponderosa pine), *Quercus gambelii* (Gambel oak), and *Pseudotsuga menziesii* (Douglas fir) to outcompete piñon and juniper.

Precipitation patterns alone do not drive vegetation patterns. A heterogeneous vegetational mosaic also arises out of the complex dissection of the region into canyons and mesas, hills and valleys, and south-facing and north-facing slopes. Even if the land were completely flat, recurring natural disturbances such as fire and disease would create a mosaic of habitats. Differences in soils arising out of various parent rocks, depositional processes, and weathering also create diverse microhabitats. Long-term climatic variability and local anomalies such as the rain shadow in the lee of Sleeping Ute Mountain prevent the establishment of a single uniform type of vegetation over any large portion of the area.

As a result of this physical heterogeneity, our area of interest supports a variety of plant communities in addition to the woodlands that are the focus of this book. At low elevations near the San Juan River on the southern border of Mesa Verde Country is a high desert scrubland dominated by *Atriplex canescens, A. gardneri,* and *A. obovata* (shadscale), as well as numerous forbs (herbaceous plants other than grasses, sedges, and rushes), including *Eriogonum corymbosum* (buckwheat), *Sphaeralcea coccinea* (globe mallow), and grasses such as *Hilaria jamesii* (galleta grass) and *Sporobolus aeroides* (alkali sacaton). Bare, rocky soils are common, as vegetation cover is low; microorganisms form well-developed soil crusts (Chapter 5). Scattered junipers also may dot the landscape.

At slightly higher elevations, *Artemisia tridentata* (big sagebrush), *Gutierrezia sarothrae* (snakeweed), and *Chrysothamnus nauseosus* (rabbitbrush) sometimes form dense stands on deeper soils. Junipers also may intermingle with these tall shrub species.

From elevations greater than 1,700 m to the upper reaches of the Mesa Verde cuesta at 2,500 m, piñon and Utah juniper share dominance, and the density of the woodland increases. Patches of young, open woodland with a dense shrub understory are interspersed with old-growth stands. Mountain shrublands abut the woodlands at the highest elevations on the cuesta. These shrublands share many floristic affinities with the piñon-juniper woodlands but have very different physiognomy. Gambel oak, *Amelanchier utahensis* (serviceberry), and *Fendlera rupicola* (fendlerbush) form dense stands that resprout vigorously after fire and dominate where fire frequency is relatively high.

Along streams and rivers and around ponds and wetlands is distinctive riparian vegetation composed of plant species that require a permanent water source: *Populus fremontii* (cottonwood), *Salix* species (willows), *Typha latifolia* (cattails), and a diversity of sedges, rushes, and other herbs (Chapter 14).

This simple introduction to the landscape and vegetation patterns sets the stage for the chapters to come, which elucidate special characteristics of piñon-juniper woodlands and their nearby riparian or shrubland neighbor communities of Mesa Verde Country.

SUMMARY AND ROADMAP

We begin by developing a general concept of old-growth conditions for the piñon-juniper vegetation that is so well developed in Mesa Verde Country (Part 1). Chapter 2 focuses on the trees' age and structural complexity, which supports a dense biological network. In Chapters 3 to 10, various experts describe the multitude of plants, microorganisms, mammals, birds, reptiles, amphibians, and insects that depend on the old-growth woodland in Mesa Verde Country.

Focus then turns to the physical parameters that make this biological reality possible (Part 2). Underlying the piñon-juniper woodland is a fascinating substrate:

Chapters 11 to 13 examine the geologic and soil-forming processes that sustain old-growth piñon-juniper woodlands. Specific climatic conditions, especially those involving available water (Chapter 14), are needed to support the old woodland and its biological inhabitants. Chapter 15 describes the climate of the last hundred years.

We recognize that change is part of all landscapes, and the old woodland at Mesa Verde is no exception. Part 3 examines factors that have helped shape today's ancient woodland, including fire (Chapters 16 and 17) and the role of humans, both prehistoric and historic (Chapters 18 to 21).

Finally, we venture into the future, identifying potential threats to the ecological integrity of Mesa Verde Country and discussing what can be done to manage this landscape for a healthy future (Part 4).

Each part opens with prefatory remarks in which we outline the structure of the following chapters. Remember that these chapters are the product of years of study and reflection by different experts, and each tends to have its own flavor and viewpoint. All of the chapters are united, however, by their focus on this unique and complex landscape. The diversity of perspectives and writing styles in this book reflects the ecological diversity of Mesa Verde Country itself.

PART I

WHAT ARE THE BIOLOGICAL AND ECOLOGICAL CHARACTERISTICS OF OLD-GROWTH PIÑON-JUNIPER WOODLANDS IN MESA VERDE COUNTRY?

IN THE NINE CHAPTERS THAT MAKE UP THIS FIRST PART OF THE BOOK, authors of diverse backgrounds lay out a description of the old-growth piñon-juniper woodlands that are so characteristic of Mesa Verde Country. Although "old-growth" has been studied extensively in other ecosystems, notably the well-watered Douglas-fir forests of the Pacific Northwest, this concept has not yet been widely applied to piñon-juniper woodlands in the Southwest. Thus Chapter 2 offers a formal definition and characterization of old-growth conditions, based on the exceptionally well-developed old-growth piñon-juniper stands of Mesa Verde. These stands are situated in a unique geographic location that provides all of the necessary ingredients for old-growth development. In addition to the favorable geographic and climatic conditions, portions of Mesa Verde have been free of extensive severe disturbance (for example, fire) for many centuries. Chronic, small-scale, low-intensity disturbances (e.g., death of individual trees as a result of insects, pathogens, and snow damage) have created small gaps in the otherwise continuous canopy of piñon and juniper that supports high biological diversity. In the following chapters we show why the absence of large, severe disturbances is necessary for development of old-growth conditions, but small, low-severity disturbances also contribute to the key structural characteristics of old-growth stands.

The first three chapters in this part flesh out the botanical and microbial characteristics of old-growth piñon-juniper woodlands. Chapter 3 deals with vascular plants: the shrubs and herbs that grow among the trees. Some species are abundant and can be found throughout the woodlands, whereas others are rare and restricted to one or a few peculiar geologic substrates or soil types. Some lie in the soil for decades, as dormant seeds, and appear only ephemerally after local disturbance such as a fire. Others are so closely associated with piñon-juniper woodlands that they are actually endemic to the Mesa Verde cuesta. Chapter 4 examines the less conspicuous fungi, lichens, and byophytes. These fungi and nonvascular plants have a variety of remarkable adaptations that enable them to survive heat and drought. Chapter 5 celebrates an even more overlooked but equally important group of organisms: the microscopic plants, bacteria, and fungi that form soil crusts, which stabilize soil and provide fixed nitrogen and germination sites for the seeds of vascular plants.

Chapters 6 to 9 deal with the animals of the piñon-juniper woodland. It appears that no species of vertebrate is entirely restricted to old-growth piñon-juniper. Numerous species of birds and mammals spend most of their time in this habitat, however, and thus are closely associated with piñons and junipers. The physical structure of the old-growth woodlands is unappealing to some species (for example, the large hawks and eagles) but provides perfect nesting, foraging, and hiding habitat for others (such as the black-throated gray warbler). Among the most surprising denizens of the generally arid Mesa Verde Country are the amphibians, which require open water during critical stages in their life history. Frogs and toads can be found not only along the few perennial springs and streams of the region but also in isolated "potholes" on the mesa tops. These small depressions within bare rock outcrops fill with water during infrequent rains and provide a rare microhabitat for a host of water-obligate species, including not just amphibians but also numerous invertebrates and algae. Finally, Chapter 10 describes the many species of insects found in old-growth piñon-juniper woodlands and offers some fascinating stories of plant-insect interactions.

GNARLY OLD TREES: CANOPY CHARACTERISTICS OF OLD-GROWTH PIÑON-JUNIPER WOODLANDS

M. Lisa Floyd, Marilyn Colyer, David D. Hanna, and William H. Romme

Although over twenty-four million hectares of piñon-juniper woodlands are scattered across the southwestern United States and northern Mexico, old-growth piñon-juniper woodlands are rare in today's landscape. Later chapters discuss characteristics of natural and human disturbances that have reduced the chances that woodlands would attain old age and describe the biological wealth supported by the few remaining old-growth woodlands such as those of Mesa Verde Country. But before we do that, we must develop a definition of old-growth that suits piñon-juniper woodlands. The common vision of old-growth that includes towering trees and lush moss-covered logs (such as the old-growth forests of the Pacific Northwest or montane regions of the Rocky Mountains) just does not fit the lower elevations of the Southwest. Rather, piñon pines and several species of junipers—trees that rarely attain 10 m in height—are the dominant species, and low precipitation precludes a lush understory. Nonetheless, given enough time, old-growth develops in a way characteristic of this arid ecosystem. In protected pockets of Mesa Verde Country stand very old, complex, piñon-juniper woodlands. Twisted, gnarled junipers (replete with rot pockets and strips of shaggy bark) are accompanied by dark-barked and often stately piñon pines, some of which were established soon after the Ancestral Puebloans left the region in the late thirteenth century. Our research on the structure of undisturbed woodlands at Mesa Verde provides an opportunity to define old-growth. This concept of old-growth creates the basis for all subsequent chapters; for without the complex, old, woody structure provided by the trees themselves, the biological wealth of mammals, birds, amphibians, and insects described in Chapters 3 to 10 could not be sustained. Conversely, without the physical environments and ecosystem processes described in later chapters, the old-growth trees—the very structure of the woodland—could not persist.

Our definition might also provide land managers with a template on background structural attributes in piñon-juniper woodlands as they existed before European settlement to gauge and evaluate management activities. We would not suggest that Mesa Verde's woodland become a reference model beyond the region defined in Chapter 1 (Moore et al. 1999). Lower annual precipitation or shorter fire intervals than occur on Mesa Verde might support a different "potential" structure in other piñon-juniper habitats. Defining our relatively undisturbed old-growth condition at Mesa Verde will help fill in gaps in our knowledge of presettlement structures in the Southwest. Other researchers have begun to offer their definitions of old-growth piñon-juniper woodlands in the Great Basin (Kaufmann et al. 1992; Kruse and Perry 1994) and eastern Oregon's juniper woodland (Miller et al. 1999). No definition has yet been offered for our corner of the Colorado Plateau. A sense of urgency exists because resource managers are developing new management plans that will incorporate fire as a more natural ecosystem process in some piñon-juniper woodlands (Chapter 16), but the appropriate model being used as a goal in these activities may be controversial (see the following discussion and Chapter 23). The definition of old-growth that we develop in this chapter will shed light on that controversy.

Finally, a definition is needed to help identify and conserve the few remaining old-growth piñon-junipers. Much of piñon-juniper woodland is considered to be "expanding" into native grasslands (Miller and Wigand 1994) or to be in a state of "super-dominance" (West and Van Pelt 1987) or is even regarded as "degraded" (Jacobs and Gatewood 1999). While many piñon-juniper woodlands might also be characterized as "healthy" early successional woodlands, few meet the criteria of "old-growth" (Stevens 1997). Concern has been expressed at several regional conferences during the last decade that our piñon-juniper woodlands are in terrible condition, primarily because of post-European settlement activities like grazing, altered fire regimes, and mining (Shaw et al. 1994). It is, of course, impossible to have a truly undisturbed piñon-juniper landscape. In our region Ancestral Puebloans no doubt cleared woodlands for corn fields (Chapter 20). Later, Ute Indians set fires to modify their environment or assist hunting (Bonnicksen 2000), and cowboys brought grazers to the Mancos Valley in the mid-1800s (Chapter 21). Yet stands of piñon-juniper woodland in Mesa Verde Country have been relatively undisturbed for centuries, imparting to these rare woodlands a high conservation value.

DEFINING OLD-GROWTH IN PIÑON-JUNIPER WOODLANDS

Defining old-growth is problematic in any ecosystem. In all forest types a distinction is made between the structure and function in young versus old-growth stands;

that distinction often involves tree age and size, accumulations of deadwood, species composition, and measures of ecosystem functioning (Bonnicksen 2000; Mehl 1992; Miller at al. 1999).

The structure of the stand is important to any definition. Logically, density ought to increase with stand age; but it is powerfully influenced by climatic regime, soil depth and texture, intra- and interspecific competition, and the stresses that the stand has experienced. Indeed some very dense piñon-juniper stands are relatively young (Swetnam and Betancourt 1999), and some old stands have relatively low density (Floyd 1989). So density alone is an inadequate measurement for our purposes.

No doubt time is critical. Given enough time, piñons may grow to a meter in girth, while shaggy, multistemmed junipers grow much larger. Not only do individual trees appear ancient, but the very structure of the woodland is intricate and complex. Limbs have broken and become part of a complicated three-dimensional network; snags have accumulated; down wood has piled up on the woodland floor. The number of stressful events the individual tree or stand has experienced increases with its age. The most important stresses are drought, heavy clinging snowfall, wind, winterkill, disease, insect outbreaks, rockfall and landslides, floods, and grazing. Stresses result in defoliation, death of buds and branches, and rot, contributing to down wood, dead branches, and cavities.

Disturbance is also part of the normal ecosystem. The impact of humans who have lived among the piñons and junipers in Mesa Verde Country has been profound (Chapters 18–21). In addition, small fires, involving a single tree or less than 5 hectares, occur frequently, opening patches in the canopy. Eight large wildfires (thousands of hectares in extent) have occurred in the past century (Chapter 16). Outbreaks of insects and pathogens create small patches of dead piñons, and heavy snows break the crowns of old trees. As a result of these natural disturbances, the old-growth stands on Mesa Verde are only one part of a heterogenous mosaic of piñon-juniper woodlands of varying ages. Yet the woodland must be free of severe disturbances for centuries for full development of the old-growth characteristics that we describe in this chapter.

To tackle the definition of old-growth, we focus on the structure of the ancient piñon-juniper woodland on the Mesa Verde cuesta. How old are the trees? What are the key structural characteristics of the woodland? How uniform is the woodland? Finally, we begin the discussion of unique biological characteristics of ancient piñon-juniper woodlands in Mesa Verde Country that continues in the remainder of this book.

TREE AGE IN THE WOODLAND

Among perennial plants, trees are the easiest to age. Their wood, or secondary xylem, is dead when functioning in water transport, and the lignified cell walls persist for centuries in some species while other portions of the tree carry out "living" functions. With a simple core extracted from the tree, one can count annual rings within the wood and determine its age accurately and nondestructively. While it is possible to obtain increment cores from most adult piñons, it is problematic to obtain a representative increment core from a Utah juniper. Juniper trees grow in erratic radial patterns, so it is difficult to determine if a sampled radial core is truly representative of the age of the entire tree. Also, rot pockets are common in the older wood. Therefore, to determine the age structure of piñon-juniper woodlands we took advantage of recently cut piñon and juniper stumps created by management activities at Mesa Verde National Park, either by a fuel reduction program or by the removal of fire-killed trees that threatened archaeological sites. Hundreds of piñon and juniper stumps were dated, then age-diameter models were developed to predict the approximate age from the diameters of other trees. Population structure (age, diameter, height, canopy) was sampled in three woodlands in Mesa Verde National Park.

Just how old are the piñon and juniper trees in Mesa Verde Country? Our oldest sampled piñon trees from three stands on Mesa Verde became established in the 1560s (Figure 2.1). Few trees of this cohort remain. Abundant piñon trees less than 190 years old germinated on the mesa shortly after 1800. Large junipers with basal areas of 1 m and ages close to 600 years are scattered throughout the woodland (Figure 2.2). These junipers were established in the late 1300s, and only a few remnant trees remain from this cohort. Most sampled junipers were less than 400 years old. Juniper density decreases in the larger ages, reflecting patterns of establishment and mortality that tend to vary over time (in contrast to a more consistent pattern in piñon pines). Moreover, the density of large live piñon and juniper trees is similar to the density of stumps and snags in Mesa Verde National Park.

Even older trees have been discovered nearby in Mesa Verde Country. Two Utah junipers listed as "Colorado Champions" have been recorded on Mesa Verde; one tree was 152.5 cm in diameter and 1,350 years old! Old-growth piñon pine stands (the oldest piñon pines were 460 years) were documented near Dolores, Colorado, during the Dolores Archeological Project. In one side drainage (Dry Creek) a record "Shulman" piñon pine, 860 years old, was surrounded by other conifers over 600 years old (Shulman 1943).

Now let us focus on the younger trees that form the lower canopy of the piñon-juniper woodland. The pattern of young and old trees together is called the stand's structure. In Mesa Verde Country today, young piñons are more abundant than junipers (Table 2.1), and juniper dominates in basal area (Table 2.2). In fact, the structure and dynamics of old-growth piñon-juniper woodlands in Mesa Verde Country are remarkably similar to those of another old-growth forest type in the region: high-elevation spruce-fir forests. In old-growth, spruce-fir, *Picea engelmannii* (Engelmann spruce), commonly has greater basal area but lower density than subalpine fir in the canopy, and fir dominates the understory both in density and

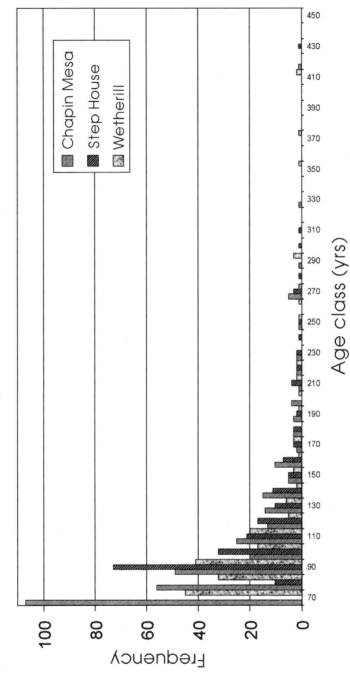

Figure 2.1. Age structure of three mature piñon pine populations, Mesa Verde National Park. Ages are estimated by using regression equations that predict age from diameter.

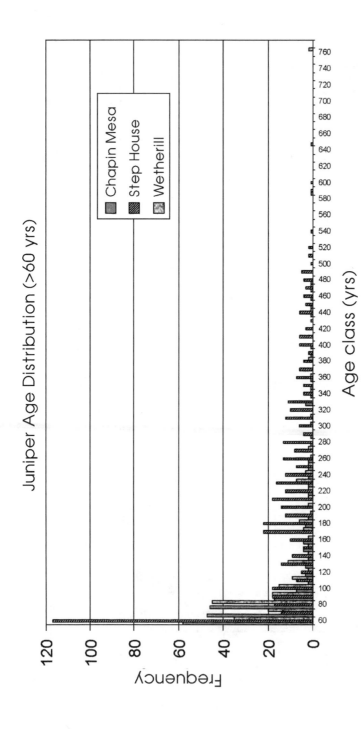

Figure 2.2. Age structure of three mature Utah juniper populations, Mesa Verde National Park. Ages are estimated by using regression equations that predict age from diameter.

Table 2.1. Density of piñon and juniper less than 5 cm in diameter, Mesa Verde National Park (number of trees per hectare)

	Wetherill Mesa	Chapin Mesa
Pinus edulis	1,000	1,360
Juniperus osteosperma	245	725

Note: This represents recruitment during the last two to three decades.

Table 2.2. Average basal area per hectare (in cm^2) of piñon and juniper, Mesa Verde National Park

	Wetherill Mesa	Chapin Mesa
Pinus edulis	385	280
Juniperus osteosperma	1,420	1,000

in basal area. Both species persist indefinitely in the stand, however, because of higher mortality rates in fir and longer life-spans in spruce (Veblen 1986a, 1986b; Veblen et al. 1991). In the old-growth piñon-juniper woodland of Mesa Verde, it is the juniper that has greater basal area and greater longevity (analogous to the spruce) and the piñon that has higher density in the understory and higher rates of mortality (analogous to the fir).

Interpreting these stand structures may become controversial. Two different explanations have been offered to account for stand structures like those described. One is that stand density may be significantly greater today than in past centuries. Clearly, the density of young trees (less than 100 years) greatly outweighs the density of older trees at Mesa Verde (Figure 2.1). Such a pattern suggests woodland "expansion" under current climatic and disturbance patterns (e.g., grazing, abnormally long fire intervals), promoting an unprecedented density of understory conifers. The second possible explanation is that this structure is due to higher mortality in the seedlings and saplings relative to adults. Survival is greater after trees reach reproductive age. This differential mortality across age classes results in the "inverse J" curve described in many mature forest stands (Silvertown 1984).

Such alternative explanations fuel a debate among range managers, resource managers, and ecologists as to whether piñon-juniper woodlands are "thickening" due to recent changes in the fire history (Chapter 16) or land-use practices (Miller and Wigand 1994). Historic photographs document that density has indeed increased in some areas. But this may not be universally true in piñon-juniper woodlands, particularly those that formed under long fire return intervals (Chapter 16), where today's high density may represent the "natural" condition. How can we decide if piñon and juniper reproduction in Mesa Verde Country is changing (the "increasing density" hypothesis) or if much of the current seedling/sapling cohort will die off and the adult population structure in several hundred years will look much as it does today (the "differential mortality" hypothesis)? The best way to approach this would be to tag and follow individual trees over their entire lifetime, but these conifers live far longer than any observant ecologist. Short of that approach, other pieces of evidence must be brought to bear on this question.

First, what evidence do we have from Mesa Verde Country concerning patterns of mortality and survival? In permanent study plots within Mesa Verde National Park, the numbers of piñon and juniper seedlings (less than 15 cm tall) have been recorded since 1975. The study sites span the mesa, from Far View at the highest elevation, to Cedar Tree Tower at the mid-elevation, and Glade at the lower elevation. Each site shows considerable annual variation in seedling densities. For example, high piñon densities were recorded in 1976, 1977, and 1978, but lower densities occurred after 1980 (Table 2.3). Relatively high juniper seedling densities were recorded in 1975 in Cedar Tree Tower plots, whereas 1978 was the big year for juniper seedlings at Far View. Density of juniper seedlings has been fairly uniform in Glade since 1980. So annual variation is the theme in Mesa Verde Country.

Two things may affect seedling density in any given year: seed supply and microclimate for establishment. Seed supply fluctuates from year to year. Piñon pines "mast," producing large seed crops at intervals usually ranging from five to ten years. The large seeds, which are typically not viable in the soil for longer than one year, either germinate quickly or are carried off and eaten or cached by birds, especially piñon jays and scrub-jays (Balda and Bateman 1971; Ligon 1978), and small mammals like *Peromyscus maniculatus* (the deer mouse) and *P. truei* (the piñon mouse). Junipers also produce large crops at irregular intervals, and their soft cones are popular foods. Juniper seeds show up in feces after passing through the digestive system of birds, coyotes, or bears. Most of the seeds left behind by these vectors find themselves in better shape to germinate and become established than if left on top of the soil, fully exposed to the sun.

After germination, what affects the seedlings' chance of survival? Weather patterns may hold an answer. An important climatic shift occurred throughout the Southwest around 1976, when the Southern Oscillation Index (SOI)—a measure of the strength of oceanic-atmospheric temperature and pressure anomalies in the tropical Pacific Ocean (Diaz and Markgraf 2000; Fagan 1999)—began to be dominated by strong negative values. These are associated with warm, moist El Niño episodes. Increased winter and spring precipitation characterized the years 1991–1995. High precipitation may promote successful seedling establishment in both piñon and juniper; however, we observed a slight lowering of densities of both species during this period. In sum, there has been no significant change in overall stand density in these study plots since 1975, despite the vagaries of seed production, seedling establishment (Table 2.3), or precipitation patterns. We have

Table 2.3. Density of piñon pines in Mesa Verde National Park (number of individuals per 688 m² plot)

Year	Average Density	Standard Error	Sample Size	Year	Average Density	Standard Error	Sample Size
1975	106	25	4	1987	80	16	2
1980	81	11	5	1989	68	1	2
1981	99	13	6	1990	74	10	6
1982	92	14	6	1991	85	6	6
1983	93	15	6	1992	88	1	2
1984	100	17	6	1993	79	8	6
1985	77	9	6	1994	75	7	4
1986	120	16	4	1994	93	7	4

Note: Three locations were sampled: Far View and Cedar Tree Tower were sampled annually from 1975 to 1995; Glade was sampled from 1980 to 1995. Sample size indicates the number of 688 m² plots sampled in a given year.

collected similar long-term data (not shown) in two other woodlands in Mesa Verde Country. The conclusion that emerges from these studies over twenty-five years is that survival patterns of the young trees smooth out over time, indicating that no net increase in density has occurred in Mesa Verde's piñon-juniper stands. Evidently these stands are naturally dense, in part because of their great age and relative freedom from disturbance. This is quite a different pattern than has been documented for some piñon-juniper woodlands elsewhere, such as in central New Mexico (Swetnam and Betancourt 1999), in western Oregon (Miller and Wigand 1994), and in portions of the Great Basin (Shaw et al. 1994; West and Van Pelt 1987).

Mature tree mortality is also influenced by climatic changes. Decade-scale periods of relatively wet or dry climates may have influenced woodland structure. Two dry periods, from 1575 to 1595 and from 1951 to 1956, are well documented in the Southwest (Swetnam and Betancourt 1999). The paucity of trees older than 400 years in Mesa Verde and elsewhere in the Southwest might reflect the drought of the late sixteenth century. The 1950s drought was less severe in Mesa Verde Country than elsewhere in the Southwest, however; average annual precipitation was 46 cm during the 1950s, and the average annual precipitation for the period from 1922 to 1997 was 45.2 cm. Also, we do not see remnants of a tree "die-off" in Mesa Verde analogous to those in Arizona and New Mexico (Swetnam and Betancourt 1999).

Nonetheless, scattered stands of dead trees are a common sight punctuating the woodland canopy, and most are caused by death of piñon pines. A lethal fungal pathogen, *Leptographium wageneri* Engelmann (the black-stain root disease), enters the xylem and kills trees within two years of infection. It has destroyed small patches (10–50 trees, often covering more than 1 hectare) of mature piñon pines during the last sixty years across our study region and has also caused the death of seedlings and saplings in its path before dampening out. Peaks in infestation occurred in 1976–1978 in Mesa Verde National Park, but a few trees have died each year since 1964 (Russel Wagner, 1990 unpublished report). A similar pattern of tree mortality was recorded in the 1980s in southwest La Plata County and elsewhere in Mesa Verde Country. Also, 2002 was marked by a widespread death of piñons in southwestern Colorado caused by an infestation of the beetle *Ips confusus* (Chapter 10). Such infestations and pathogenic outbreaks leave behind a series of canopy gaps, openings with relatively high light penetration.

Canopy gaps are a critical structural component of the old-growth piñon-juniper woodland. Within the small openings, understory herbaceous species are more numerous and diverse than directly under the shade of the canopy, thus contributing to the high plant diversity in the old-growth piñon-juniper woodlands (Chapter 3). Piñon and juniper seedlings/saplings (less than 10 cm in diameter) also were four times more abundant in the gaps than in adjacent closed woodland. A nitrogen-fixing shrub, *Purshia tridentata* (bitterbrush), thrives in the openings, possibly enhancing the total nitrogen of soils and thereby promoting forb and conifer seedling growth.

In sum, Mesa Verde Country is characterized in at least some places by old-growth piñon-juniper stands that probably have not changed much in density from the recent past. Our data and observations do not support the "increasing density" hypothesis. We suggest that the density and structure seen today probably characterized the woodland of past centuries as well. High—and highly variable—patterns of mortality reduce each successive seedling and sapling cohort to produce a mature canopy much like the one we see today. Historic photographs taken early in the 1900s that show a dense canopy with many piñon and juniper saplings (Figure 2.3) support this interpretation. Moreover, we find no evidence (e.g., fire scars or charcoal) for a low-intensity surface fire regime in past centuries that might have thinned the understory and created a low-density, open stand. We return to this idea in a more thorough look at fire history in Chapter 16.

DOWN WOOD, UNTURNED TREES, AND CAVITIES: THE WOODY NETWORK OF OLD-GROWTH

If Mesa Verde's piñon-juniper woodlands are (conservatively) four or even five centuries old, surely a mass of woody limbs, snags, and downed woody debris should have accumulated. Down wood (limbs lying on the woodland floor) is an outstanding feature of the old-growth piñon-juniper woodland in Mesa Verde (Table 2.4). Although similar measurements have not been made in other old-growth piñon-juniper woodlands, we can compare coarse woody debris in Mesa Verde with old-growth ponderosa-pine forests at higher elevations in the nearby San Juan Mountains. The average mass of down woody material was quite com-

Figure 2.3. Old-growth piñon-juniper woodland on Wetherill Mesa, Mesa Verde National Park, circa 1934. Courtesy, National Park Service, Mesa Verde Archive.

parable in the two vegetation types—9.3 tons/acre in the piñon-juniper woodlands of Mesa Verde and 10.6 tons/acre in an old-growth ponderosa-pine forest at Corral Mountain, about 100 miles from Mesa Verde (Romme et al. 1997).

In addition to down wood on the ground, old-growth is characterized by cavities, broken attached limbs, flaky bark, and a plethora of holes throughout the forest. This intricate web of lignin creates favorable habitat for wildlife by providing suitable areas for foraging, reproduction, and protection from predators. A few examples are discussed in the following definition of old-growth. Chapters 3 to 10 elaborate on each of the biotic components of the woodland.

Mature junipers can be literally riddled with rot pockets that have endured in live trees for decades (Figure 2.4). The cavities may eventually consume much of the heartwood, leaving only a living shell that may become a dwelling for birds and mammals for several decades. Old-growth piñon trees that die of black-stain root rot provide nest sites, since the soft wood is readily excavated. Piñon cavities, however, persist for less than one decade. Social insects that require large nesting spaces (such as honeybees, bumblebees, some wasps, carpenter bees, and minute black ants) also utilize the cavities. Bats use these tree cavities for maternity wards

Table 2.4. Characteristics of deadwood in old-growth piñon-juniper woodlands, Mesa Verde National Park

Deadwood Type and Size	Average Weight (tons/acre)
0–.25 in	.11 (.02)
.25–1 in	.58 (.17)
1–3 in	1.23 (.65)
Rotten coarse woody debris > 3 in	1.39 (.67)
Sound coarse woody debris > 3 in	5.92 (4.14)
Depth of coarse woody debris	1.10 (.3)
Standing deadwood: fine	21.7 (15.3)
Standing deadwood: foliage	16.9 (12.1)
Standing deadwood: cordwood	36.4 (29.6)
Standing deadwood: slash	42.6 (31.8)

Note: Coarse woody debris values are averages of 48 planar intercept transects. Standing deadwood is averaged over 20 line transect samples.

and for hibernacula (Chapter 8). Cavities may be enlarged and cleaned by woodpeckers and later become homes to many other birds (Chapter 6).

Occasionally heavy snows accumulate to 45 cm depths, piling up on conifer limbs and causing widespread breakage. At Mesa Verde National Park such storms were recorded in 1987, 1992, and 1995, leaving a belt of broken trees at 2,150 m, 2,333 m, and 2,275 m elevations, respectively. The broken limbs, which usually remain alive and attached to the tree for decades, are immediately put to use by other forms of life. Cottontail rabbits nest under the portions lying on the ground. Deer mice and Mexican voles nest in the tangled vegetation. Mosses, bryophytes, and lichens are encouraged by this new source of shade.

Trees are often uprooted by such snowstorms, when coupled with high wind and saturated ground. In these cases the ground around the tangle of exposed roots is dug out and excavated into well-protected nest mounds. The gray fox, spotted skunk, striped skunk, rock squirrel, Mexican woodrat, and several species of mice utilize these root houses.

It is not uncommon to see dead treetops or upper limbs protruding from the canopy. Like snags, dead tops serve an important function for raptors, flycatchers, mountain bluebirds, and other insect-feeding birds. These species utilize such perches as launches during hunting. Perches are especially effective on slopes when insects are driven upward by winds and thermals.

The stout limbs (especially on Utah juniper), spanning from the ground to the treetop, afford a ladder for wildlife (Figure 2.5). That is partly why gray foxes and spotted skunks, both species of piñon-juniper woodland, are able to climb up the trees. The striped whipsnake also navigates the tree structure. Small birds, especially during vulnerable periods of winter and fledging, dive into the dense branches

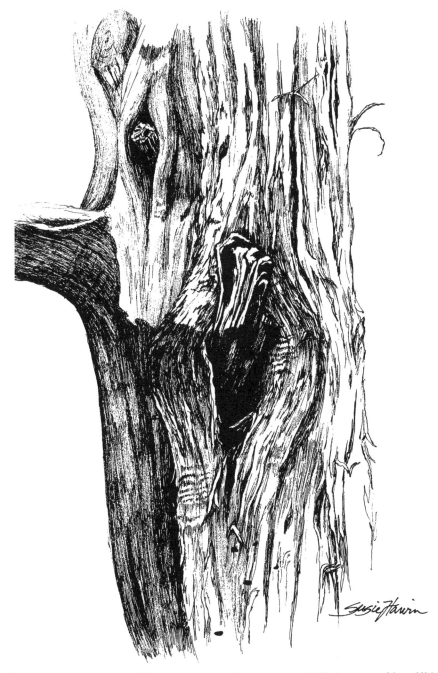

Figure 2.4. *A close look at a Utah juniper trunk shows crevices and folds that are used by wildlife. Illustration by Susie Harvin.*

of a full piñon tree—and disappear! If they are pursued, they can flee out of the other side of the tree. Birds take shelter in these shaded, thick branches during the inactivity of winter months. The squatty piñon or juniper, with a plethora of crooked limbs, provides excellent hiding shelter. Limb networks, where the branches make a nearly solid curtain, also reduce the windchill factor. This is important to wildlife for day and night bedding.

Parasitic mistletoe forms tight branch clusters on junipers. These proliferate during wet years and decrease during dry years when the tree limbs, weakened from the parasitic growth, may die. Mistletoe fruits are an important winter food for western bluebirds and Townsend's solitaires.

In summary, we add to the structural definition of old-growth this array of coarse woody debris, broken, attached limbs, flaky bark, exposed roots, cavities, and rot pockets. The old-growth piñon-juniper woodland is much more than old trees and dense saplings. It is truly an intricate web of lignin, forming the network upon which a highly diverse biota relies (Chapters 3–10).

VARIABILITY IN WOODLAND STRUCTURE

No one structure characterizes old-growth, even in as small a region as Mesa Verde Country. Old-growth stand structure, tree density, and stature vary considerably across different sites and depend on both elevation and soil characteristics. Here we consider the range of natural variation in woodland structures attained in various elevation and soil conditions. Soil types are described fully in Chapter 13.

Silty loess covers the mesa tops, where it has collected on Cliff House Sandstone. Loess-covered mesa tops of Mesa Verde can be divided for convenience into a series of bands, each supporting a different type of stand and each defined by a characteristic soil depth and elevation. While a large proportion of these are old-growth, patches of younger stands occur as well, in response to fine-scale disturbances. From 2,303 to 2,390 m, loess about 1.8–2 m in depth supports a "transition zone" between the mountain shrublands that dominate the higher elevations and the lower-elevation piñon-juniper woodlands. Stands in the transition zone are characterized by an average cover of 20% piñon and 20% juniper. At a slightly lower elevation, which occurs to the south of this transition area on the mesa, silty loess accumulates to over 3 m. Here a dense growth of relatively tall, large-diameter piñon and juniper trees develops. Where loess accumulates to its greatest depths at this elevation, and precipitation reaches 45 cm, we find the tallest piñon and juniper trees. The trees grow to spectacular heights with straight boles. Woodland cover is 30% piñon, 30% juniper, higher than in most woodlands on the cuesta and surrounding region.

In the middle-elevation range of the loess-covered mesas, from 1,848 m to 2,151 m, trees tend to attain relatively shorter heights but comparable cover, and growth is slower. Annual precipitation is only about 0.4 cm lower, but hotter temperatures increase evapotranspiration. Archaeologists call this the "corn layer," believing that the temperature, precipitation, and soil type were well suited for the corn agriculture of the Ancestral Puebloans.

The lower reaches of the loess-covered mesas are dramatically different from the higher elevations already described. Mesa Verde Resource Management staff recorded soil temperature and moisture between 1986 and 1999; soils are drier and considerably warmer most of the year on the lower elevations of the mesas. Trees do not attain the heights that they reach in the higher elevations, and even the very old trees have smaller diameters due to slower growth rates. Effective reproduction is lower here as well. The canopy tends to be more open than at higher elevations.

Mancos Shale, Menefee Shale, and Morrison Shale Formations support characteristically short and bushy trees. Canopy ranges from 10% piñon and 10% juniper on Menefee shales to 50% piñon and 15% juniper on Mancos shales. On Mancos shales grow rare, exceptionally tall, robust trees that produce unusually large cone crops, often each year.

Loamy soils, a mixture of clay and sand, occur on the expansive landslides that encircle the Mesa Verde cuesta, Weber Mountain, and Menefee Mountain. These landslides are composed of a mixture of Point Lookout Sandstone and Mancos Shale. The woodland on the resultant benches is primarily composed of piñon and a small proportion of *Juniperus osteosperma* and *J. scopulorum* and as such differs from other piñon-juniper stands in the region. Total canopy cover averages 20%.

A unique piñon-juniper woodland occurs on oyster beds, where trees tend to "crawl" along the surface. Live trees that never attain great heights tip and recline on elbow-like projections. Trees over 900 years old have been dated from these sites, yet the largest trees reach only 5 m in height and are less than 1 m in diameter because of their slow growth. Tree longevity is probably influenced by extremely infrequent fires. Lack of understory plants reduces fuel potential in these oyster-bed derived soils. Piñon trees in these locales mast at unusually long intervals of about ten years. Junipers, which are more common than piñons, show considerable retrenchment of growth.

Sandy soils are found in the lower-elevation extreme of our study area, near Hovenweep National Monument. Here old piñon and juniper trees are very short relative to trees on the Mesa Verde cuesta. Juniper tends to dominate in density, and the total tree canopy may be less than 10% cover.

These structural attributes help to characterize old-growth piñon-juniper in Mesa Verde Country. Natural variation is also important in understanding the potential development of the woodland. More research is needed to characterize the full range of developmental trajectories and potential in the woodlands of this region. Even within Mesa Verde National Park, where most research has been conducted, we have shown that tall, straight trees develop on some sites during

Figure 2.5. As Utah junipers age, the bark becomes stringy and peels off in sheets. These sheets are used by wildlife. Illustration by Susie Harvin.

long, disturbance-free periods, whereas short, leaning trees are all that can be expected on other sites. Old-growth is regulated by site conditions, and much of the beauty of the piñon-juniper woodland lies in its diversity.

SUMMARY

Given the variability in old-growth described here, we propose a working definition of old-growth piñon-juniper woodlands in Mesa Verde Country that expands on those previously outlined by other researchers (Kruse and Perry 1994; Mehl 1992; Miller et al. 1999). Old-growth piñon-juniper woodlands require at least several centuries to develop. Although junipers dominate in basal area, piñon has greater density, in part because canopy openings promote the piñon reproduction. There is considerable variation among old-growth sites, depending on elevation (hence precipitation and temperature) as well as soil characteristics. Fine woody debris is present but is greatly exceeded by coarse woody debris. Standing dead trees are common; and dead limbs persist on live trees, forming a complex array of live and dead material. This complex canopy structure is punctuated by frequent gaps, so that total tree cover rarely exceeds 50%. The density and sizes of stumps and snags are similar to those of live canopy trees. This intricate and complex woody structure supports an exceptionally diverse biological network in the old-growth piñon-juniper woodlands of Mesa Verde Country. The following chapters describe the plants, fungi and bacteria, birds, bats and other mammals, reptiles, and insects that depend upon and add to these remarkable old woodlands.

REFERENCES CITED

Balda, R., and G. C. Bateman. 1971. Flocking and annual cycle of the Piñon Jay *Gymnorhinus cyanocephalus*. *Condor* 78: 287–302.

Bonnicksen, T. M. 2000. *America's ancient forests: From the Ice Age to the age of discovery*. John Wiley and Sons, Inc., New York.

Brown, James K. 1974. *Handbook for inventorying downed woody material*. USDA Forest Service Publication General Technical Report INT-16.

Davenport, D. W., D. D. Breshears, B. P. Wilcox, and C. D. Allen. 1998. Viewpoint: Sustainability of piñon-juniper ecosystems—a unifying perspective of soil erosion thresholds. *Journal of Range Management* 51: 231–240.

Diaz, H. F., and V. Markgraf (eds.). 2000. *El Niño and the Southern Oscillation: Multiscale variability and global and regional impacts*. Cambridge University Press, Cambridge.

Dillinger, Kenneth C. 1970. Evaluation of mule deer habitat in Mesa Verde National Park. Master of Science thesis. Department of Fisheries and Wildlife Biology, Colorado State University, Fort Collins.

Everett, R. L., and S. Koniak. 1981. Understory vegetation in fully stocked pinyon-juniper stands. *Great Basin Naturalist* 41: 467–475.

Fagan, B. 1999. *Floods, famines, and emperors: El Niño and the fate of civilizations*. Basic Books, New York.

Fleischner, T. K. 1994. Ecological costs of livestock grazing in western North America. *Conservation Biology* 8: 629–644.

Floyd, M. L. 1989. Inter- and intra-specific variations in stand structure in *Pinus edulis*. *Botanical Gazette* 147: 180–188.

Floyd, M. L., and T. Kohler. 1990. Current productivity and prehistoric use of piñon (*Pinus edulis*, Pinaceae) in the Dolores Archeological Project Area, southwestern Colorado. *Economic Botany* 44: 141–156.

Huber, A. S. Goodrick, and K. Anderson. 1999. Diversity with successional status in the pinyon-juniper/mountain mahogany/bluebunch wheatgrass community types near Dutch John, Utah. In S. Monsen and R. Stevens (comps.), *Proceedings: Ecology and management of pinyon-juniper communities within the interior West*, September 15–18, 1997, Provo, Utah. Proc. RMRS-P-9. USDA Forest Service, Ogden, Utah.

Jacobs, B. F., and R. G. Gatewood. 1999. Restoration studies in degraded pinyon-juniper woodlands of north-central New Mexico. In S. Monsen and R. Stevens (comps.), *Proceedings: Ecology and management of pinyon-juniper communities within the interior West*, September 15–18, 1997, Provo, Utah. Proc. RMRS-P-9. USDA Forest Service, Ogden, Utah.

Kaufmann, M. R., W. H. Moir, and W. W. Covington. 1992. Old-growth forests: What do we know about their ecology and management in the southwest and Rocky Mountain Regions. In *Old-growth forests in the Rocky Mountains and Southwest Conference*, Portal, Ariz., March 9–13, 1992, pp. 1–12. USDA General Technical Report RM-258.

Kruse, W. H., and H. M. Perry. 1994. Ecosystem management research in an "old-growth" piñon-juniper woodland. In D. W. Shaw, E. F. Aldon, and C. LoSapio (eds.), *Desired future conditions for piñon-juniper ecosystems*. USDA Forest Service General Technical Report RM-258.

Ligon, J. David. 1978. Reproductive interdependence of piñon pines and piñon jays. *Ecology* 48: 111–126.

Little, E. L. 1948. *Food analysis of piñon nuts*. USDA Forest Service, Southwestern Forest and Range Experiment Station. Note 49.

Meeuwig, Richard O., and Jerry Budy. 1981. *Point and line intersect sampling in pinyon-juniper woodland*. USDA Forest Service General Technical Report INT-104.

Mehl, M. S. 1992. Old-growth descriptions for the major forest cover types in the Rocky Mountain region. In *Old-growth forests of the Rocky Mountains and Southwest Conference*, Portal, Ariz., March 9–13, 1992, pp. 106–121. USDA Forest Service General Technical Report RM-213.

Miller, R., R. Tausch, and W. Waichler. 1999. Old-growth juniper and pinyon woodlands. In S. Monsen and R. Stevens (comps.), *Proceedings: Ecology and management of pinyon-juniper communities within the interior West*, September 15–18, 1999, Provo, Utah. Proc. RMRS-P-9. USDA Forest Service, Ogden, Utah.

Miller, R. F., and R. E. Wigand. 1994. Holocene changes in semiarid pinyon-juniper woodlands. *BioScience* 44: 465–474.

Moore, M. M., W. W. Covington, and P. Z. Fule. 1999. Reference conditions and ecological restoration: A southwestern ponderosa pine perspective. *Ecological Applications* 9: 1266–1277.

Naillon, D., K. Memmott, and S. B. Monsen. 1999. A comparison of understory species at three densities in a pinyon-juniper woodland. In S. Monsen and R. Stevens (comps.), *Proceedings: Ecology and management of pinyon-juniper communities within the interior West,* September 15–18, 1999, Provo, Utah. Proc. RMRS-P-9. USDA Forest Service, Ogden, Utah.

Peet, R. K. 1981. Forest vegetation of the Colorado Front Range. *Vegetatio* 45: 3–75.

Romme, W. H., D. W. Jamieson, J. Redders, G. Bigsby. J. Page Lindsey, D. Kendall, R. Cowen, T. Kreynes, A. Spencer, and J. C. Ortega. 1997. Old-growth forests of the San Juan National Forest in southwest Colorado. In *Old-growth forests in the Rocky Mountains and Southwest Conference,* Portal, Ariz., March 9–13, 1992. USDA Forest Service General Technical Report RM-258.

Roovers, L. M., and A. J. Rebertus. 1993. Stand dynamics and conservation of an old-growth Engelmann spruce–subalpine fir forest in Colorado. *Natural Areas Journal* 13: 256–267.

Shaw, D. W., E. F. Aldon, and C. LoSapio (eds.). 1994. *Desired future conditions for piñon-juniper ecosystems.* USDA Forest Service General Technical Report RM-258.

Shulman, E. 1943. Over-age drought conifers of the Rocky Mountains. *Journal of Forestry* 41: 422–425.

Silvertown, J. W. 1984. *Introduction to plant population ecology.* Longman Press, London and New York.

Swetnam, T. W., and J. L. Betancourt. 1999. Mesoscale disturbance and ecological response to decadal climatic variability in the American Southwest. *Journal of Climatology* 11: 3128–3147.

Vander Wall, S. B. 1997. Dispersal of singleleaf pinyon pine by seed caching rodents. *Journal of Mammalogy* 78: 181–191.

Veblen, T. T. 1986a. Age and size structure of subalpine forests in the Colorado Front Range. *Bulletin of the Torrey Botanical Club* 113: 225–240.

———. 1986b. Treefalls and the coexistence of conifers in subalpine forests of the central Rockies. *Ecology* 67: 644–649.

Veblen, T. T., K. S. Handley, and M. S. Reid. 1991. Disturbance and stand development of a Colorado subalpine forest. *Journal of Biogeography* 18: 707–716.

West, N. E., and N. S. Van Pelt. 1987. Successional patterns in pinyon-juniper woodlands. In R. L. Everett (comp.), *Proceedings: Pinyon juniper conference,* January 13–16, 1986, Reno, Nev., pp. 43–52. USDA Forest Service General Technical Report INT-215.

BENEATH THE TREES: SHRUBS, HERBS, AND SOME SURPRISING RARITIES

M. Lisa Floyd and Marilyn Colyer

Each season hundreds of shrubs, grasses, and wildflowers germinate, flower, produce seeds, and then disappear in the piñon-juniper woodlands of Mesa Verde Country. May's landscape is characterized by deep orange globe mallows, intense purple lupines, bright white flowers of the fendlerbush, and yellow lomatiums. In midsummer the prevalent lavender flowers of the low penstemon, the creamy yucca flowers, and the cryptic commandras dot the woodland floor. By September white and purple asters and deep yellow balsamroots catch the eye. The less conspicuous but not always less colorful lichens, mosses, and cyanobacteria cling to bark, soil, and rock year-round. Myriad understory forbs (herbaceous annuals and perennials) and grasses thrive, sometimes at higher densities in the spaces between canopies than under the trees themselves. Others persist in the shade of old-growth trees, while still others are restricted to the edges where woodland meets meadow or to a particular soil. Many understory plants depend on fires and canopy disturbances. Biological diversity in the understory contributes in no small way to the beauty of the area. Here we examine the breadth of these plants that make their home under the conifers on Mesa Verde. Well-established common names are presented with species names (Weber and Wittman 1996) as the species are introduced and are used thereafter.

Among piñon-juniper stands on Mesa Verde, canopy closure varies with soil type and depth, exposure, and stand age. At the upper-elevation ecotone on the Mesa Verde cuesta, piñon-juniper woodlands are gradually supplanted by *Pseudotsuga menziesii* (Douglas fir) forests and mountain-shrub communities. These ecotones are similar floristically to the dense piñon-juniper woodlands that occupy lower elevations but are very different structurally, as they are dominated by *Quercus gambelii* (Gambel oak) and *Amelanchier utahensis* (Utah serviceberry). In addition, conifers may be fairly rare. Elsewhere in the region surrounding the Mesa Verde cuesta, upper-elevation ecotones are characterized by overlap of piñon-juniper woodlands and ponderosa-pine forests. In still other areas, ecotones consist of dense stands of *Seriphidium (Artemisia) tridentata* (big sagebrush), accompanied by scattered piñon and junipers. Ponderosa pine is distinctly uncommon on the Mesa Verde cuesta, for reasons not yet understood, so the ecotone between piñon-juniper woodlands and ponderosa-pine forests that is so common elsewhere on the Colorado Plateau is not found on Mesa Verde.

Part of the architectural diversity among woodland sites has to do with stand age and with the frequency of canopy "gaps." As the woodland ages, the canopy may become dense; yet the very nature of these short, squatty trees allows ample light to penetrate the canopy and reach the soil. As the woodland matures, trees that succumb to insect and fungal pathogens and natural senescence create canopy openings. Small, often single-tree fires are a natural annual occurrence, leaving openings in the woodland canopy. The resultant heterogenous canopy structure supports an exciting assortment of herbaceous species.

This type of structural diversity is less common elsewhere on the Colorado Plateau, where closed canopy piñon-juniper woodlands are characterized by few plants in the understory, at least partially because of dwindling light and water (Huber et al. 1999; Naillon et al. 1999). At Mesa Verde National Park, however, we have found no evidence that even the densest canopy precludes species richness in the understory. In a comprehensive vegetation study involving nearly 400 sampling areas, the number of understory forbs and shrubs was the same in open and dense canopies (averaging 12 species per 100 m^2), with a slightly greater number (16 species per 100 m^2) in the intermediate canopies. This may be in part because the woodland becomes more and more open as it ages, with an average tree canopy cover of less than 50%. The intercanopy spaces favor shrub and herbaceous growth. Thus long periods without any large or severe disturbances, combined with soil stability, have allowed piñon-juniper woodland to age, develop an open canopy structure, and maintain a rich understory flora. A similar pattern has been reported in deep, well-drained soils in juniper ecosystems in Oregon (Miller et al. 1999) and in northeastern Utah (Austin 1999).

Indeed, most of Mesa Verde's piñon-juniper woodlands support a rich diversity of understory species. While *Purshia tridentata* (bitterbrush) is often the dominant understory shrub, other species include Gambel oak (Figure 3.1), Utah serviceberry (Figure 3.2), big sagebrush (Figure 3.3), *Seriphidium novum* (= *Artemisia nova*, black sagebrush), *Peraphyllum ramosissimum* (squawapple), and *Fendlera rupicola* (fendlerbush). The forb layer varies considerably across elevations and among

Figure 3.1. The most common tall shrub on the Mesa Verde cuesta is Quercus gambelii *(Gambel oak), a species that resprouts vigorously after fire. Illustration by Walt Anderson.*

microsites, but the common forbs are *Penstemon linarioides coloradoënsis, Pedicularis centranthera, Petradoria pumila, Cryptantha bakerii, Eriogonum racemosum, Lupinus ammophilus, Astragalus scopulorum, Calochortus nuttallii, Commandra umbellata, Cymopterus bulbosus, C. purpureus, Lomatium triternatum, L. grayii,* and *Yucca baccata.*

An abundant bunchgrass layer is rather uncommon for piñon-juniper woodlands, many of which have been so heavily affected by cattle and sheep grazing that the native understory grasses are no longer present (Fleischner 1994; Orodho et al. 1990). We are fortunate that dense grasses have returned to the Mesa Verde National Park woodland, in contrast to woodlands in parts of northern New Mexico's Jemez Mountains, where bunch grasses are uncommon and soils are subject to extensive erosion (Davenport et al. 1998). On Mesa Verde dense patches of the common *Poa fendleriana* (muttongrass) attain 80% cover of the ground, providing soil stability through its intricate root system. Whereas muttongrass is by far the most common perennial bunch grass on the Mesa Verde cuesta, it is accompanied by *Achnatherum (Oryzopsis) hymenoides* (Indian ricegrass) (Figure 3.4), *Elymus elymoides* (= *Sitanion hystrix*, squirreltail grass), *Pascopyrum smithii* (= *Agropyron smithii*, western wheat), *Koeleria macrantha* (junegrass), *Hesperostipa comata* (= *Stipa comata*, needle-and-thread grass), and *Agropyron trachycaulum* (slenderwheat). These are but a few of the hundreds of species associated with piñon-juniper woodlands in Mesa Verde Country; a more complete listing is provided in Appendix 3.1.

The arrangement of these hundreds of plants is anything but random. Unique groups of species occur on particular soils (red loess, shales, limestones, alluvial soils, and colluvial soils) and within specific elevation belts on the mesa and in the surrounding Mesa Verde Country (Figure 1.3). To begin to see floristic patterns, these communities have been documented on thirty transects over many years in Mesa Verde National Park. For example, Mesa Verde's prominent red loess soil, when found at the upper elevation of the mesa (2,363–2,484 m) where temperatures are cool and precipitation relatively high, supports tall piñon and juniper trees and vigorous growth of understory forbs and grasses. At the middle elevations (2,273–2,364 m) dense stands of bitterbrush, muttongrass, and *Penstemon linarioides* are interspersed with *Phlox hoodii* (Hood's phlox), *Polygonum sawatchense* (piñon knotweed), and *Eriogonum racemosum* (slender buckwheat). At low elevations on the cuesta (2,060–2,273 m) drier conditions prevail, and exposure of sandstone caprock is common. Many of the same species already listed occur here too, but generally at lower densities. *Euphorbia fendleri* (fendler's spurge) and *Astragalus calycosus* (calcite milk-vetch) become more common.

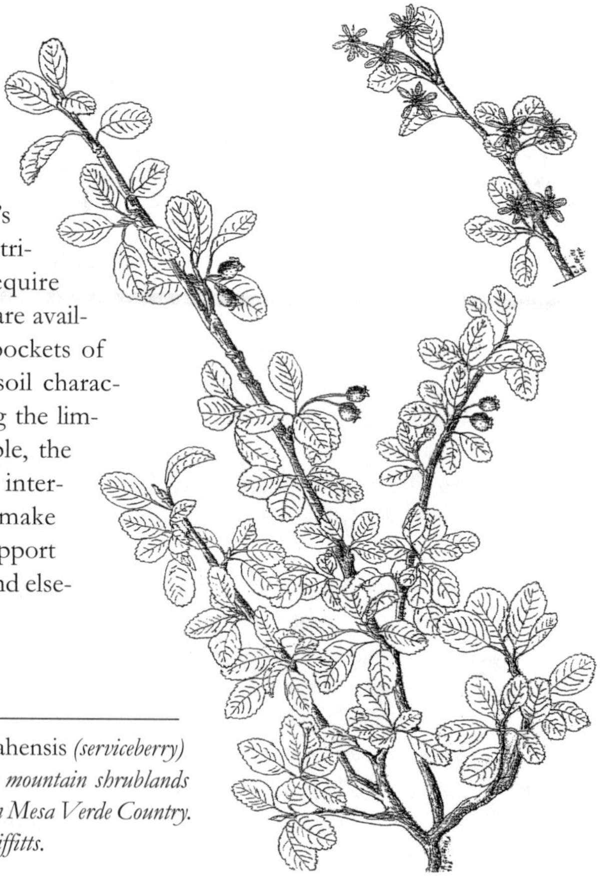

RESTRICTED AND RARE SPECIES

Many of Mesa Verde's plants are limited in distribution because they require specific conditions that are available only in restricted pockets of the landscape. Unique soil characteristics are often among the limiting factors. For example, the shales, sandstones, and intermittent bentonite that make up the Menefee Shale support plants that are rarely found else-

Figure 3.2. Amelanchier utahensis *(serviceberry) is a common tall shrub of the mountain shrublands and piñon-juniper woodlands in Mesa Verde Country. Illustration by Elizabeth Griffitts.*

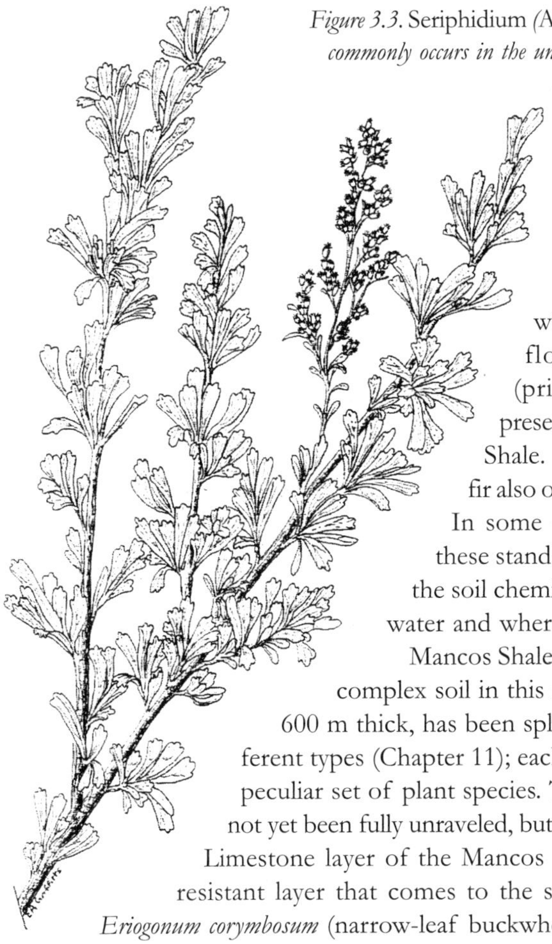

Figure 3.3. Seriphidium *(*Artemisia*)* tridentata *(big sagebrush) commonly occurs in the understory of old-growth piñon-juniper woodlands and in clearings, where it can become the dominant shrub. Illustration by Elizabeth Griffitts.*

where. With its showy yellow inflorescence, *Stanleya pinnata* (prince's plume) advertises the presence of selenium in the Menefee Shale. Localized stands of Douglas fir also occur on Menefee-derived soils. In some cases geologists suspect that these stands occur along fault lines where the soil chemistry is altered by passing rainwater and where moisture is relatively high.

Mancos Shale undoubtedly fosters the most complex soil in this region. This formation, up to 600 m thick, has been split by geologists into eight different types (Chapter 11); each layer often supports its own peculiar set of plant species. The soil-flora relationship has not yet been fully unraveled, but we can discuss the Greenhorn Limestone layer of the Mancos Shale, a 10-m-thick weather-resistant layer that comes to the surface on Mesa Verde. Here *Eriogonum corymbosum* (narrow-leaf buckwheat) and *Frasera speciosa* (green gentian) are found in sparse stands of Utah serviceberry, piñon, juniper, *Juniperus scopulorum* (Rocky Mountain juniper), and scattered ponderosa pines.

Cliff House Sandstone is usually buried by the red loess soils discussed above, but unique pockets of Cliff House–derived sand form along the mesa rims and in larger slices of exposed Cliff House Sandstone on flat mesa tops. These pockets support restricted plant species. For example, *Astragalus deterior* (Cliff Palace milkvetch) is found only on a narrow "beach" deposit derived from the upper reach of the Cliff House Sandstone, which is overlain by a fossiliferous red-brown layer.

A cobbly mixture of sandstone pieces may form when caprock and shales fall down steep slopes. This colluvium on rocky slopes is a frequent habitat for *Ephedra viridis* (Mormon tea) and big sagebrush. *Elymus cinereus* (Basin wildrye) has survived on rough, colluvial slopes where cattle grazing has been prevented by natural barriers. Indeed these Basin wildrye communities may possibly represent a

Figure 3.4. Achnatherum (Oryzopsis) hymenoides *(Indian ricegrass) is a common grass of the piñon-juniper woodland. Illustration by Elizabeth Griffitts.*

once-common flora that formerly dominated the lower elevations of Mesa Verde Country.

The old-growth piñon-juniper woodland supports hundreds of plants, but we cannot say with conviction that any plant species *requires* the complex conditions presented by mature piñon-juniper woodlands. We do know that growth of many rare plants is favored by well-developed woodland overstory and that they are less common or absent in young piñon-juniper woodlands (Appendix 3.2). Several endemic species are known only from Mesa Verde.

The first three plants we consider have the distinction of official federal or state "Threatened or Endangered" status. *Astragalus humillimus* (Mancos milkvetch) is considered endangered by the U.S. Fish and Wildlife Service (Spackman et al. 1997). It forms low, tight mats on Point Lookout Sandstone. Townshend Brandegee (1843–1925), a botanist with the Hayden Surveys who made important collections in Mesa Verde Country, collected Mancos milkvetch over a century ago; but it has not been found again. *Astragalus deterior* (Cliff Palace milkvetch) is a Colorado endemic found only on the Mesa Verde cuesta, restricted to cracks and depressions in the white beach sand of the Upper Cliff House Sandstone. While it has no known locations on the Mesa Verde cuesta, the federally threatened *Sclerocactus mesae-verdae* (Mesa Verde cactus) is known from sparsely vegetated piñon-juniper stands on Mancos shales to the southwest (Spackman et al. 1997).

More *Astragalus* species are Mesa Verde endemics. *Astragalus schmolliae* (Schmoll's milkvetch) is considered "critically endangered throughout its range" (G1/S1) by the Colorado Natural Heritage Program and is now reportedly being considered for state and federal threatened status (L. Hudson and J. Coles, personal communication, 1999). This species, known only from the Mesa Verde cuesta, is restricted to sandy loess on Cliff House Sandstone in old-growth piñon-juniper woodlands. While it is locally abundant on Chapin Mesa, a small population was recently discovered in the perimeter of the 1996 Chapin 5 fire, where it apparently is both resprouting from rootstocks and germinating from seeds. The older Chapin Mesa population consists primarily of resprouting older plants that may have established long ago.

Several other plant species are found primarily in the old piñon-juniper woodlands, but with local abundance that precludes state or federal status. *Pedicularis centranthera*, with its fernlike leaves and deep-purple bilaterally symmetrical flowers, is a common sight under the dense piñon-juniper canopies. Although only occasional elsewhere, *Pedicularis centranthera* occurred at over three-quarters of the piñon-juniper sampling points in a vegetation survey of Mesa Verde National Park (Floyd-Hanna et al. 1993). *Aletes macdougalii subsp. breviradiatus* is a bright yellow–flowered, pungent perennial that forms large clusters where seeps support moist habitats within the piñon-juniper woodland. This subspecies is considered "vulnerable throughout its range or found locally in a restricted range" (G3/S1) by the Colorado Natural Heritage Program.

Large wildfires ravaged sections of the Mesa Verde cuesta in 1934, 1959, 1972, 1989, 1996, 2000, and most recently during the summer of 2002. Although devastating in many ways, these fires have allowed several species to spread, perhaps by promoting germination of buried seeds awaiting fire-induced signals or perhaps because of ephemeral boosts in soil nutrients. Extensive populations of *Nicotiana attenuata* (wild tobacco) germinated after the 1989, 1996, and 2000 fires but only in severely burned old-growth piñon-juniper woodlands. Thousands of wild tobacco plants germinated from seeds stored in the seed bank, possibly responding to smoke-induced germination signals (Brown and Van Staden 1997; Keeley et al. 1995; Preston and Baldwin 1999) or light-enhanced cues. Wild tobacco was very likely utilized by the Ancestral Pueblo people who lived on the Mesa Verde cuesta prior to A.D. 1300, and they may have set fires to encourage its growth. The tiny round black seeds may survive in the soil for perhaps a century (J. Winter, personal communication, 1996). Since the abandonment of the mesa by Ancestral Puebloans, it is likely that periodic fires have encouraged populations, allowing replenishment of tobacco seeds in the soil.

One of the most surprisingly beautiful sights after fires in Mesa Verde is the contrast of small ice-blue flowers against the blackened landscape. Normally found only in small isolated populations, the endemic borage *Hackelia gracilenta* (Mesa Verde stickseed) quickly invades burned piñon-juniper woodlands (Floyd-Hanna et al. 1999). *Hackelia* is a species of concern for Mesa Verde National Park and is ranked "vulnerable throughout its range or found locally in a restricted range" (G3/S1) by the Colorado Natural Heritage Program. After the 1996 Chapin 5 fire we recorded over thirty-five Mesa Verde stickseed locations, all of which were in burned piñon-juniper woodlands. It persists for several years after fires.

Collomia grandiflora (collomia) is rare on the Mesa Verde cuesta except after fires. Otherwise, it is known from two widely separated populations in Colorado. This small forb—with an unusual salmon-colored flower—was common after the 1996 fire, recorded throughout piñon-juniper and piñon-juniper/oak woodland habitats. It is ephemeral, however; by 1998 only very few plants could be found (Floyd-Hanna et al. 1999).

Iliamna rivularis (wild hollyhock), although not officially listed, is a proposed Colorado State Species of Concern. Several decades ago wild hollyhock invaded small fires on the Mancos Shale slopes on the north escarpment of Mesa Verde; although we have searched for this species after the fires of the last decade, it has not been found.

Many other plant species are locally rare in our Mesa Verde Country, are restricted to particular local substrates, or are considered "Park Sensitive" by Mesa Verde National Park (Appendix 3.2). While not all are recognized by the federal or state species of concern lists, each locally rare population or restricted species is important. They provide food for insect and bird pollinators as well as homes for insect pests, stabilize the soils, and add richness to floristic biodiversity.

Our listings demonstrate the breadth of diversity of plant species in the old-growth piñon-juniper woodland and provide a nascent checklist that will most certainly expand as interest in the flora of the piñon-juniper woodlands grows.

SUMMARY

The richness of the flora of Mesa Verde Country is illustrated by the variety of its common grass and forb species as well as the rare, often restricted, components. We know of 259 understory plants that occur in the old-growth piñon juniper woodlands of Mesa Verde Country, typifying the breadth of species that might grow in untouched woodlands. By contrast, in southwestern landscapes where livestock grazing has been prominent for over a century, it is unlikely that many piñon-juniper stands support such a rich biodiversity. Mesa Verde Country, and particularly the relatively protected Mesa Verde National Park, allow us a glimpse into the complex and diverse habitats associated with old-growth piñon-juniper woodlands. Chapters 4 and 5 describe nonvascular plants and plantlike organisms of the old woodlands.

APPENDIX 3.1. Old-growth piñon woodland understory species (nomenclature according to Weber and Wittman 1996)

	Loess upper	Loess mid	Loess low	Menefee Shale	Mancos Shale	Greenhorn	Cliff House Sandstone	Pt. Lookout Sandstone	Dakota Sandstone	Alluvium upper	Alluvium low	Colluvium	Minette	Morrison
OVERSTORY TREES														
Juniperus (Sabina) osteosperma	20%	30%	30%	10%	15%	15%	15%	5%	15%	<5%	<5%	15%	10%	—
Juniperus (Sabina) scopulorum	—	—	—	—	—	—	—	10%	10%	20%	20%	—	—	—
Pinus edulis	20%	30%	30%	10%	50%	30%	15%	15%	—	<5%	<5%	15%	10%	—
UNDERSTORY SHRUBS AND FORBS														
ACERACEAE														
Negundo aceroides	—	—	—	—	—	—	—	—	—	L	—	—	—	—
AGAVACEAE														
Yucca baccata	U	U	C	U	R	R	C	?	C	—	—	U	?	—
ALISMATACEAE														
Sagittaria latifolia	—	—	—	—	—	—	—	—	—	—	—	—	—	—
ALLIACEAE														
Allium acuminatum	—	—	—	U	—	—	C	C	C	U	U	U	—	—
Allium macropetalum	—	—	—	—	Ab	—	—	—	—	—	—	—	—	—
Allium textile	—	—	—	—	Ab	—	—	—	—	—	—	—	—	—
AMARANTHACEAE														
Amaranthus blitoides	D	D	—	—	—	—	—	—	—	D	D	D	—	—
Amaranthus graezicans	—	D	—	—	—	—	—	—	—	D	D	—	—	—
ANACARDIACEAE														
Rhus aromatica subsp. *trilobata*	R	—	—	—	U	—	R	—	L	C	C	U	—	—
APIACEAE														
Aletes macdougalii	—	—	—	—	—	—	—	—	—	—	—	—	—	—
Cymopterus bulbosus	U	C	C	U	C	—	—	—	—	R	R	C	—	—
Cymopterus constancei	U	C	C	U	C	—	—	—	—	—	—	C	—	—

Species	1	2	3	4	5	6	7	8	9	10	11	12
Cymopterus purpureus	—	—	—	—	R	—	—	—	—	—	—	—
Lomatium dissectum	L	—	—	—	—	Rm	Rm	—	—	—	Rm	—
Lomatium grayi	C	—	—	—	—	L	L	—	—	—	L	—
Lomatium triternatum subsp. Platycarpum	—	—	—	—	—	C	—	—	—	—	C	—
ASCLEPIADACEAE												
Asclepias subverticillata	—	U	—	—	—	—	U	—	—	U	U	—
ASTERACEAE												
Achillea lanulosa	C	U	—	C	C	—	C	C	C	C	U	U
Antennaria dimorpha	—	—	—	—	—	—	—	—	—	—	—	—
Antennaria parvifolia	—	—	—	—	—	—	—	—	—	—	—	—
Antennaria rosea	—	L	—	—	—	—	—	—	L	L	—	—
Artemisia frigida	C	U	L	C	—	—	—	—	—	—	—	U
Artemisia ludoviciana	—	—	—	C	—	—	D	D	U	U	D	—
Artemisia spinescens	C	—	—	—	—	—	—	—	—	—	—	—
Balsamorhiza sagittata	—	—	—	—	—	—	D	D	U	U	D	D
Brickellia californica	—	—	—	—	—	—	C	C	—	—	—	—
Chaenactis douglasii	C	—	U	C	C	?	—	—	—	—	—	—
Cirsium undulatum	C	C	C	C	C	—	—	—	C	C	C	C
Cyclachaena xanthifolia	D	—	—	—	—	—	—	—	—	—	—	—
Erigeron compactus	C	C	C	C	—	—	—	—	—	D	—	—
Erigeron divergens	C	U	C	C	—	—	—	D	U	U	U	U
Erigeron speciosus	C	C	C	—	—	—	—	—	U	U	U	U
Erigeron utahensis	—	—	—	Ab	—	—	—	—	—	—	—	—
Grindelia arizonica var. stenophylla	D	D	D	D	D	R	R	D	C	C	—	C
Gutierrezia sarothrae	D	—	D	D	—	—	—	—	—	—	—	—
Helianthus rigidus var. subrhomboideus	—	U	U	C	U	U	U	U	U	C	C	?
Heliomeris multiflora	C	—	—	C	C	—	—	—	—	—	—	—

continued on next page

APPENDIX 3.1—continued

	Loess upper	Loess mid	Loess low	Menefee Shale	Mancos Shale	Green-horn	Cliff House Sandstone	Pt. Lookout Sandstone	Dakota Sandstone	Alluvium upper	Alluvium low	Colluvium	Minette	Morrison
Heterotheca villosa	D	D	D	D	D	U	U	U	U	C	C	D	U	—
Hymenopappus filifolius	—	—	—	—	U	—	—	—	U	—	—	C	—	—
Hymenoxys richardsonii subsp. *floribunda*	—	U	U	—	—	—	—	—	—	—	—	—	—	—
Iva axillaris	—	—	—	—	—	—	—	—	—	C	C	—	—	—
Lygodesmia grandiflora	—	—	—	—	R	—	—	—	—	—	—	—	—	—
Machaeranthera gracilis	—	—	—	—	—	—	—	—	U	—	—	U	—	—
Oligosporus dracunculus subsp. *glaucus*	U	—	—	—	—	—	—	—	—	Ab	—	—	—	—
Packera multilobata	C	C	C	—	—	—	C	—	—	C	C	C	—	—
Petradoria pumila	C	C	C	—	—	—	—	—	C	C	C	C	—	—
Solidago velutina	—	—	—	U	U	—	—	—	—	—	—	—	—	—
Stenotus armerioides	Rc	—	—	—	—	—	Rc	—	U	—	—	—	—	—
Stephanomeria exigua	—	—	—	—	—	?	—	—	—	—	—	—	—	—
Stephanomeria tenuifolia	C	C	C	C	C	U	C	U	C	—	—	C	—	—
Tetraneuris ivesiana	—	U	U	U	C	—	C	—	C	—	—	U	U	—
Townsendia incana	C	U	D	C	—	—	U	—	—	U	U	U	U	—
Wyethia arizonica	C	—	—	—	—	—	—	—	—	—	—	—	—	—
Xanthium strumarium	—	—	—	—	—	U	—	—	—	—	—	—	—	—
BERBERIDACEAE														
Mahonia repens	R	—	—	U	U	U	—	Dr	Dr	U	—	Dr	—	—
BORAGINACEAE														
Cryptantha crassisepala var. *elachantha*	—	—	—	—	—	—	—	—	—	—	—	—	U	U
Cryptantha gracilis	—	—	—	—	—	—	—	—	—	—	—	C	—	—
Cryptantha minima	—	—	—	—	—	—	—	—	—	—	—	—	U	U
Descurainia pinnata	D	D	D	D	D	D	D	D	D	—	—	D	—	U
Draba cuneifolia	—	—	—	—	—	—	—	—	—	C	—	—	—	—

Species	1	2	3	4	5	6	7	8	9	10	11	12
Draba reptans	—	—	—	C	—	C	—	C	C	C	—	—
Lithospermum incisum	C	U	C	C	—	—	—	—	C	C	—	—
Mertensia brevistyla	—	—	—	—	—	—	—	U	—	U	—	C
Oreocarya bakeri	—	—	—	—	U	—	—	—	—	U	—	—
Oreocarya flavoculata	—	—	—	—	U	—	—	—	—	U	—	—
Oreocarya fulvocanescens	—	—	—	—	—	—	—	—	—	U	—	—
Oreocarya humilis subsp. *nana*	—	—	—	—	—	—	—	—	—	—	—	—
BRASSICACEAE												
Arabis hirsuta	—	—	—	—	—	—	—	—	L	—	—	—
Boechera drummondii	—	C	—	—	—	—	—	—	L	—	—	—
Boechera fendleri	U	U	—	—	U	—	—	—	—	—	—	—
Boechera lignifera	—	—	Rc	—	—	—	—	—	—	—	—	—
Boechera pulchra	—	U	U	U	—	U	—	U	—	U	U	—
Boechera retrofracta	—	C	—	C	—	—	—	—	—	—	—	—
Boechera selbyi	C	C	U	C	—	—	—	—	—	—	—	—
Erysimum inconspicuum	U	U	—	C	—	C	C	C	—	U	—	—
Lepidium montanum	U	C	—	C	—	C	C	C	—	C	—	—
Lesquerella rectipes	—	C	—	C	—	C	C	—	—	C	C	—
Physaria acutifolia	C	C	—	U	—	U	—	—	—	—	—	—
Schoenocrambe linifolium	—	—	—	—	U	U	—	—	—	—	—	—
Stanleya pinnata	—	—	—	C	—	—	—	—	—	C	—	—
Streptanthus cordatus	—	—	—	C	—	—	—	—	C	C	—	—
Thelypodium laxiflorum	—	—	—	—	L	—	?	—	—	U	U	—
Thlaspi arvense	—	—	—	—	—	—	—	—	—	—	—	—
Ximenesia encelioides	—	D	—	D	D	—	—	—	—	U	—	—
CALOCHORTACEAE												
Calochortus gunnisonii	C	U	R	C	—	—	—	—	—	—	—	—
Calochortus nuttallii	C	U	R	C	—	—	—	—	—	U	—	—
CAPRIFOLIACEAE												
Symphoricarpos rotundifolius	C	—	—	—	U	U	—	—	—	—	—	—

continued on next page

APPENDIX 3.1—continued

	Loess upper	Loess mid	Loess low	Menefee Shale	Mancos Shale	Green-horn	Cliff House Sandstone	Pt. Lookout Sandstone	Dakota Sandstone	Alluvium upper	Alluvium low	Colluvium	Minette	Morrison
CELASTRACEAE														
Atriplex canescens	U	U	U	—	U	U	U	—	U	U	U	U	—	—
Atriplex confertifolia	—	—	U	U	—	—	—	—	—	—	U	—	—	—
Paxistima myrsinites	—	—	—	—	—	—	—	—	C	—	—	—	—	—
Sarcobatus vermiculatis	—	—	U	L	—	—	—	—	L	—	U	—	U	—
CHENOPODIACEAE														
Chenopodium berlandieri	P	P	—	—	—	—	—	—	—	R	—	—	—	—
Chenopodium fremontii	P	P	U	D	D	U	U	U	U	U	—	—	—	—
Chenopodium incanum	—	—	—	—	—	—	—	—	—	—	—	C	—	C
CONVALLARIACEAE														
Maianthemum amplexicaule	—	—	—	U	C	—	—	—	—	—	—	U	—	—
Maianthemum stellatum	—	—	—	U	C	—	—	—	—	—	—	—	—	—
CYPERACEAE														
Carex deweyana	—	—	—	—	—	—	—	—	—	C	—	—	—	—
Carex festivella	C	C	—	C	—	—	—	C	—	U	—	—	U	—
Carex geophyla	—	—	?	C	C	U	—	C	—	—	—	C	—	—
Carex geyeri	—	—	—	Ru	Ru	—	—	—	—	—	—	—	—	—
Carex occidentalis	—	—	—	—	—	—	—	—	—	C	—	—	—	—
Carex pensylvanica subsp. *heliophila*	—	—	—	—	—	—	—	—	—	U	—	—	—	—
Carex praegracilis	—	—	—	—	—	—	—	—	—	C	—	—	—	—
Eleocharis macrostachya	Ri	—	—	—	—	—	—	—	—	Ri	Ri	—	—	—
Scirpus pallidus	—	—	—	—	—	—	—	—	—	—	R	—	—	—
Scirpus pungens	—	—	—	—	—	—	—	—	—	Ri	Ri	—	—	—
EUPHORBIACEAE														
Tithymalus robustus	—	—	—	—	—	—	R	—	—	—	—	R	—	—

Taxon	1	2	3	4	5	6	7	8	9	10	11	12	13
FABACEAE													
Astragalus bisulcatus	—	—	—	—	—	—	—	—	U	—	—	—	—
Astragalus calycosus var. *scapus*	—	R	Rc	—	—	—	—	—	U	—	—	—	—
Astragalus ceramicus	—	—	—	—	—	—	C	—	—	—	—	—	C
Astragalus flexuosus	—	—	C	—	—	—	C	—	—	—	C	—	—
Astragalus lentiginosus var. *palans*	—	—	Rc	?	—	—	—	—	—	—	—	—	C
Astragalus missouriensis var. *amphibolus*	—	—	—	—	—	—	—	—	C	—	—	—	—
Astragalus missouriensis var. *missouriensis*	—	—	—	—	—	—	—	—	—	—	—	—	—
Astragalus newberryi	—	—	—	—	—	—	—	—	—	—	—	—	U
Astragalus nuttalliana var. *micranthemiformis*	—	—	—	—	—	—	—	C	—	C	—	—	—
Astragalus pattersonii	—	—	—	—	L	—	—	—	—	—	—	—	—
Astragalus scopulorum	C	—	C	C	U	—	—	—	C	C	C	—	—
Astragalus tenellus	—	U	—	—	C	—	—	—	—	—	—	—	U
Astragalus wingatanus	—	U	U	U	C	R	—	—	U	—	—	—	—
Hedysarum boreale	C	—	—	—	—	—	—	—	—	—	—	—	—
Lathyrus leucanthus	C	—	—	—	—	—	—	—	—	—	—	—	—
Lupinus ammophilus	U	—	C	C	—	—	—	—	—	—	C	U	—
Lupinus caudatus	C	—	U	C	U	—	—	—	—	—	—	—	—
Lupinus kingii	—	—	—	—	—	—	—	—	—	—	—	—	—
Pediomelum megalanthum	—	—	—	—	—	—	—	—	—	—	C	—	D
Trifolium longipes	—	—	—	—	—	—	—	—	—	—	C	—	R
Vicia americana	—	—	—	—	—	—	—	—	C	—	—	—	U
FAGACEAE													
Quercus gambelii	U	R	—	C	C	U	U	U	U	U	L	U	—
GENTIANACEAE													
Centaurium exaltatum	—	—	—	—	Rm	—	—	—	R	—	—	—	—
Frasera albomarginata	—	—	Rc	—	—	—	—	—	—	—	—	—	—

continued on next page

APPENDIX 3.1—continued

	Loess upper	Loess mid	Loess low	Menefee Shale	Mancos Shale	Green-horn	Cliff House Sandstone	Pt. Lookout Sandstone	Dakota Sandstone	Alluvium upper	Alluvium low	Colluvium	Minette	Morrison
GROSSULARIACEAE														
Ribes aureum	—	—	—	—	—	—	—	—	—	—	—	—	—	—
Ribes leptanthum	—	—	—	—	—	—	—	—	—	C	U	U	—	—
HELLEBORACEAE														
Delphinium nuttallianum	C	C	C	C	U	U	—	—	—	C	U	C	—	—
HYDRANGEACEAE														
Fendlera falcata	U	R	R	U	L	U	U	U	R	—	—	U	U	—
Philadelphus microphyllus	—	—	—	—	—	—	U	U	—	—	—	—	—	—
HYDROPHYLLACEAE														
Ellisia nyctelea	U	—	—	—	—	—	—	—	—	—	—	—	—	—
Phacelia heterophylla	C	U	—	—	—	—	—	—	—	—	—	U	—	—
JUNCACEAE														
Juncus bufonius	—	—	—	—	—	—	—	—	—	U C	C	—	—	—
Juncus dudleyi	—	—	—	—	—	—	—	—	—	C	—	—	—	—
Juncus longistylis	—	—	—	—	—	—	—	—	—	U	—	—	—	—
Juncus saximontanus	C	—	—	—	—	—	—	—	—	—	—	—	—	—
LAMIACEAE														
Dracocephalum parviflora	D	D	—	D	D	—	—	—	—	—	—	U	R	—
Mentha arvensis	—	—	—	—	—	—	—	—	—	C	—	—	—	—
LINACEAE														
Adenolinum lewisii	D	D	—	D	D	—	—	—	—	U	U	D	D	—
LOASACEAE														
Nuttallia cronquistii	—	—	—	D	D	—	—	—	—	—	—	—	—	U
Nuttallia multiflora	—	—	—	D	D	—	—	—	—	—	—	—	—	—
MALVACEAE														
Iliamna rivularis	—	—	—	E	R	—	—	—	—	—	—	—	—	—
Sphaeralcea coccinea subsp. dissecta	D	D	D	D	D	L	—	—	—	U	U	D	D	—
Sphaeralcea parvifolia	—	—	—	—	C	—	—	—	—	C	—	C	—	—

MELANTHIACEAE												
Toxicoscordion paniculatum	C	U	—	—	—	—	—	U	—	—	—	—
MONOTROPACEAE												
Pterospora andromedea	—	—	—	—	—	—	R	R	—	—	—	—
NYCTAGINACEAE												
Abronia fragrans	—	—	—	U	—	—	—	—	U	U	—	—
Mirabilis multiflora	—	—	—	R	C	—	—	—	—	—	—	—
Oxybaphus linearis	—	R	—	—	U	—	—	—	—	—	—	—
OLEACEAE												
Forestiera pubescens	—	—	—	—	—	—	L	L	—	—	—	—
Fraxinus anomala	—	—	—	—	—	—	L	L	U	U	—	—
ONAGRACEAE												
Epilobium brachycarpum	C	—	—	C	—	U	—	—	C	U	D	—
Oenothera caespitosa	D	D	U	C	—	—	—	—	U	U	—	—
Oenothera coronopifolia	—	—	—	U	C	C	—	—	C	—	—	—
ORCHIDACEAE												
Epipactis gigantea	—	—	—	—	—	R	O	—	—	—	—	—
Pinus ponderosa	—	—	—	L	C	L	L	—	—	—	—	—
PLANTAGINACEAE												
Plantago patagonica	—	—	—	C	—	—	—	—	C	C	—	—
POACEAE												
Achnatherum hymenoides	D	D	C	C	U	C	L	C	U	C	U	—
Bouteloua curtipendula	—	—	—	—	—	L	—	R	C	C	R	—
Bromopsis porteri	—	—	—	—	—	C	C	C	C	C	—	—
Chondrosum gracile	—	—	U	—	—	U	C	U	U	U	C	—
Critesion jubatum	—	—	—	C	—	C	U	D	C	C	—	—
Elymus elymoides	C	C	U	C	C	C	U	U	C	U	C	—
Elymus trachycaulus subsp. *andinus*	—	—	—	U	C	C	L	U	U	C	—	—
Hesperostipa comata	U	—	—	U	U	U	C	C	C	C	C	—

continued on next page

APPENDIX 3.1—continued

	Loess upper	Loess mid	Loess low	Menefee Shale	Mancos Shale	Green-horn	Cliff House Sandstone	Pt. Lookout Sandstone	Dakota Sandstone	Alluvium upper	Alluvium low	Colluvium	Minette	Morrison
Hesperostipa neomexicana	—	—	—	—	—	—	—	—	—	—	—	—	—	—
Hilaria jamesii	C	U	—	—	C	C	—	—	C	C	C	C	—	—
Koeleria macrantha	—	—	—	C	—	—	—	—	—	C	C	—	—	—
Muhlenbergia asperifolia	—	—	—	—	C	—	—	—	—	—	C	—	—	—
Muhlenbergia richardsonis	—	—	—	—	—	—	—	—	—	—	—	—	—	R
Panicum capillare var. occidentale	U	U	—	U	C	C	U	C	—	C	C	—	U	—
Pascopyrum smithii	C	C	C	C	C	C	—	—	C	C	C	C	U	—
Piptatherum micranthum	—	U	—	U	U	U	—	—	—	U	U	U	C	—
Poa fendleriana subsp. longiligula	C	C	C	C	U	—	C	—	—	U	U	C	C	—
Poa pratensis	Dr	Dr	Dr	Dr	Dr	Dr	—	—	—	Dr	—	Dr	—	—
Poa secunda	—	U	—	—	—	—	—	—	—	C	C	U	—	—
Sporobolus contractus	—	—	—	—	—	—	—	—	—	—	—	—	—	—
Sporobolus cryptandrus	—	—	—	—	C	C	C	C	C	C	C	C	C	—
POLEMONIACEAE														
Aliciella haydenii	U	—	—	—	—	—	—	—	—	—	—	—	—	—
Collomia linearis	—	C	U	C	C	U	—	—	—	—	—	—	—	—
Eriastrum diffusum	—	—	—	—	—	—	—	—	C	—	—	U	C	—
Gilia haydenii	—	—	U	—	C	U	—	—	C	U	C	C	U	—
Gilia sinuata	C	C	C	—	—	—	—	—	—	—	—	—	—	—
Ipomopsis aggregata	—	—	—	—	—	—	—	—	—	—	—	—	—	—
Ipomopsis congesta	—	—	—	—	—	—	—	—	—	—	—	—	—	—
Ipomopsis polycladon	—	—	—	—	—	—	—	C	C	—	—	U	—	—
Ipomopsis pumila	—	—	U	—	C	—	C	—	C	—	U	C	—	—
Leptodactylon pungens	C	C	C	C	C	—	C	—	—	—	—	—	—	—
Microsteris gracilis	—	—	—	C	C	L	C	C	L	C	—	C	C	—
Phlox austromontana	—	C	C	C	C	—	C	—	C	—	—	C	C	—
Phlox hoodii subsp. canescens	C	C	—	U	C	—	C	C	C	C	C	C	C	—
Phlox longifolia	—	—	—	—	—	—	—	—	C	U	—	U	C	—

Family / Species	1	2	3	4	5	6	7	8	9	10	11	12	13	14
POLYGONACEAE														
Androsace septentrionalis	—	—	—	—	—	—	—	—	—	Rm	L	Rm	U	—
Eriogonum corymbosum	—	U	—	—	—	L	—	—	—	—	—	—	—	—
Eriogonum jamesii	C	C	C	—	—	C	C	—	—	C	—	—	—	—
Eriogonum lonchophyllum	C	C	C	O	C	—	C	C	—	U	—	—	C	—
Eriogonum racemosum	C	C	C	C	C	C	—	C	—	C	U	—	U	—
Eriogonum umbellatum	L	L	L	L	L	—	L	L	—	L	C	L	C	—
Polygonum douglasii	—	—	—	—	—	—	—	L	—	R	R	L	—	—
Pterogonum alatum	—	—	—	—	R	—	—	—	—	—	R	—	—	—
Rumex triangulivalvis	—	—	—	—	—	—	—	—	—	—	—	—	—	—
PORTULACACEAE														
Claytonia lanceolata	—	—	—	—	—	—	Dr	—	—	C	C	—	—	—
RANUNCULACEAE														
Atragene pseudoalpina	—	—	Rm	—	R	—	—	—	—	—	Rm	—	—	—
Halerpestes cymbalaria subsp. *saximontana*	—	—	—	—	—	—	—	—	—	—	—	—	—	—
RHAMNACEAE														
Ceanothus fendleri	—	—	—	—	—	—	—	R	L	—	—	—	—	—
ROSACEAE														
Amelanchier utahensis	C	U	R	C	C	C	U	U	U	U	U	R	C	—
Cercocarpus montanus	U	R	R	L	L	L	D	U	D	U	U	—	L	—
Padus virginiana subsp. *melanocarpa*	—	—	—	U	—	—	—	—	—	—	U	—	—	—
Peraphyllum ramosissimum	R	—	—	—	L	—	—	—	—	L	L	—	—	—
Purshia stansburiana	—	—	—	—	—	R	R	—	—	—	R	—	—	—
Purshia tridentata	C	C	C	U	—	—	—	U	U	U	U	U	D	—
Rosa woodsii	R	—	—	U	—	U	—	—	—	C	C	—	U	—
RUBIACEAE														
Galium coloradoense	—	—	Rc	C	C	—	—	—	—	—	—	—	—	—

continued on next page

APPENDIX 3.1—continued

	Loess upper	Loess mid	Loess low	Menefee Shale	Mancos Shale	Green-horn	Cliff House Sandstone	Pt. Lookout Sandstone	Dakota Sandstone	Alluvium upper	Alluvium low	Colluvium	Minette	Morrison
SALICACEAE														
Lycium pallidum	—	—	—	—	C	—	—	—	—	U	U	—	—	—
Populus acuminata	—	—	—	R	—	—	R	—	—	—	—	—	—	—
Populus angustifolia	—	—	—	—	—	—	—	—	—	C	U	U	—	—
Populus deltoides subsp. *wislizenii*	—	—	—	R	—	R	R	U	C	C	C	—	—	—
Populus tremuloides	—	—	—	—	—	—	—	R	—	—	—	—	—	—
Salix exigua	—	—	—	—	—	—	—	—	—	C	C	—	—	—
Salix lucida subsp. *lasiandra*	—	—	—	—	—	—	—	—	—	U	—	—	—	—
SANTALACEAE														
Commandra umbellata	C	C	U	U	U	L	—	—	—	U	U	C	U	—
SAXIFRAGACEAE														
Heuchera parvifolia	—	—	—	L	R	—	Rm	Rm	Rm	—	—	—	R	—
Pseudotsuga menziesii	—	—	—	—	—	—	—	—	—	—	—	—	—	—
SCROPHULARIACEAE														
Castilleja chromosa	U	—	—	C	C	—	U	—	—	—	—	C	—	—
Castilleja linariifolia	—	U	C	C	C	U	C	C	C	C	U	—	—	—
Colinsia parviflora	U	U	U	U	C	—	U	?	U	U	U	U	—	—
Cordylanthus wrightii	C	—	—	C	C	—	C	—	—	—	—	U	U	U
Pedicularis centranthera	—	—	—	—	—	—	—	—	—	—	—	—	—	—
Penstemon angustifolius var. *vernalensis*	—	—	—	C	C	—	—	—	C	—	—	C	—	—
Penstemon barbatus	—	—	—	—	—	—	—	—	—	—	—	—	—	—
Penstemon crandallii	—	—	—	C	C	—	C	—	C	—	—	C	—	—
Penstemon eatonii	—	—	—	C	C	—	C	—	C	—	—	C	C	U
Penstemon linarioides subsp. *coloradoensis*	U	C	U	C	C	U	U	U	C	R	R	U	U	—

Species												
Penstemon rostriflorus	C	—	—	—	—	—	—	—	—	—	—	—
Penstemon strictus	—	C	—	U	—	U	U	C	—	—	C	—
SOLANACEAE												
Chamaesaracha coronopus	—	—	T	—	R	—	—	—	R	—	—	—
Physalis virginiana	—	—	—	—	—	—	—	—	D	—	U	—
TYPHACEAE												
Typha domingensis	—	—	—	—	—	L	—	L	—	—	—	—
Typha latifolia	—	—	—	—	—	—	—	—	C	—	C	—
VISCACEAE												
Arceuthobium divaricatum	—	—	L	—	L	—	—	—	—	—	—	—
Phoradendron juniperinum	C	C	U	R	R	—	—	—	—	U	—	—

Key: Definition of abundance and distribution terms applied to plants of Mesa Verde Country

Plants with less than 1% cover

R = Rare: very few stands, or one or few known locations
L = Localized: found in specific soil types
E = Extirpated: no longer found in previously known locality

Plants with 1% to 5% cover

Rm = Restricted moist: confined to moist drainages or hydric sites
Rc = Restricted calcareous: confined to calcareous substrates
Ru = Restricted upper: confined to higher elevations of the geologic type
O = Restricted oyster beds: confined to oyster bed substrates
Ri = Restricted riparian: confined to riparian habitats
Dr = Disturbed, restricted: confined to disturbed sites

Plants with greater than 5% cover

U = Uncommon: 5 to 25% cover
C = Common: 25 to 50% cover
Ab = Abundant: > 50% cover
D = Disturbed: high cover in disturbed habitats
P = Postfire: high cover following fires, otherwise uncommon

APPENDIX 3.2. Plants of concern, Mesa Verde National Park

Species by Family	Common Name	Number of Known Locations	Status
ACERACEAE			
Acer grandidentatum	Bigtooth maple	1	G4, S1
ADIANTACEAE			
Adiantum aleuticum	Maidenhair fern	1	G5, S2
AGAVACEAE			
Yucca harrimanae	Narrow leafed yucca	21	
ALISMATACEAE			
Sagittaria cuneata	Arrowleaf	5	SQ, Park
ALLIACEAE			
Allium cernuum	Nodding lavender onion	2	Park
Androstephium breviflorum	Funnel lily	2	SQ, Park
ALSINACEAE			
Moehringia macrophylla	Big leaf sandwort	2	Park
AMARANTHACEAE			
Amaranthus retroflexus	Creeping pigweed	2	Park
ANACARDIACEAE			
Toxicadendrons rydbergii	Poison ivy	30	SQ, Park
APOCYNACEAE			
Amsonia jonesii	Pink ansomia	2	SQ
Apocynum androsaemifolium	Spreading dogbane	2	SQ, rare in SW CO
Apocynum cannabinum	Indian hemp	7	SQ, Park
ASCLEPIADACEAE			
Asclepias asperula	Yerba de inmortal	7	SQ, Park
Asclepias speciosa	Showy milkweed	10	Park
ASPIDIACEAE			
Dryopteris filix-mas	Male fern	1	Relic, SQ, Park
ASTERACEAE			
Acrolasia albicaulis	Golden blazing star	1	Park, SQ
Ageratina herbacea	Western joe pye weed	3+	SQ, Park
Agoseris aurantiaca	Orange mountain dandelion	2	Park
Agoseris glauca	Yellow mountain dandelion	1	Park
Ambrosia tomentosa	Skeletonleaf bursage	3	S1
Antennaria microphylla	Midget pussytoes	1	Park
Bahia dissecta	Cutleaf bahia	2	Park
Brickellia brachyphylla	Short-leaf bricklebush	2	Park
Brickellia grandiflora	Grand bricklebush	10	SQ, Park
Brickellia microphylla subsp. scabra	Littleleaf bricklebush	8	SQ, Park
Brickellia oblongifolia	Narrowleaf bricklebush	1+	SQ, Park
Lactuca tatarica subsp. pulchella	Purple wild lettuce	3	Park
Lygodesmia grandiflora	Purple skeleton lettuce	4	Park
Malacothrix sonchoides	False dandelion	1	Park
Oreocarpa humilis	Mat miner's candle	1	
Packera neomexicana	Early ragwort	2	Park

continued on next page

APPENDIX 3.2—*continued*

Species by Family	Common Name	Number of Known Locations	Status
Picrothamnus desertorum	Spiny sagebrush	4	Park
Pseudognaphalium viscosum	Cudweed	4	SQ, Park
Psilochenia modocensis	Hawksbeard	1	Park
Rudbeckia ampla	Big coneflower	3	Park
Senecio ambryoides	Montane ragweed	1	Park
Senecio longiflobus	Desert groundsel	1	Park
Seriphidium pygmaeum	Pygmy sagebrush	1	Park
Stephanomeria pauciflora	Pink desert skeletonweed	1+	S2
Suffruticosi flaccidus var. douglasii	Narrowleaf ragwort	2+	Park
Tetradymia spinosa	Cottonhorn	1	Park
Thelesperma subnudum	Navajo tea	2	SQ
Triangulares eremophilus	Tall ragweed	3	Park
Virgulaster ascendens	Montane aster	7	Park
Virgulus falcatus	White fall aster	1	Park
Wyethia scabra	Rough mule's ear daisy	1	SQ, Park, S2
ATHYRIACEAE			
Cystopteris fragilis	Brittle fern	1	SQ, Park
BERBERIDACEAE			
Berberis fendleri	Barberry	6	SQ, Park; rare in W Co.
BETULACEAE			
Alnus incana	Alder	7	SQ, Park
Betula fontinalis	River birch	3	State plant communiy of concern
BORAGINACEAE			
Cryptantha confortifolia	Tufted miner's candle	1	
Cryptantha fendleri	Fendler's miner's candle	1	SQ, Park
Hackelia gracilenta	Mesa Verde stickseed	5	G2, S2
Lithospermum multiflorum	Yellow grommel	1	SQ, Park
Mertensia lanceolata	Narrowleaf bluebell	1	Park
Myosotis asiatica	Wild forget-me-not	1	Park
Oreocarya flava	Yellow miner's candle	various	SQ, Park
BRASSICACEAE			
Camelina rumelica	False flax	1	Park
Descurainia pinnata	Short pod tansy mustard	1	Park
Rorippa teres	Cress	1	SQ, Park
Thelypodium integrifolium	Giant thelypodium mustard	3	SQ, Park
CACTACEAE			
Coryphantha missouriensis	Nipple cactus	1	SQ, rare in SW CO
Cylindropuntia whipplei	Rattail cactus	7	SQ, Park
Echinocereus triglochidiatus	Hedgehog cactus	20+	SQ, Park
Opuntia fragilis var. barchyarthra	Brittle cactus	4	SQ, Park

continued on next page

APPENDIX 3.2—*continued*

Species by Family	Common Name	Number of Known Locations	Status
Pediocactus simpsonii	Simpson's hedgehog cactus	3	SQ
Sclerocactus mesae-verdae	Mesa Verde cactus	2	G2, S2
CALOCHORATACEAE			
Calychortus flexuosus	Flexing sego lily	4+	SQ, park
CANNABACEAE			
Humulus lupulus subsp. americanus	Wild hops	4	SQ, rare in SW CO
CAPPARACEAE			
Cleome lutea	Yellow beeweed	1	Park
Cleome serrulata	Rocky Mountain beeplant	3	Park
CAPRIFOLIACEAE			
Sambucus coerulea	Blue elderberry	1	Park
Sambucus microbotryus	Elderberry	2	Park
CARYOPHYLLACEAE			
Anotites menziesii	Douglas fir campion	1	SQ, Park
Silene antirrhina	Catchfly	1	SQ, Park
CHENOPODIACEAE			
Atriplex grayi	Spiny hopsage	4	Park
Atriplex patula	Saline orache	1	Park
Chenopodium glaucum	White goosefoot	1	Park
Monolepis nuttalliana	Poverty weed	1	Park
Proatriplex pleiantha	Mancos saltbrush	1	G4
Suaeda moquinii	Suaeda	3	S1
CONVALLARIACEAE			
Maianthemum amplexicaule	Tall Solomon's seal	7	SQ, Park
CORNACEAE			
Swida sericea	Spreading dogwood	4	Park
CRASSULACEAE			
Amerosedum lanceolatum	Stonecrop	3	Park, SQ; vulnerable
CROSSOSOMATACEAE			
Forsellesia meionandra	Spiny greasebush	3	S1, Park
ELAEAGNACEAE			
Shepherdia argentea	Silverberry	3	Park
EQUISETACEAE			
Equisetum arvense	Field horsetail	5	SQ, Park
Hippochaete hyemalis var. affinis	Jointgrass	1	Park
Hippochaete laevigata	Jointgrass	2	Park
EUPHORBIACEAE			
Chamaesyce fendleri	Fendler's spurge	3+	S1, Park
Tithymalus brachyceras	Robust spurge	7	SQ, Park
FABACEAE			
Astragalus deterior	Cliff Palace milkvetch	various	G2, S2
Astragalus humillimus	Mancos milkvetch	2+	G2, S2

continued on next page

APPENDIX 3.2—continued

Species by Family	Common Name	Number of Known Locations	Status
Astragalus lonchocarpus	Long pod milkvetch	1	Park
Astragalus molissimus var. thompsoniae	Milkvetch	1	Park
Astragalus schmolliae	Schmoll's milkvetch	various	G1, S1
Astragalus tortipes	Ute Mountain milkvetch	30	
Glycyrrhiza lepidota	Wild liquorice	various	SQ, Park
Lathyrus eucosmus	St. John's sweetpea	2	SQ, Park
Lotus wrightii	Butter and eggs	2	SQ, Park
Trifolium gymnocarpum	Dwarf clover	3	S1, Park
Vicia ludoviciana var. texana	Little white vetch	1	SQ, Park
FAGACEAE			
Quercus ajoensis	Ajo oak	1	S1
Quercus havardii tuckeri	Havard's oak	3	SQ, Park
Quercus turbinella	Canyon live oak	1	SQ, Park
Quercus undulatum	Wavy leaf oak	2	SQ, Park
FRANKENIACEAE			
Frankenia jamesii	Frankenia	3	S2, Park
FUMARIACEAE			
Corydalis aurea	Golden smoke	3	SQ, Park
GERANIACEAE			
Geranium caespitosum subsp. atropurpurem	Magenta geranium	6	Park
GROSSULARIACEA			
Ribes aureum	Golden currant	4	Park
HELLEBORACEAE			
Aquilegia elegantula	Red columbine	3	SQ, Park
Aquilegia mancosana (a spurless mutation of A. micrantha)	Spurless columbine	?	G2, S2
Delphinium andersonii scaposum	Bicolored delphinium	8	Park
HYPOLEPIDACEAE			
Pteridium aquilinum subsp. lanuginosum	Bracken fern	3	SQ, Park
IRIDACEAE			
Sisyrinchium montanum	Mountain blue-eyed grass	5	Park
JUNCACEAE			
Juncus arcticus subsp. ater	Baltic rush	1	
Juncus interior	Desert rush	1+	Park
Juncus nevadensis	Nevada rush	1	
Juncus nodosus	Nodular rush	1	
JUNCAGINACEAE			
Triglochin maritima	Arrowgrass	1+	Park
LAMIACEAE			
Hedeoma drummondii	Pennyroyal	4	S1, Park
Monarda fistulosa	Beebalm	3	S1, Park
Salvia reflexa	Lanceleaf sage	1	SQ, Park

continued on next page

APPENDIX 3.2—*continued*

Species by Family	Common Name	Number of Known Locations	Status
LILIACEAE			
Fritillaria atropurpurea	Fritillary lily	3	S2, Park
LINACEAE			
Mesynium puberulum	Copper flax	1	SQ, Park?
LOASACEAE			
Acrolasia albicaulis	Golden blazing star	2	Park
Acrolasia humilis	Adobe blazing star	1	
Mentzelia pumila	Little blazing star	1	SQ, Park
MALVACEAE			
Iliamna grandiflora	Wild hollyhock	2+	GQ, S1
Sidalcea candida	White checkers mallow	2	S1, Park
Sidalcea neomexicana	New Mexico mallow	2	S1
MONOTROPACEAE			
Pterospora andromedea	Pinedrop	2	Park
NYCTAGINACEAE			
Mirabilis oxybaphoides	Trailing four o'clocks	4	SQ, Park
Tripterocalyx carneus var. wootonii	Narrowleaf sand verbena	1+	SQ, Park
OLEACEAE			
Forestiera pubescens	Desert adelia	3	Park
Fraxinus anomala	Single leaf ash	4	S2, Park
ONAGRACEAE			
Calylophus lavandulifolius	Changing color calylophus	1	SW CO 1, Park
Camissonia scapoidea	Little evening primrose	1	SQ, Park
Chamerion danielsii	Fireweed	2	Park
Gaura coccinea	Scarlet butterfly weed	1	SQ, Park
Gaura mollis	Tall guara	1	Park
Oenothera albicaulis	Low white evening primrose	3	Park
Oenothera elata subsp. hirsutissima	Tall evening primrose	3	Park
Oenothera flava	Low yellow evening primrose	1+	Park
Oenothera pallida	Whitepole evening primrose	variable	SQ, Park
Oenothera villosa	Villus evening primrose	1	Park
ORCHIDACEAE			
Calypso bulbosa	Calypso orchid	3	S1, Park
Corallorhiza maculata	Coral root (Spotted coral root)	4	Park
Epipactis gigantea	Helleborine	3	G4, S2
Spiranthes diluvialis	Ute Indian tresses	?	G2, S2
OROBANCHACEAE			
Orobanche multiflora	Big broomrape	3	Park
PINACEAE			
Pinus strobiformis	Mexican white pine	6	S1
POACEAE			
Achnatherum speciosum	Desert needlegrass	3	Park
Agrositanion scaxicola	Squirreltail wheat grass	1	SQ, Park
Distichlis stricta	Desert saltgrass	4	Park

continued on next page

APPENDIX 3.2—*continued*

Species by Family	Common Name	Number of Known Locations	Status
Leymus salina	Salina wildrye	6+	SQ, Park
Muhlenbergia montana	Montane muhly grass	3+	Park
Phragmites australis	Common reed	6	Park
POLEMONIACEAE			
Aliciella micromeria	Little blue gilia	1	Park, SQ
Collomia grandiflora	Salmon collomia	variable-fire	G5, S2
Collomia tinctoria	Sticky collomia	1	New CO record in 1997
Gilia polycladon	Yellow gilia	1	
Ipomopsis gunnisonii	Gunnison gilia	3+	S1, SQ, Park
Ipomopsis longiflora	Longleaf gilia	1	SQ, Park
Polemonium foliosissimum	Jacob's ladder	4	SQ, Park
POLYGONACEAE			
Eriogonum divaricatum	Red annual buckwheat	1	
Eriogonum hookeri	Hooker's buckwheat	2	SQ
Eriogonum microthecum var. simpsonii	Simpson's buckwheat	2	SQ, Park
Eriogonum microthecum	Adobe buckwheat	3	SQ?, Park
Eriogonum ovalifolium var. ovalifolium	Round leaf buckwheat	5+	Park
Eriogonum salsulgiuose	Low buckwheat	1	
Persicaria coccinea	Bladderpod smartweed	2	SQ, Park
Rumex hymenosepalus	Canaigre	4	Park
PORTULACACEAE			
Portulaca oleracea	Purslane	3	Park
Talinum parviflorum	Depression purslane	1	S2, Park
PRIMULACEAE			
Primula specuicola	Cliff primrose	1	G2
PYROLACEAE			
Pyrola rotundiflora subsp. asarifolia	Round shinleaf	2	SQ, Park
RANUNCULACEAE			
Batrachium trichophyllum	Water crowfoot	1	SQ, Park
Clematis ligusticifolia	Virgin's bower	12	Park
Coriflora hirsutissima	Leatherflower	4	SQ, Park
Pulsatillo patens	Pasque flower	1	SQ
RHAMNACEAE			
Rhamnus smithii	Smith's buckthorn	1	SQ, Park
ROSACEAE			
Cercocarpus intricatus	Littleleaf mountain mahogany	1	SQ, Park
Geum macrophyllum	Avens	1	Park, SQ
Holodiscus discolor	Rock spirea	1	Park, rare in SW CO
Potentilla hippiana	Many fingered cinquefoil	2	Park
Rosa nutkana	Nutkan's rose	3	Park

continued on next page

APPENDIX 3.2—*continued*

Species by Family	Common Name	Number of Known Locations	Status
Rubus idaeus melanolasius	Wild raspberry	1	Park
RUBIACEAE			
Galium aparine	Cleavers	2	Park, SQ
Galium septentrionale	Northern bedstraw	2	SQ, Park
Galium trifidum subsp. brevipes	Three part bedstraw	1	SQ, Park
SALICACEAE			
Salix lucida subsp. lasiandra	Montane willow	3	Park, SQ
Salix lutea	Bicolored willow	3	Park, SQ
SAXIFRAGACEAE			
Lithophragma tenellum	Woodland nymph	4	Park
SCROPHULARIACEAE			
Orthocarpus luteus	Yellow owl clover	4	Park
Penstemon breviculus	Little penstemon	4	G3, S2
Penstemon comarrhenus	Late blue penstemon	1	Park
Penstemon ophianthus	Lavender penstemon	1	Park
Penstemon parviflorus	Three leaf penstemon	1	G2, S2
Penstemon utahensis	Utah penstemon	1	SQ, Park
SINOPTERIDACEAE			
Cheilanthes feei	Lip fern	12	SQ, Park
Pellaea glabella subsp. simplex	Cliff brake fern	1	S2, Park
SOLANACEAE			
Chamaesaracha coronopus	Chaemasaracha	2	S1, Park
Datura wrightii	Moon flower	various	SQ, Park
Lycium barbarum	Matrimony vine	1	SQ, Park
Nicotiana attenuata	Coyote tobacco	various-fire	Park, SQ
Physalis hederifolia	Big leaf ground cherry	2	SQ, Park
Solanum jamesii	Wild potato	3	SQ
Solanum triflorum	Cutleaf night shade	5	Park
THALICTRACEAE			
Thalictrum fendleri	Meadow rue	4	Park
TYPHACEAE			
Typha angustifolia	Narrow-leaved cat-tail	3	S1, Park
UMBELLIFERAE			
Aletes macdougalii breviradiatus	Mesa Verde aletes	8	G3, S2
Cymopterus fendleri	Yellow biscuitroot	1+	Park
Ligusticum porteri	Osha	4	Park
Osmorhiza depauperata	Sweet cicely	3	Park
URTICACEAE			
Urtica gracilis	Stinging nettle	3	Park
VALERIANACEAE			
Valeriana edulis	Valerian	5	SQ, Park
VIOLACEAE			
Viola adunca	Purple violet	2	Park
Viola nuttallii	Dwarf yellow violet	2	SQ, Park

continued on next page

APPENDIX 3.2—*continued*

Species by Family	Common Name	Number of Known Locations	Status
Viola rydbergii	Marsh violet	1	SQ, Park
Viola sororia	Mottled violet	1	SQ, Park
VITACEAE			
Parthenocissus vitacea	Virginia creeper	3	Park, SQ
WOODSIACEAE			
Woodsia oregana subsp. *cathcartiana*	Woodsia fern	2	SQ, Park

Key:
Relic: few individuals, once common
Park: park-sensitive, Mesa Verde National Park
S1: critically imperiled in state (Colorado)
S2: imperiled in state because of rarity
S3: vulnerable in state
SQ: state status uncertain
G1: critically imperiled globally because of extreme rarity
G2: imperiled globally because of rarity
G3: vulnerable throughout range or local in restricted range
G4: secure globally but rare in parts of range
G5: secure globally but quite rare in parts of range, especially periphery
GQ: uncertainty about taxonomic status

REFERENCES CITED

Austin, D. D. 1999. Changes in plant composition within a pinyon-juniper woodland. In S. B. Monsen and R. Stevens (comps.), *Proceedings: Ecology and management of pinyon-juniper communities within the interior West*, September 15–18, 1997, Provo, Utah. Proc. RMRS-P-9. USDA Forest Service Rocky Mountain Research Station, Ogden, Utah.

Brown, N.A.C., and J. Van Staden. 1997. Smoke as a germination cure: A review. *Plant Growth Regulation* 22: 115–124.

Davenport, D. W., D. D. Breshears, B. P. Wilcox, and C. D. Allen. 1998. Viewpoint: Sustainability of piñon-juniper ecosystems—A unifying perspective of soil erosion thresholds. *Journal of Range Management* 51: 231–240.

Fleischner, T. K. 1994. Ecological costs of livestock grazing in western North America. *Conservation Biology* 8: 629–644.

Floyd-Hanna, L., D. Hanna, C. Funicelli, M. Colyer, and G. San Miguel. 1999. The role of fire and noxious weed infestations on rare plants. Mesa Verde National Park, Colorado Natural History 1999 Small Grants Program. Final Report on file at Mesa Verde National Park and Colorado Natural History Association.

Floyd-Hanna, L., W. H. Romme, and D. Hanna. 1993. Vegetation of Mesa Verde National Park. Final Report on file at Mesa Verde National Park.

Huber, A., S. Goodrich, and K. Anderson. 1999. Diversity with successional status in the pinyon-juniper/mountain mahogany/bluebunch wheatgrass community type near

Dutch John, Utah. In S. B. Monsen and R. Stevens (comps.), *Proceedings: Ecology and management of pinyon-juniper communities within the interior West,* September 15–18, 1997, Provo, Utah. Proc. RMRS-P-9. USDA Forest Service Rocky Mountain Research Station, Ogden, Utah.

Keeley, J. E., B. A. Morton, A. Padres, and P. Trotter. 1995. Role of allelopathy, heat and charred wood in the germination of chaparral herbs and suffrutescents. *Journal of Ecology* 73: 445–458.

Miller, R., R. Tausch, and W. Waichler. 1999. Old-growth juniper and pinyon woodlands. In S. Monsen and R. Stevens (comps.), *Proceedings: Ecology and Management of pinyon-juniper communities within the interior West,* September 15–18, 1997, Provo, Utah. Proc. RMRS-P-9. USDA Forest Service Rocky Mountain Research Station, Ogden, Utah.

Naillon, D., K. Memmott, and S. B. Monsen. 1999. A comparison of understory species at three densities in a pinyon-juniper woodland. In S. B. Monsen and R. Stevens (comps.), *Proceedings: Ecology and management of pinyon-juniper communities within the interior West,* September 15–18, 1997, Provo, Utah. Proc. RMRS-P-9. USDA Forest Service Rocky Mountain Research Station, Ogden, Utah.

Orodho, A. B., M. J. Trlica, and C. D. Bonham. 1990. Long-term heavy-grazing effects on soil and vegetation in the Four Corners region. *Southwestern Naturalist* 35: 9–14.

Preston, C. A., and I. T. Baldwin. 1999. Positive and negative signals regulate germination in the post-fire annual, *Nicotiana attenuata. Ecology* 80: 481–494.

Spackman, S., B. Jennings, J. Coles, C. Dawson, M. Minton, A. Kratz, and C. Spurrier. 1997. *Colorado rare plant field guide.* Prepared for the Bureau of Land Management, U.S. Forest Service, and the U.S. Fish and Wildlife Service by the Colorado Natural Heritage Program, Denver.

Weber, William A., and Ronald C. Wittman. 1996. *Colorado Flora: Western Slope.* Rev. ed. University Press of Colorado, Niwot.

AT THE GROUND LEVEL: FUNGI AND MOSSES

Jayne Belnap and J. Page Lindsey

Under the canopy of ancient piñon and juniper trees thrives a rich understory community. Flowering forbs, grasses and small shrubs, and the vascular plants described in Chapter 3 are only part of the woodland floor. The nonvascular biota—fungi, lichens, and bryophytes—are equally well developed in the ancient woodlands. We know less about the nonvascular species that occur, yet the importance of these groups for decomposition and nutrient cycling (fungi), nitrogen fixation (lichens), and understory diversity (bryophytes) cannot be overstated. In this chapter, we provide necessary background information on the natural history and biological uniqueness of the relatively unknown nonvascular plants, introduce adaptations to arid conditions, and describe the fungi and bryophytes known from Mesa Verde Country.

INTRODUCING FUNGI, LICHENS, AND BRYOPHYTES

Fungi are ubiquitous, occurring in almost all environments across the globe. We generally see only the fruiting bodies (such as mushrooms or woody conks) of fungi, because most of their body mass is hidden inside plants or in the soil. Free-living fungi are called saprotrophs. Fungi that live in close relationships with other organisms are either mutualistic or parasitic, depending on the nature of the relationship.

Fungi cannot produce their own energy from the sun but instead rely on their ability to decompose plant material (thus they are heterotrophs). Fungi obtain their nutrients by absorbing solubilized organic and inorganic compounds from their environment. Most fungi are composed of highly branched, cylindrical filaments called hyphae, which collectively make up a mat called the mycelium. These hyphae weave throughout the substrate, encountering and absorbing nutrients. Although hyphae of different species may superficially appear as similar thin-walled, highly branched tubes, they differ greatly in cell wall composition and structure; in the number, type, and position of their nuclei; and in the type of septa (cross-walls) present.

Fungi secrete enzymes to break down the surrounding substrate, releasing nutrients that are then absorbed. These enzymes may soften the surrounding tissue, making it easier for the fungi to penetrate farther. The long, narrow hyphae have a large surface:volume ratio, making them very efficient at absorbing nutrients. As the hyphal tip grows, the older parts die when nutrients around them are exhausted and toxic waste products build up. Once dead, these hyphal sections are sealed off by septa.

All fungi are an integral part of the decomposition biota. Of particular interest are the mycorrhizal fungi, which form associations with many vascular plants. The fungal hyphae mine the surrounding soils for water and nutrients, which they provide to the host plant. Piñon and juniper are known to have mycorrhizal associations: among the most common mycobionts are *Amanitopsis vaginata* (Fr.) Vitt and various species of *Inocybe* and *Russula*.

Some fungi form large "conks" on living trees; others are located on the underside of slash or the inside of stumps. Many of the slash-rotting fungi are inconspicuous but quite active decomposers of wood. There are two main groups of wood-rotting fungi: the brown rot fungi (which only remove cellulose from the wood) and the white rot fungi (which remove cellulose and lignin from the wood). Wood-rotting fungi have been surveyed in Mesa Verde National Park (Appendix 4.1 and the following discussion). In addition to their critical value as decomposers, these fungi are of special interest in a park like Mesa Verde, where wood fragments on the ground are often very important in dating remains or defining other characteristics of archaeological sites. As is true across the Colorado Plateau, on Mesa Verde conditions for growth of fungi are limited to the times of year when moisture is available and temperatures are favorable (usually late July through September). Most of the rest of the year the conspicuous fleshy fungi such as the mushrooms are simply not visible. Wood-rotting fungi can make use of moisture held in pieces of wet wood for longer periods, and many of them can actually fruit at near-freezing temperatures. Thus decomposition of wood may occur almost year-round or under snow in some situations, albeit slowly.

Lichens are fungi that capture and cultivate green algae and/or cyanobacteria to produce carbohydrates for the lichens photosynthetically. The association with green algae creates a phycolichen, whereas the partnership with cyanobacteria produces a cyanolichen. In addition to producing carbohydrates, the cyanobacteria "fix" nitrogen in the cyanolichen by transforming atmospheric nitrogen (N_2) into biologically usable forms (ammonium, nitrates, and nitrites). The lichen fungus, in return, is thought to provide protection from desiccation and thus benefits the alga or cyanobacterium, although benefits to the algal partner have not been demonstrated definitively. In addition, the free-living photosynthetic partner can almost always be found on substrates adjoining the lichen. Therefore, many scientists believe that the lichen relationship is not a symbiotic one as previously believed but instead is a parasitic one, with the fungi being the parasite (Purvis 2000). In any case, the association between a fungus and a photosynthetic partner results in a new morphological entity and is named for the fungal partner, as the same algae or cyanobacteria can occur with a variety of fungi. There are over 20,000 known species of lichens, which frequently thrive in extreme environments, where they often represent most of the living biomass.

Lichens come in a wide variety of shapes, sizes, and colors. They generally have an outer fungal layer and in cross section reveal a layer of algal cells. Lichens can cover the soil surface like a layer of skin (called a crustose or squamulose growth form) or can be three-dimensional and leaflike in appearance (called a fruticose or foliose life-form). They can be green, yellow, pink, red, orange, brown, white, or black. Sexual reproduction is limited to the fungal partner in this relationship. The reproductive structures are generally round, dark-colored disks on the lichen surface. Lichens can also reproduce asexually and produce either small or large fragments that easily break off from the main lichen thallus. Lichens with asexual reproduction can colonize disturbed sites much more rapidly than those species that lack asexual reproduction, because the green algal or cyanobacterial partners may not be readily available for capture by the fungi.

Lichens can play many important roles in the ecosystems in which they occur, especially when they are part of biological soil crusts (Chapter 5). All lichens contribute carbon to the surrounding soils, which can then be utilized by microbes; as noted, cyanolichens contribute nitrogen to ecosystems as well. Lichens also promote soil stability by gluing soil particles together (Appendix 4.2).

Mosses, liverworts, and hornworts are ancient plants that are collectively called bryophytes. In comparison with vascular plants, bryophytes are less complex in structure, are much smaller, and do not usually possess specialized tissue to transport water, nutrients, and sugars throughout the plant. All bryophytes have a similar life cycle, which is an alternation between asexual and sexual generations. Because the sexual structures are often small and hidden, these organisms are called cryptogams ("crypto" means hidden and "gams" represents gametophytes).

Liverworts are the simplest of all land plants and occur in two general forms: thalloid and leafy. The thalloid form has a greenish black–colored thallus or flat, narrow ribbon of dichotomously branching material. Thalloid liverworts are pressed close to the soil surface and can fold in half, almost disappearing from view; they look like thin black lines on the soil surface when dry. Leafy liverworts are rare in arid environments. They can look like mosses but are smaller, and the leaves are very black when dry. Liverworts have rhizoids that anchor them to the soil surface.

Mosses are generally easy to identify because they have greenish-colored leaves when moist and look just like miniature vascular plants with stems and leaves. There are different types of mosses in Mesa Verde Country: most are either "tall" or "short" mosses, which, as their description implies, have long or short stems, respectively (Appendix 4.3). Like vascular plants, mosses can be either annuals or perennials. Unlike vascular plants, their rootlike rhizoids have a limited ability to conduct water to the leaves. Mosses generally reproduce by spore capsules that rise above the leaves. In an arid environment, however, mosses often lack sexual reproductive structures and usually propagate asexually by simple or specialized fragmentation. Mosses often curl up as they dry, exposing a heavily protected underside to the sun that can resist the damaging effects of solar radiation.

HOW DO THESE ORGANISMS ADAPT TO DESERTS?

Fungi, lichens, and bryophytes share several important features that allow them to live in deserts and semiarid woodlands like those found in Mesa Verde Country. Unlike vascular plants that either die or must regrow new tissue, these nonvascular forms are able to dry out completely without any negative effects and are therefore referred to as poikilohydric. Poikilohydric organisms have small cells with many small water vacuoles scattered throughout each cell. Most poikilohydric organisms equilibrate their water content with atmospheric humidity or soil surface moisture content. These organisms are only active when wet and are able to become metabolically active very quickly upon wetting. This is important in semiarid woodlands, where wet times can be very short. Bryophytes and lichens can begin photosynthesizing almost immediately after wetting, often producing usable products within minutes.

Whereas fungi live mostly below the soil surface when they are not reproductive and thus are protected from solar radiation, bryophytes and lichens are often constantly exposed to harmful solar rays. They have several adaptations to deal with this. First, they generally have heavy ultraviolet-protective pigmentation in the tissue that is most exposed to the sun. In lichens this is their top layer. In liverworts and mosses, which both fold up as they dry, this is generally the back of the leaves. Second, dry tissue is much less susceptible to damage than hydrated tissue. Thus the ability to dry completely protects these species from literally boiling to death during hot summer days. When their surfaces are wetted on hot days, convective air cooling drops their temperature quickly and reduces tissue damage. There is a point during drying, however, when tissue damage frequently occurs. In addition, quick drying at high air temperatures often results in little carbon gain for these organisms. For these reasons, most of these organisms can be actually harmed rather than benefited by summer rains. Luckily, even in deserts where summer rains are frequent, a given spot on the ground generally does not receive many rain events, and thus these organisms can survive in this challenging environment.

FUNGI, BRYOPHYTES, AND LICHENS OF MESA VERDE COUNTRY

Fungal species were systematically collected and identified in Mesa Verde National Park by Dr. J. Page Lindsey and others (Appendix 4.1), and the species that occur in the old-growth piñon-juniper woodland are summarized in Table 4.1.

Table 4.1. Fungal species identified in Mesa Verde National Park from old-growth piñon-juniper woodlands of Chapin Mesa, Wetherill Mesa, or Bobcat Canyon

Species	Family	Common Habitat	Ecological Role
Acanthophysium cerrusatum (Bres.) Boidin	Corticiaceae	juniper slash	wr
Acanthophysium mesaverdense (Lindsey) Boidin	Corticiaceae	underside of live piñon limbs	wr
Amanitopsis vaginata (Fr.) Vitt	Agaricaceae	soil under piñon	m
Antrodia albida (Fr.) Donk	Polyporaceae	juniper slash	br
Athelia bombacina Pers.	Corticiaceae	piñon bark	wr
Calvatia booniana A. H. Smith	Lycoperdaceae	sandstone under piñon	s
Clitocybe sp.	Agaricaceae	decayed piñon wood	m or s
Geaster coronatus (Schaeff.) Schroet.	Lycoperdaceae	soil under piñon	s
Gloeophyllum sepiarium (Wulf ex Fr.) Karst	Polyporaceae	piñon slash	br
Gymnosporangium nelsonii	Pucciniaceae	live juniper branch	op
Gymnosporangium speciosum	Pucciniaceae	live juniper branch	op
Inocybe sp.	Agaricaceae	soil under piñon	m or s
Lactarius barrowsii Hesler and Smith	Agaricaceae	soil and duff under piñon	m
Mycena sp.	Agaricaceae	soil under piñon and juniper	s
Ophiostoma wagneri Goheen and Cobb	Ophiostomataceae	live piñon butt and roots	p
Paxina acetabulum (L.) Kuntze	Pezizaceae	soil under piñon	s
Pyrofomes demidoffii (Lev.) Kotl. et Pouz.	Polyporaceae	live juniper trunk	wr
Pyronema domesticum	Pyronemataceae	burned soils under piñon	s
Russula sp.	Agaricaceae	soil under piñon	s
Russula xerampelina Fr.	Agaricaceae	soil under piñon/oak	s
Suillis americanus (Pk.) Snell	Boletaceae	soil under piñon	m or s
Trechispora mollusca (Pers.:Fr.) Liberta	Corticiaceae	piñon log	wr
Tubulicrinus glebulosis (Bres.) Donk	Corticiaceae	piñon slash	wr?
Tubulicrinus medius (Bourd. & Galzin) Oberw.	Corticiaceae	wood in soil under piñon/juniper	wr
Tulostoma sp.	Tulostomatales	soil under piñon	s

Note: Collections by M. Colyer and J. Page Lindsey. Identifications by J. Page Lindsey and others as listed in Mesa Verde herbarium. Abbreviations used for ecological role: wr = white rot; br = brown rot; m = mycorrhizal; s = saprophyte; op = obligate parasite; p = parasite.

Several other surveys of fungi have been conducted in the Colorado Plateau region (Gilbertson et al. 1974; Lindsey 1985, 1986a, 1986b, 1988, 1999; States 1990), but none has focused on species restricted to the woodland.

A new species, *Aleurodiscus mesaverdensis* Linds., was initially described from the dripline of live large piñon branches within Mesa Verde National Park (Lindsey 1987). The fungus is not parasitic but seems to be restricted to larger, older piñons. It has been found in several other locations as well.

Following the Chapin 5 fire in Mesa Verde National Park in August 1996, a bright red prostrate fungus was observed in severely burned soils, exclusively under piñon logs or at the base of piñon snags. It was quite widespread in these conditions for several weeks. This species was identified as the ascomycete *Pyronema domesticum* (Fr.) Sacc. It is common on burned soil and greenhouse sterilized soil. *Pyronema* reappeared throughout the first two years of postfire succession, confined to wet microsites such as the undersurface of decaying logs.

The activity of one fungal pathogen in particular has allowed periodic gaps to occur in the old-growth woodland, enhancing the richness of habitats for birds and other foragers. These small gaps create openings in the canopy, which are characteristic of Mesa Verde's old-growth forest. *Leptographium wageneri* (black-stain root disease) has destroyed small patches (10–50 trees, often more than 1 hectare) of mature piñons during the last sixty years (Harrington and Cobb 1988; Russel Wagner, unpublished data). Mostly large trees are affected, but saplings may also die from black-stain root disease. After the tree is killed, wood-rotting fungi invade the intact wood and weaken it, making it a favorable site for nest preparation by woodpeckers. Woodpeckers are known to "sound out" wood and avoid drilling directly into intact wood. Thus the wood-rotting fungi play an extremely important role in preparing snags for use by these birds and by other animals that may subsequently benefit from their abandoned nest sites (Gilbertson 1980). Within the openings created by the black-stain root disease, the understory thrives on increased light and moisture.

Other fungi that are common on piñon but have not been reported from Mesa Verde include *Pleurotus ostreatus* Jacq.:Fr., *Phellinus pini* (Thore.:Fr.) A. Ames, *Cryptoporus volvatus* (Pk.) Shear, *Pycnoporus cinnabarinus* (Jacq.:Fr.) Karst., *Trichaptum albietinum* (Dicks.:Fr.) Ryv., and *Hyphodontia arguta* (Fr.) J. Erikss., all white rot fungi, and *Neolentinus ponderosus* (O.K. Miller) Redhead & Ginns, a brown rot fungus. *Phellinus pini* causes a white trunk heartrot of live piñon, making older trees become hollow. *Cryptoporus volvatus* invades and fruits abundantly on the trunk of recently killed piñons, especially beetle kills. Apparently it is an early pioneer species in the decay of the piñon wood, producing small, egglike white conks that last only one season.

Other fungi on juniper that are common in the area but have not been reported from Mesa Verde are *Cylindrobasidium corrugum* (Burt) Ginns and *Ceriporia tarda* (Berk.) Ginns, both white rot fungi, and *Antrodia juniperina* (Murr.) Niem. & Ryv., a brown rot fungus.

SUMMARY

The nonvascular plants and fungi are critical components of ecosystems. Thorough studies of the fungi in Mesa Verde National Park (Appendix 4.1) document the species in Mesa Verde Country. Two lichen surveys have been completed in Mesa Verde National Park, so the list of lichens in Appendix 4.2 is fairly complete. Unfortunately, the bryophytes included in Appendix 4.3 probably represent an incomplete listing for the region.

APPENDIX 4.1. Fungi species from Mesa Verde National Park (identifications by P. Lindsey and others listed in Mesa Verde Herbarium)

Species	Common Habitat
Agaricus arvensis	soil
Agrocybe dura	soil in oak thicket under ponderosa pine
Aleurodiscus cerussatus	juniper slash
Aleurodiscus mesaverdensis	underside branches of live piñon
Amanitopsis vaginata	soil under piñon
Antrodia albida	juniper slash
Antrodia juniperina	juniper
Antrodia serialis	juniper
Armillaria ostoyae	live oak, juniper
Athelia bombacina	piñon bark
Boletus chrysenteron	soil under oak
Calvatia booniana	sandstone under piñon
Camarophylllus spp.	soil in piñon-juniper forest
Cercospora sequoiae	juniper
Ceriporia tarda	oak slash
Clitocybe sp.	soil and well-decayed wood under piñon/juniper
Coleosporium jonesii	on *Ribes aureus* in piñon-juniper
Coprinus micaceus	soil in oak thicket
Cortinarius sp.	soil in piñon-juniper forest
Cronartium occidentale	specfic habitat unknown
Crucibulum levis	soil in piñon-juniper forest drainage
Cumminsiella sanguinea	parasite on Oregon grape, under oak
Cylindrobasidium corrugum	juniper
Didymascella tetramicrospora	juniper
Diplocarpon mespili	on host *Amelanchier utahensis*
Diplomitoporus rimosus	juniper
Erysiphe cichoracearum	on host *Gutierrezia sarothrae*
Exidia glandulosa	oak slash
Fibulomyces mutabilis	pine slash
Fomitopsis cajanderi	juniper
Fuligo spp.	soil duff in piñon-juniper forest
Geaster coronatus	soil under piñon/juniper
Gloeophyllum sepiarium	piñon slash

continued on next page

APPENDIX 4.1—*continued*

Species	Common Habitat
Gymnosporangium betheli	juniper, on *Crataegus*, globose galls
Gymnosporangium exiguum	juniper, on *Crataegus*
Gymnosporangium globosum	juniper, on *Crataegus*, globose galls
Gymnosporangium harknessianum	juniper, on *Amelanchier*
Gymnosporangium inconspicuum	juniper, on *Amelanchier, Peraphyllum*
Gymnosporangium juvenescens	juniper, on *Amelanchier*, witches' brooms
Gymnosporangium kernianum	juniper, on *Amelanchier*, witches' brooms
Gymnosporangium multiporum	juniper
Gymnosporangium nelsoni	juniper branch, on *Amelanchier, Peraphyllum*, globose galls
Gymnosporangium nidus-avis	juniper, on *Amelanchier*, witches' brooms
Gymnosporangium speciosum	fusiform swellings on juniper, also on *Fendlera, Philadelphus*
Hebeloma crustuliniforme	soil under oak
Herpotrichia juniperi	juniper
Heterobasidion annosum	juniper
Holmiella sabina	juniper
Hyphodontia crustosa	oak slash
Hyphodontia hastata	oak slash
Hyphodontia subalutacea	pine slash
Inocybe sp.	soil under piñon/juniper
Kabatina juniperi	juniper
Knieffiella fibrosa	pine slash
Lactarius barrowsii	piñon, soil and duff
Leocarpus fragilis	soil in oak thicket under ponderosa pine
Lophodermium juniperinum	juniper
Lycoperdon sp.	soil in oak thicket under ponderosa pine
Microsphaera calocladophora	on leaves of *Quercus gambelii*
Mycena sp.	soil under piñon/juniper
Panus rudis	oak slash
Paxina acetabulum	soil under piñon/juniper
Peniophora pseudoversicolor	specific habitat unknown
Perenniporia ellisiana	on host *Shepherdia argenta*
Pestalotiopsis funerea	juniper
Phanerochaete avellanea	oak slash
Phellinus texanus	juniper
Phomopsis juniperovora	juniper
Phyllactinia angulata	on host *Lupinus* sp.
Phyllosticta sp.	on host *Lupinus* sp.
Phyllosticta solidaginis	on host *Solidago petradoria*
Phymatotrichopsis omnivora	juniper
Polyporus arcularius	oak slash
Puccinia enceliae	on host *Viguiera multiflora*
Puccinia grindeliae	on host *Solidago petradoria*
Puccinia similis	specific habitat unknown
Puccinia sp.	on host *Sitanion hystrix*
Puccinia tumidipes	on host *Lycium pallidum*

continued on next page

APPENDIX 4.1—*continued*

Species	Common Habitat
Pycnoporus cinnabarinus	soil in piñon-juniper forest
Pyrofomes demidoffii	live juniper trunk
Pyronema domesticum	burned soils, under piñon
Rhizoctonia solani	juniper
Russula sp.	soil under piñon-juniper and oak thicket under ponderosa pine
Russula xerampelina	soil under piñon/oak
Sepedonicum sp.	soil under oak
Sistotrema brinkmannii	oak slash
Sphaeropsis sapinea	juniper
Sthughesia juniperi	juniper
Suillis americanus	soil under piñon
Tomentella terrestris	oak slash
Trechispora mollusca	piñon log
Trechispora vaga	oak slash
Tubulicrinis glebulosus	piñon/juniper slash
Tubulicrinis medius	wood in soil under piñon/juniper
Tulostoma sp.	soil under piñon
Verticicladiella wagneri	specfic habitat unknown
Volvariella sp.	soil in piñon-juniper forest
Xeromphalina sp.	soil in oak thicket under ponderosa pine

APPENDIX 4.2. Lichen species from Mesa Verde National Park (identifications by Nash and Sayce 1980)

Species	Common Habitat
Acarospora fuscata	on Cliff House Sandstone
Acaraspora cf. schleicheri	on Cliff House Sandstone
Acaraspora stapfiana	on *Caloplaca trachyphylla*
Acaraspora strigata	on Cliff House Sandstone
Acaraspora cf. utahensis	on Cliff House Sandstone
Agrestia cyphellata	calcareous soils in the piñon-juniper forest
Agrestia hispida	soil
Aspicilia caesiocinerea	on Cliff House Sandstone
Aspicilia calcarea	on Cliff House Sandstone
Bacaida beckhausii	on *Quercus gambelii*
Buellia alboatra	on Cliff House Sandstone
Buellia punctata	on *Quercus gambelii, Pseudotsuga menziesii,* and exposed wood
Buellia retrovertens	on Cliff House Sandstone
Buellia triphragmioides	on juniper wood
Buellia zahlbruckneri	on *Pseudotsuga menziesii*
Caloplaca cf. approximata	on Cliff House Sandstone
Caloplaca arizonica	on juniper bark and wood, and *Quercus gambelii*
Caloplaca aurantiaca	on *Juniperus osteosperma*

continued on next page

APPENDIX 4.2—*continued*

Species	Common Habitat
Caloplaca cerina	on wood and *Quercus gambelii* bark
Caloplaca cerina var. *muscorum*	on pleurocarpous moss
Caloplaca cf. *cinnamomea*	on *Grimmia* (moss)
Caloplaca decipiens	on Cliff House Sandstone
Caloplaca epithallina	on *Dimelaena oreina* and *Rhizoplaca melanophthalma*
Caloplaca ferruginea	on *Pseudotsuga menziesii*
Caloplaca cf. *festiva*	on Cliff House Sandstone
Caloplaca flavovirescens	on Cliff House Sandstone
Caloplaca holocarpa	on *Pseudotsuga menziesii* and *Quercus gambelii*
Caloplaca cf. *modesta*	on Cliff House Sandstone
Caloplaca saxicola	on Cliff House Sandstone
Caloplaca tominii	soil
Caloplaca trachyphylla	on Cliff House Sandstone
Candelariella aurella	on mosses, limey sandstone, and wood
Candelariella deflexa	on *Quercus gambelii*
Candelariella rosulans	on Cliff House Sandstone
Candelariella spraguei	on Cliff House Sandstone
Candelariella vitellina	data missing
Candelariella vitellina var. *assericola*	on *Pseudotsuga menziesii* twigs
Catapyrenium squamulosum	component of the soil lichen crusts in the piñon-juniper forest
Catillaria herrei	on *Peltigera rufescens*
Catillaria kansuensis	on Cliff House Sandstone
Catillaria sp.	on Cliff House Sandstone
Cladonia balfourii	soil and decaying wood
Cladonia coniocraea	piñon-juniper forest
Cladonia fimbriata	mosses and decaying wood
Cladonia pyxidata	soil and mosses
Collema coccophorum	soil in all habitats
Collema conglomeratum	on *Juniperus osteosperma* and *Quercus gambelii*
Collema cristatum	on mosses
Collema furfuraceum	on *Quercus gambelii*
Collema polycarpon	on Cliff House Sandstone
Collema subflaccidum	on sandstone, *Pseudotsuga menziesii*, and *Quercus gambelii*
Collema tenax	soil in all habitats
Collema tuniforme	on mosses over sandstone and soil
Collema undulatum	on soil mosses
Collema undulatum var. *granulosum*	soil and sandstone
Dermatocarpon lachneum	on soil
Dermatocarpon cf. *leptophyllum*	on Cliff House Sandstone
Dermatocarpon miniatum	on Cliff House Sandstone
Dermatocarpon moulinsii	on Cliff House Sandstone
Dermatocarpon plumbeum	on Cliff House Sandstone
Dimelaena oreina	on Cliff House Sandstone
Diploschistes actinostomus	on Cliff House Sandstone
Diploschistes muscorum	soil

continued on next page

APPENDIX 4.2—*continued*

Species	Common Habitat
Diploschistes scruposus	soil, in piñon-juniper forest
Endocarpon pusillum	soil
Fulgensia bracteata	soil
Fulgensia desertorum	soil
Glypholecia scabra	on Cliff House Sandstone
Heppia lutosa	soil, under overhangs
Huilia sp.	on Cliff House Sandstone
Lecania erysibe	on Cliff House Sandstone
Lecania cf. nylanderiana	on Cliff House Sandstone
Lecanora sp.	on Cliff House Sandstone
Lecanora alphoplaca	on Cliff House Sandstone
Lecanora carpinea	on *Pseudotsuga menziesii*
Lecanora cenisia	on Cliff House Sandstone
Lecanora cf. chlarona	on *Pseudotsuga menziesii*, *Quercus gambelii*, and wood
Lecanora christoi	on Cliff House Sandstone
Lecanora dispersa	on Cliff House Sandstone
Lecanora dispersa cf. parasitans	on *Dermatocarpon miniatum*
Lecanora frustulosa	on Cliff House Sandstone
Lecanora garovaglii	on Cliff House Sandstone
Lecanora hageni	on *Quercus gambelii*
Lecanora muralis	on Cliff House Sandstone
Lecanora cf. novomexicana	on Cliff House Sandstone
Lecanora piniperda	on *Pinus edulis*, *Quercus gambelii*, and shrub twigs
Lecanora cf. saligna	on *Pseudotsuga menziesii*
Lecidea atrobrunnea	on Cliff House Sandstone
Lecidea berengeriana	on decaying *Peltigera*
Lecidea tessellata	on Cliff House Sandstone
Lecidella carpathica	on Cliff House Sandstone
Lecidella glomerulosa	on juniper wood, *Pseudotsuga menziesii*, and *Quercus gambelii*
Lecidella sp.	on Cliff House Sandstone
Lecidella stigmatea	on Cliff House Sandstone
Lecidella viridans	on Cliff House Sandstone
Lepraria incana	on Cliff House Sandstone
Leptogium furfuraceum	on *Quercus gambelii*
Leptogium lichenoides	on mosses over sandstone and *Quercus gambelii*
Leptogium tenuissimum	on mosses
Leptogium teretiusculum	on moss over bark
Microthelia sp.	on *Quercus gambelii*
Pachyspora mutabilis	on *Juniperus osteosperma*, *Pseudotsuga menziesii*, and *Quercus gambelii*
Parmelia elegantula	on Cliff House Sandstone, *Pinus edulis*, *Pseudotsuga menziesii*, and *Quercus gambelii*
Parmelia exasperatula	on juniper wood and *Pseudotsuga menziesii*
Parmelia soredica	on juniper wood
Parmelia subolivacea	on juniper wood, *Pinus edulis*, *Pseudotsuga menziesii*, and *Quercus gambelii*

continued on next page

APPENDIX 4.2—*continued*

Species	Common Habitat
Parmelia substygia	on Cliff House Sandstone
Peltigera canina	soil in piñon-juniper forest
Peltigera rufescens	soil
Peltula bolanderi	soil
Pertusaria saximontana	on wood
Phaeophyscia cernohorskyi	soil over sandstone, mosses, juniper wood, and oak
Phaeophyscia ciliata	on *Pseudotsuga menziesii* and *Quercus gambelii*
Phaeophyscia decolor	on mosses
Phaeophyscia orbicularis	on Cliff House Sandstone and *Quercus gambelii*
Phaeophyscia sciastra	on Cliff House Sandstone
Physcia adscendens	on *Quercus gambelii*
Physcia biziana	on *Pseudotsuga* bark and wood
Physcia caesia	on mosses
Physcia dubia	on Cliff House Sandstone, juniper wood, *Pseudotsuga menziesii*, and *Quercus gambelii*
Physcia stellaris	on *Pseudotsuga menziesii* and *Quercus gambelii*
Physconia detersa	on mosses, sandstone, and pine
Placynthium nigrum	on Cliff House Sandstone
Protoblastenia rupestris	on limey sandstone
Psora cerebriformis	soil
Psora crenata	soil
Psora decipiens	soil in piñon-juniper forest
Psora globifera	soil
Psora nipponica	soil
Psora rubiformis	soil, piñon-juniper forest
Psora russellii	soil
Psorotichia cf. schaereri	on Cliff House Sandstone
Rhizocarpon disporum	on Cliff House Sandstone
Rhizocarpon cf. intersitum	on Cliff House Sandstone
Rhizoplaca chrysoleuca	on Cliff House Sandstone
Rhizoplaca melanophthalma	on Cliff House Sandstone
Rinodina cf. ascociscana	on Cliff House Sandstone
Rinodina nimbosa	soil
Rinodina cf. novoconfragosa	on Cliff House Sandstone
Rinodina pyrina	on juniper wood and *Pseudotsuga menziesii*
Rinodina violascens	on Cliff House Sandstone
Sarcogyne privigna	on Cliff House Sandstone
Sarcogyne simplex	on Cliff House Sandstone
Staurothele clopima	on Cliff House Sandstone
Toninia aromatica	soil
Toninia caeruleonigricans	soil
Toninia candida	soil
Toninia cf. lobulata	soil
Toninia sedifolia	calcareous sandstone
Toninia tristis	soil
Usnea fulvoreagens	on *Pseudotsuga menziesii*

continued on next page

APPENDIX 4.2—*continued*

Species	Common Habitat
Usnea hirta	on *Pseudotsuga menziesii*
Usnea subfloridana	on *Pseudotsuga menziesii*
Verrucaria sp.	on Cliff House Sandstone
Verrucaria cf. virens	on Cliff House Sandstone
Xanthoparmelia cf. arseneana	on Cliff House Sandstone
Xanthoparmelia lineola	on Cliff House Sandstone
Xanthoparmelia mexicana	on Cliff House Sandstone
Xanthoparmelia cf. taractica	on Cliff House Sandstone
Xanthoria elegans	on Cliff House Sandstone
Xanthoria fallax	on *Cercocarpus, Pinus,* and *Quercus gambelii*
Xanthoria polycarpa	on *Quercus gambelii*
Xanthoria sorediata	on Cliff House Sandstone

APPENDIX 4.3. Bryophyte species from Mesa Verde National Park

Species	Common Habitat
Amblystegium compactum	soil in moist caves
Bryum argenteum	in soil
Bryum caespiticium	soil
Bryum cuspidatum	under brush thickets
Bryum sp.	with lichen crusts in piñon-juniper forest
Bryum turbinatum	moist coal outcrops
Ceratodon purpureus	under oak thickets
Crossidium aberrans	
Didymodon rigidulus	
Encalypta vulgaris	
Grimmia anodon	sandstone rock
Grimmia apocarpa	bedrock
Grimmia calyptrata	talus slopes
Gymnostomum calcareum	calcareous sandstone seep walls
Hypnum revolutum	data missing
Leskea tectorum	on *Juniperus osteosperma*
Orthotrichum alpestre	on *Juniperus osteosperma*
Orthotrichum hallii	shaded, moist ledges
Orthotrichum jamesianum	under calcareous ledges
Pterygoneurum ovatum	soil
Timmia bavarica	on shaded ledges
Tortula obtusifolia	sandstone and in piñon-juniper forest
Tortula ruralis	in piñon-juniper forest
Tortula subulata	calcareous streambed below an extensive spring
Weisia controversa	on decaying *Juniperus osteosperma*

REFERENCES CITED

Flowers, S. 1973. *Mosses: Utah and the West.* Ed. A. Holmgren. Brigham Young University Press, Provo.

Gilbertson, R. L. 1980. Wood-rotting fungi of North America. *Mycologia* 72: 1–49.

Gilbertson, R. L., K. J. Martin, and J. P. Lindsey. 1974. Annotated check list and host index for Arizona wood-rotting fungi. *University of Arizona Agricultural Experiment Station Technical Bulletin* 209: 1–48.

Harrington, T. C., and F. W. Cobb. 1988. Leptographium *root diseases on conifers.* APS Press, American Phytopathological Society, St. Paul, Minn.

Lindsey, J. P. 1985. Basidiomycetes that decay Gambel oak in southwestern Colorado. *Mycotaxon* 22: 327–362.

———. 1986a. Basidiomycetes that decay Gambel oak in southwestern Colorado: II. *Mycotaxon* 25: 67–83.

———. 1986b. Basidiomycetes that decay Gambel oak in southwestern Colorado: III. *Mycotaxon* 27: 325–345.

———. 1987. A new species of *Aleurodiscus* from Colorado. *Mycotaxon* 30: 433–437.

———. 1988. Annotated check-list with host data and decay characteristics for Colorado wood-rotting Basidiomycotina. *Mycotaxon* 33: 265–278.

———. 1999. Addendum to annotated check-list with host data and decay characteristics for Colorado wood-rotting Basidiomycotina. *Mycotaxon* 70: 99–102.

Nash, T. H., III, and K. Sayce. 1980. *Preliminary study of the lichens of Mesa Verde National Park.* Arizona State University, Department of Botany and Microbiology, Tempe.

Purvis, W. 2000. *Lichens.* Smithsonian Institution Press, Washington, D.C.

States, J. S. 1990. *Mushrooms and truffles of the Southwest.* University of Arizona Press, Tucson.

Welsh, S. L., and J. A. Erdman. 1964. *Annotated checklist of the plants of Mesa Verde, Colorado.* Brigham Young University, Provo.

MAGNIFICENT MICROBES: BIOLOGICAL SOIL CRUSTS IN PIÑON-JUNIPER COMMUNITIES

Jayne Belnap

In the piñon-juniper woodlands of Mesa Verde the open spaces between higher plants are usually covered by biological soil crusts (also referred to as cryptogamic, cryptobiotic, microbiotic, or microphytic soil crusts). Cyanobacterial, green algal, and microfungal filaments weave throughout the top few millimeters of soil, gluing loose soil particles together and forming a coherent skin on the soil surface (Figure 5.1). Once soil surfaces are stabilized, lichens and mosses colonize as well. These crusts are not unique to piñon-juniper woodlands but occur in all hot, cool, and cold arid and semiarid regions of the world. They have only recently been recognized as having a major influence on terrestrial ecosystems (Harper and Marble 1988).

SOIL CYANOBACTERIA: BUILDING BLOCKS OF BIOLOGICAL SOIL CRUSTS

Soil cyanobacteria, one of the major components of biological soil crusts, are prokaryotic oxygenic phototrophic bacteria that often fix nitrogen. Soil cyanobacteria occur on all continents of the world and in most ecosystems, ranging from extreme deserts to tropical rain forests. These organisms, one of the oldest life-forms on earth, are found as 3.5-billion-year-old fossils. They were the dominant life-form on earth for almost 2 billion years (from 2.5 billion to 570 million years ago) and are considered responsible for oxygenating the atmosphere, thus enabling the evolution of oxygen-breathing life-forms (Schopf 1983). As prokaryotes,

Figure 5.1. A scanning electron micrograph (SEM) depicts the components of biological soil crusts (see the text for details on microorganisms).

cyanobacteria have cell walls but no internal membrane-bound structures (e.g., nuclei, chloroplasts), although many species have a thickened mass of material called a central body. Cyanobacteria are either unicellular or filamentous. Most cyanobacteria have a surrounding mucilaginous sheath that protects them from desiccation and sharp sand grains, also binding and chelating nutrients. They move through moist soils by gliding or being mechanically expelled when their surrounding polysaccharide sheath swells with water. These sticky, mucilaginous sheaths also help bind soil particles together, fostering a more stable soil environment for mosses and lichens to become established and thus creating biological soil crusts.

SPECIES COMPOSITION AND GROWTH FORMS

Globally, biological soil crusts have many similarities in species composition, despite occurring in unconnected and seemingly dissimilar environments. Most of the dominant cyanobacteria, lichens, and moss species and genera found in Mesa Verde piñon-juniper woodlands have a cosmopolitan distribution. The cyanobacteria found at Mesa Verde are typical of those found in deserts where rainfall during

Figure 5.2. Microbial crusts develop over many decades, forming thick crusts in the old-growth piñon-juniper woodland, such as these at Mesa Verde National Park.

cool seasons is dominant and hot-season rainfall does not exceed 35% of the total annual rainfall. In these areas *Microcoleus vaginatus,* a large, highly mobile filamentous species, heavily dominates the cyanobacterial species. Also present but smaller and less abundant are the cyanobacterial genera *Scytonema, Lyngbya, Phoridium, Oscillatoria,* and *Nostoc.* Green algae are very low in biomass (less than 1%) but can have a fairly high species diversity (Grondin and Johansen 1993). Common lichens found in Mesa Verde soil crusts include *Fulgensia bracteata, F. desertorum, Squamarina lentigera, Diploschistes muscorum, Psora decipiens, Catapyrenium squamulosum, Toninia sedifolia, Collema tenax, Peltigera rufescens,* and several *Cladonia* species. Common mosses include *Tortula ruralis, Bryum* spp., *Encalypta vulgaris,* and *Ceratodon purpureus.*

Crusts occur in four general growth forms. Hot deserts that lack frost-heaving are generally characterized by smooth cyanobacterial crusts or rugose lichen-moss crusts. In cool deserts where frost-heaving is present, soil crusts with low lichen cover are often pinnacled, due to frost-heaving upward and differential erosion downward. Crusts in cool deserts with a heavy lichen-moss cover have a gently rolling appearance, as frost-heaving and erosion are mitigated by the extensive lichen-moss cover. Both pinnacled and rolling growth forms are found in the piñon-juniper woodlands of Mesa Verde. In the larger plant interspaces found in less productive sites, the crusts are dominated by cyanobacteria, with moss/lichen cover generally less than 20%. Here frost-heaving and erosion result in soil pinnacles up to 10 cm high with a highly complex and greatly increased soil surface area. This is the growth form most commonly found in low-elevation sites on the Colorado Plateau. In the more productive piñon-juniper stands, where plant interspaces are smaller, the crusts are generally dominated by mosses and lichens and are of the rugose form. This growth form is typical of higher-elevation sites of the Colorado Plateau as well as those of the northern Great Basin.

Because the dominant components of biological soil crusts are photosynthetic organisms, they require sunlight. When soils are dry, the bulk of the crustal biomass is up to 0.5 mm below the soil surface, with some individuals found down to 4 mm (Garcia-Pichel and Belnap 1996). While mosses and lichens have UV pigments or heavy coloration to protect them from UV, only some cyanobacteria have such protection. Large filamentous species such as *Microcoleus*, *Lyngbya*, *Phoridium*, and *Oscillatoria* do not have UV-protective pigments and thus are seldom found on the soil surface except on cloudy days when soils are moistened. Cyanobacteria such as *Scytonema* and *Nostoc* do manufacture UV-screening pigments and can be found on the soil surface even when soils are dry.

ECOLOGICAL ROLES

Carbon Fixation

Biological soil crusts are an important source of fixed carbon on sparsely vegetated areas common throughout the West (Beymer and Klopatek 1991). While vascular plants provide organic matter to soils directly beneath them, large spaces between plants have little opportunity to receive such input. Where biological soil crusts are present in these interspaces, carbon contributed by these organisms helps keep plant interspaces fertile and aids in supporting other microbial populations. This is less important in areas with high plant productivity, and more important as plants get farther apart and soils get shallower and/or less fertile.

Nitrogen Fixation

Nitrogen levels are low in desert ecosystems relative to other ecosystems. Surveys in cold deserts have revealed only a few nitrogen-fixing plants (Wullstein 1989). Because nitrogen can limit plant productivity (Ettershank et al. 1978; Nobel et al. 1988), maintaining normal nitrogen cycles is critical to maintaining the fertility of semiarid soils. Most soil crusts in deserts are dominated by complexes of organisms capable of fixing nitrogen, including *Microcoleus*, *Scytonema*, *Nostoc*, and the soil lichen *Collema*. In desert areas where rainfall events are too insignificant to promote plant growth, such rainfall still can stimulate crustal and microbial community activity; thus soil crusts may often be active. As with carbon, crusts contribute nitrogen to soils both underneath plants and in plant interspaces, counteracting the tendency of nutrients to concentrate around plants. From 5 to 88% of nitrogen fixed by crusts has been shown to leak into the surrounding soils (Stewart

1967). Nitrogen leaked from these organisms is available to nearby vascular plants and microbial communities (Mayland and McIntosh 1966). Vascular plants growing in crusted areas show higher leaf concentrations of nitrogen when compared with plants in uncrusted soils. Leaked nitrogen has also been found in associated fungi, actinomycetes, and bacteria (Belnap and Lange 2001).

Soil crusts can be the dominant source of nitrogen for desert shrub and grassland communities (Evans and Belnap 1999; Evans and Ehleringer 1993), where nitrogen-fixing plants generally are locally scarce. Input estimates range from 1 to 365 kg ha^{-1} annually (reviewed in Harper and Marble 1988), with estimates for the Colorado Plateau at about 20 kg ha^{-1} annually. Nitrogen inputs are highly dependent on temperature, moisture, and crustal species composition (Belnap and Lange 2001); thus timing, extent, and type of climatic regimes and past disturbance are critical in determining fixation rates (Belnap 1995, 1996).

Dust Trapping

Dust can be an essential component of desert soil fertility, and soil crusts are effective in capturing eolian dust deposits. Recent work in southeastern Utah shows that dust input significantly increases levels of all major and minor soil nutrients in the tested soils. The bioessential macronutrients (nitrogen, potassium, phosphorus, and magnesium) and micronutrients (copper, iron, molybdenum, and manganese) were enriched up to two to thirteen times the levels in surrounding bedrock (Reynolds et al. 1999).

Effects on Germination and Establishment of Vascular Plants

Soil crusts can influence the location of safe sites for seeds and the germination and establishment of vascular plants. The rough microtopography of both pinnacled and rolling crusts found in Mesa Verde provides many safe sites for seeds. Crusts can also influence seed germination. While small soil cracks can provide favorable conditions for small seeds to germinate, most large-seeded plants need soil or litter cover (or an increase in humidity similar to what litter and soil cover can provide) to germinate. Most native seeds have self-burial mechanisms (such as hygroscopic awns) or are cached by rodents. Many exotic species lack such adaptations, and crusts appear to inhibit their germination. Once seeds germinate, pinnacled or rolling crusts have not been shown to constitute a barrier to root penetration. It should also be kept in mind that seedling germination per se has not been shown to limit species density in desert plant communities. Rather, studies suggest that vascular plant cover in arid lands worldwide is controlled by water and nutrient availability rather than by other site factors (Belnap and Lange 2001).

In cool deserts survival of vascular plants is generally much higher or unaffected in crusted soils compared with uncrusted soils (Belnap and Lange 2001; Harper and Marble 1988; Lesica and Shelley 1992). No studies have shown crusts to decrease vascular plant survival. Many studies have correlated crust cover with vascular plant cover, and results have been variable, with negative, positive, and no relationships found between crust and vascular plant cover (Harper and Marble 1988; Ladyman and Muldavin 1996). At more arid sites correlations between vascular plant cover and cover of crustal components are generally positive, suggesting that plants aid survival of crustal components, especially mosses and lichens, perhaps due to microclimate conditions associated with perennial vegetation (such as decreased soil surface temperatures and increased surface moisture). At higher elevations and/or with higher plant cover it appears that plants inhibit crust cover by restricting the amount of light reaching the soil surface. No study has demonstrated a negative influence of crusts on overall plant cover.

EFFECTS ON NUTRIENT LEVELS IN VASCULAR PLANTS

Plants growing on crusted soils generally show higher concentrations and/or greater total accumulation of various bioessential nutrients when compared with plants growing in adjacent, uncrusted soils, including nitrogen, potassium, sodium, calcium, iron, and magnesium. Dry weight of plants in pots with cyanobacteria are up to four times greater than in pots without cyanobacteria (Harper and Pendleton 1993). Dry weight of plants in untrampled areas can be two times greater than in trampled areas (Belnap 1995; Belnap and Harper 1995; Brotherson and Rushforth 1983; Shields and Durrell 1964).

Several mechanisms have been postulated to explain this effect. Cyanobacterial sheath material is sticky and negatively charged, binding positively charged macronutrients and thus preventing their leaching (Belnap and Gardner 1993; Black 1968). Cyanobacteria secrete chelators that keep iron, copper, molybdenum, zinc, cobalt, and manganese more available in high-pH soils (Lange 1974). Nutrient differences may also result from a thermal effect. Dark-colored crusts warm soils, thus increasing nutrient uptake rates.

SOIL HYDROLOGY AND STABILIZATION

The effect of biological soil crusts on soil water relations is heavily influenced by soil texture, soil structure, and the growth form of the crusts. In hot deserts the presence of the mucilaginous cyanobacteria and surface smoothness can decrease water infiltration. In cool deserts such as Mesa Verde, however, where frost-heaving is common, increased surface roughness can increase water pooling and residence time. As a result, here the presence of soil crusts generally increases the amount and depth of rainfall infiltration (Brotherson and Rushforth 1983; Harper and Marble 1988; Johansen 1993; Loope and Gifford 1972).

Crusts have been shown to reduce soil loss by wind and water erosion in all types of deserts. Polysaccharides extruded by the cyanobacteria, green algae, and microfungi, in combination with lichen and moss rhizomes, entrap and bind soil particles together. As soil aggregates get larger, they are heavier, have a greater surface area, and are more difficult for wind or water to move, thus reducing both wind and water erosion. When wetted, cyanobacterial sheath material swells and covers the soil surface even more extensively than when dry, protecting soils from both raindrop erosion and overland water flow during rainstorms. Resistance to wind erosion parallels biological crust development (Belnap and Gardner 1993; McKenna-Neuman et al. 1996). Soils in arid regions are often highly erodible and soil formation extremely slow, taking 5,000 to 10,000 years or more (Dregne 1983a). Consequently, reducing soil loss is very important in these regions. Soil aggregates are also important for increasing infiltration and as microenvironments for soil biota.

EFFECTS OF DISTURBANCE

Species Composition

Crushing of crusted surfaces by people, livestock, and/or off-road vehicles generally results in a decrease in crustal species present. Untrampled crusts in Mesa Verde often have four to ten species of soil lichens and/or four to six species of cyanobacteria. In contrast, trampled areas often have no lichens and only one species of cyanobacteria (Belnap 1995).

Water Erosion

Crustal components are brittle when dry and easily crushed (Belnap 1993; Campbell et al. 1989; Harper and Marble 1988), so the soil aggregates formed by the presence of soil crusts are disrupted when trampled (Dregne 1983b; Stolzy and Norman 1961). When the roughened microtopography of undisturbed cool desert crusts is flattened, velocity of surface water flows is increased. Thus suspended sediments do not settle out and surfaces are subjected to sheet erosion (Harper and Marble 1988). Surface disturbance also reduces the depth to which abandoned cyanobacterial sheath can accumulate, thereby reducing resistance to water erosion at depth. At many disturbed sites sheath material is often not observed below 1 mm depth, in contrast to crusts up to 10 cm thick in untrampled areas (Belnap 1995). Buried sheath material is still capable of binding soil particles together and still increases nutrient and moisture retention of associated soil. Damage to such abandoned sheath material is nonrepairable, however, since living cyanobacteria are no longer present at these depths to regenerate filament and sheath materials. Consequently, trampling can greatly accelerate desertification processes through increased soil loss and water runoff (Alexander and Calvo 1990; Beymer and Klopatek 1992; Eldridge 1993a, 1993b; Eldridge and Greene 1994; Foth 1978; Harper and Marble 1988; Ladyman and Muldavin 1996).

Wind Erosion

Wind is a major erosive force in deserts, where there is little soil surface protection by organic matter or vegetative cover (Goudie 1978). Experiments have demonstrated that while well-developed undisturbed crusts protect soil surfaces from wind erosion, any compressional disturbance to these crusts leaves surfaces vulnerable to wind erosion (Belnap and Gillette 1997, 1998; Leys 1990; Williams et al. 1995).

Decrease in soil wind resistance is directly associated with increased sediment movement (Leys 1990; Williams et al. 1995). Because soil formation is slow, soil loss can have long-term consequences. In addition, nearby biological soil crusts can be buried by blowing sediment, resulting in the death of the photosynthetic organisms. Because over 75% of the photosynthetic biomass, and almost all photosynthetic productivity, are from organisms in the top 3 mm of these soils, very small soil losses can dramatically reduce site fertility and further reduce soil surface stability (Garcia-Pichel and Belnap 1996). In addition, many plants have relatively inflexible rooting depths and often cannot adapt to rapidly changing soil depths (Belnap 1995, 1996).

Nutrient Cycles

Nitrogenase activity in crusts shows short-term and long-term reductions in response to all types of experimentally applied disturbance, including human feet, mountain bikes, four-wheel-drive vehicles, tracked vehicles (tanks), and shallow and deep raking. Consequently, crust disturbance can result in large decreases in soil nitrogen through a combination of reduced biological nitrogen input and elevated gaseous loss of nitrogen and soil loss. Short-term reduction (two years) in nitrogen inputs ranges up to 100% (Belnap 1996), while long-term studies in southeastern Utah have demonstrated a 42% decrease in soil nitrogen in the twenty-five years following disturbance. The greatest long-term impact of disturbance may be on the soil microbial pool: production of plant-available nitrogen by soil microbes has been found to decrease almost 80% following disturbance (Evans and Belnap 1999; Evans and Ehleringer 1993). Fire can also decrease nitrogen input by killing the nitrogen-fixing organisms (see the section on "Fire").

Albedo

Trampled surfaces show up to a 50% increase in reflectance of wavelengths from 0.25 to 2.5 µm (Belnap 1995) when compared with untrampled crusted surfaces. This represents a change in the surface energy flux of approximately 40 watts/m^2. Large acreages of trampled areas, combined with lack of urban areas to offset this energy loss, may lead to changes in regional climate patterns in many semiarid regions (Sagan et al. 1979).

Because of changes in albedo, trampled surfaces have significantly lower surface temperatures than untrampled surfaces. Midday temperatures in southeastern Utah in June and July show air temperatures averaged 39° C and bare sand 52° C, while dark-crusted surfaces averaged 62° C. In the winter, surface temperatures of well-developed crusts are up to 14° C higher than ambient air temperatures (Belnap 1995).

Surface temperatures can regulate many ecosystem functions. Nitrogen and carbon fixation are heavily temperature dependent, with lower temperatures resulting in lowered activity levels (Belnap and Lange 2001). Altered soil temperatures affect microbial activity, plant nutrient uptake rates, and soil water evaporation rates. Soil temperatures affect seed germination time and seedling growth rates for vascular plants. Timing of these events is often critical in deserts, and relatively small delays can reduce species fitness and seedling establishment, which may eventually affect community structure (Bush and Van Auken 1991). Food and other resources are often partitioned among ants, arthropods, and small mammals on the basis of surface temperature–controlled foraging times (Crawford 1991; Doyen and Tschinkel 1974; Wallwork 1982). Many small desert animals are weak burrowers, and soil-surface microclimates are of great importance to their survival (Larmuth 1978). Consequently, altering surface temperatures can affect nutrient availability and community structure for many desert organisms, thus increasing susceptibility to desertification.

Fire

The main effect of fire on soil crusts is loss of cover, biomass, and species diversity (Johansen et al. 1993). Damage to (and recovery of) soil crusts depends on the prefire composition and structure of the vascular plant community and distribution of fuel, fire intensity, and fire frequency. Hotter and more frequent fires result in more damage to the crusts. Studies following a major fire at Mesa Verde (Belnap, unpublished data) showed that crusts in severely and moderately burned areas were completely killed. Thus these fires reduced species biomass, species diversity, soil stability, carbon input, and nitrogen input of the crusts. Recovery from fire depends on the extent of the damage to the soil crusts (see the section on "Recovery from Disturbance"). Frequent fires prevent the recovery of lichens and mosses, resulting in crusts containing only a few species of cyanobacteria.

RECOVERY FROM DISTURBANCE
Natural Recovery Rates

Soil and plant characteristics of low- and mid-elevation western United States desert ecosystems such as Mesa Verde suggest that most habitats probably evolved with low levels of soil surface disturbance by ungulates (Mack and Thompson 1982; Martin 1975; Stebbins 1981). The dominant native bunchgrasses lack adaptations to grazing, such as tillering, secondary compounds, or high tissue silica content. Dung beetles, present globally in other systems with large ungulate populations, are absent. Limited surface water and sparse forage would have kept ungulate populations small and generally limited to winter use of lower elevations, as is seen today (Parmenter and Van Devender 1995). Biological soil crusts that are intolerant to trampling are well developed and widespread, and many ecosystems appear dependent on nitrogen provided by them (Belnap 1995; Evans and Belnap 1999; Evans and Ehleringer 1993). Soils are highly compactable and slow to recover due to limited depth of soil freezes as well as low levels of soil organic matter. In addition, shallow soils and limited precipitation limit the distribution of burrowing vertebrate and invertebrate species.

Some desert surfaces in the western United States show more local disturbance by ants, termites, and burrowing mammals than other deserts, and this surface disturbance is reflected in the soil microflora. For instance, the Sonoran has much greater surface disturbance than the Colorado Plateau desert, with the Great Basin, Mojave, Colorado, and Chihuahuan deserts in between. Decomposition rates are also very different between deserts, with rates much higher in the Sonoran and Chihuahuan deserts than in the Mojave or Colorado Plateau. On the Colorado Plateau, however, standing dead and fallen trees recorded in 1898 show no visible decomposition 100 years later (Webb 1997).

Recovery from disturbance appears to parallel levels of evolutionary soil disturbance and decomposition rates. Using experimentally applied disturbances, sites in the Sonoran and Chihuahuan deserts have shown much faster recovery than those in the Mojave and Colorado Plateau deserts. It may be that surfaces that did not evolve with disturbance may depend more heavily on soil surface integrity for natural ecosystem functioning than other regions do. As a result, these deserts may be more negatively affected by soil surface disturbances than regions that evolved with higher levels of surface disturbance (Belnap, unpublished data).

Recovery rates of cyanobacterial-lichen soil crusts depend on the type and extent of disturbance, the availability of nearby inoculation material, and the temperature and moisture regimes that follow disturbance events. Estimates of time for visually assessed recovery have varied from 5 to 100 years (Harper and Marble 1988; Johansen 1993). It has been shown, however, that many components of recovery cannot be assessed visually (Belnap 1993). Assuming linear recovery rates, recovery in southeastern Utah is estimated at 15 years for cyanobacterial biomass, 45 to 85 years for lichen cover, and 200 years for moss cover in scalped 0.25 m^2 plots surrounded by well-developed crusts. Lichen recovery in some plots in the Mojave Desert, assessed after 50 years, indicates recovery times of over 1,000 years. Because recovery time is dependent on the presence of nearby inoculant, larger disturbed areas will take longer to recover.

Recovery of nitrogenase activity also appears to be quite slow. In scalped areas on the Colorado Plateau no nitrogenase activity was detectable after nine years, and the nitrogen content of soils was still much lower when compared with adjacent control plots. In areas disturbed with four-wheel-drive vehicles, no recovery could be documented after two years (Belnap 1996). Using isotopic ratios of nitrogen, soil and plant nitrogen and nitrogenase activity levels were found to be significantly lower in an area that had been released from livestock grazing for thirty years when compared with an area that was never grazed (Evans and Belnap 1999). These data suggest that negative effects on nitrogen dynamics may persist in systems for extended but variable periods after disturbance ceases.

Restoration of normal surface albedos and temperatures will depend on the restoration of cyanobacteria, lichens, and mosses. Cyanobacteria form a dark matrix in which other components are embedded; dark mosses and lichens contribute up to 40% of the cover in an undisturbed crust in southeastern Utah (Belnap 1993). Consequently, recovery of surface albedo characteristics in severely disturbed areas could take up to 250 years for even very small areas.

Inoculations of cyanobacteria have been tested for speeding crust recovery (Ashley and Rushforth 1984; Belnap 1993; St. Clair et al. 1986; Tiedemann et al. 1980). All studies found that crustal recovery was greatly enhanced with this technique.

SUMMARY

Relatively undisturbed biological soil crusts can contribute a great deal of stability to otherwise highly erodible soils. Unlike vascular plant cover, crustal cover is not reduced in drought; and unlike rain crusts these organic crusts are present year-round. Consequently, they offer stability over time and in adverse conditions that is often lacking in other soil surface protectors.

Unfortunately, disturbed crusts now cover vast areas as a result of ever-increasing recreational and commercial uses of these semiarid and arid areas. The increasing human activities in desert areas are often incompatible with the well-being of biological soil crusts. The cyanobacterial fibers that confer such tensile strength to these crusts are no match for the compressional stresses placed on them by vehicles or trampling. Crushed crusts contribute less nitrogen and organic matter to the ecosystem. Impacted soils are left highly susceptible to both wind and water erosion. Raindrop erosion is increased, and overland water flows carry detached material away.

There are now many resources available with information on biological soil crusts. A complete bibliography and further details on crusts can be obtained at www.soilcrust.org.

REFERENCES CITED

Alexander, R. W., and A. Calvo. 1990. The influence of lichens on slope processes in some Spanish badlands. In J. B. Thornes (ed.), *Vegetation and erosion,* pp. 385–398. John Wiley and Sons, Ltd., Chichester, England.

Ashley, J., and S. R. Rushforth. 1984. Growth of soil algae on topsoil and processed oil shale from the Uintah Basin, Utah, USA. *Reclamation and Revegetation Research* 3: 49–63.

Belnap, J. 1993. Recovery rates of cryptobiotic soil crusts: Assessment of artificial inoculant and methods of evaluation. *Great Basin Naturalist* 53: 89–95.

———. 1995. Surface disturbances: Their role in accelerating desertification. *Environmental Monitoring and Assessment* 37: 39–57.

———. 1996. Soil surface disturbances in cold deserts: Effects on nitrogenase activity in cyanobacterial-lichen soil crusts. *Biology and Fertility of Soils* 23: 362–367.

Belnap, J., and J. S. Gardner. 1993. Soil microstructure of the Colorado Plateau: The role of the cyanobacterium *Microcoleus vaginatus*. *Great Basin Naturalist* 53: 40–47.

Belnap, J., and D. A. Gillette. 1997. Disturbance of biological soil crusts: Impacts on potential wind erodibility of sandy desert soils in southeastern Utah. *Land Degradation and Development* 8: 355–362.

———. 1998. Vulnerability of desert biological soil crusts to wind erosion: The influences of crust development, soil texture and disturbance. *Journal of Arid Environments* 39: 133–142.

Belnap, J., and K. T. Harper. 1995. Influence of cryptobiotic soil crusts on elemental content of tissue in two desert seed plants. *Arid Soil Research and Rehabilitation* 9: 107–115.

Belnap, J., and O. L. Lange. 2001. *Biological soil crusts: Structure, function, and management.* Ecological Studies Series. Springer-Verlag, Berlin.

Beymer, R. J., and J. M. Klopatek. 1991. Potential contribution of carbon by microphytic crusts in pinyon-juniper woodlands. *Arid Soil Research and Rehabilitation* 5: 187–198.

———. 1992. Effects of grazing on cryptogamic crusts in pinyon-juniper woodlands in Grand Canyon National Park. *American Midland Naturalist* 127: 139–148.

Black, C. A. 1968. *Soil-plant relationships.* 2nd ed. John Wiley and Sons, Inc., New York.

Brotherson, J. D., and S. R. Rushforth. 1983. Influence of cryptogamic crusts on moisture relationships of soils in Navajo National Monument, Arizona. *Great Basin Naturalist* 43: 73–78.

Bush, J. K., and O. W. Van Auken. 1991. Importance of time of germination and soil depth on growth of *Prosopis glandulosa* seedlings in the presence of a C_4 grass. *American Journal of Botany* 78: 1732–1739.

Campbell, S. E., J. S. Seeler, and S. Golubic. 1989. Desert crust formation and soil stabilization. *Arid Soil Research and Rehabilitation* 3: 217–228.

Crawford, C. S. 1991. The community ecology of macroarthropod detritivores. In G. Polis (ed.), *Ecology of desert communities,* pp. 89–112. University of Arizona Press, Tucson.

Doyen, J. T., and W. F. Tschinkel. 1974. Population size, microgeographic distribution and habitat separation in some tenebrionid beetles. *Annals of the Entomological Society of America* 67: 617–626.

Dregne, H. E. 1983a. *Desertification of arid lands.* Harwood Academic Publishers, New York.

———. 1983b. Physical effects of off-road vehicle use. In R. H. Webb and H. G. Wilshire (eds.), *Environmental effects of off-road vehicles: Impacts and management in arid regions,* pp. 15–30. Springer-Verlag, New York.

Eldridge, D. J. 1993a. Cryptogam cover and soil surface condition: Effects on hydrology on a semiarid woodland soil. *Arid Soil Research and Rehabilitation* 7: 203–217.

———. 1993b. Cryptogams, vascular plants and soil hydrological relations: Some preliminary results from the semiarid woodlands of eastern Australia. *Great Basin Naturalist* 53(1): 48–58.

Eldridge, D. J., and R.S.B. Greene. 1994. Assessment of sediment yield by splash erosion in a semi-arid soil with varying cryptogam cover. *Journal of Arid Environments* 26: 221–232.

Ettershank, G., J. Ettershank, M. Bryant, and W. Whitford. 1978. Effects of nitrogen fertilization on primary production in a Chihuahuan desert ecosystem. *Journal of Arid Environments* 1: 135–139.

Evans, R. D., and J. Belnap. 1999. Long-term consequences of disturbance on nitrogen dynamics in an arid ecosystem. *Ecology* 80(1): 150–160.

Evans, R. D., and J. R. Ehleringer. 1993. A break in the nitrogen cycle in arid lands? Evidence from N^{15} of soils. *Oecologia* 94: 314–317.

Foth, H. D. 1978. *Fundamentals of soil science.* 6th ed. John Wiley and Sons, London.

Garcia-Pichel, F., and J. Belnap. 1996. The microenvironments and microscale productivity of cyanobacterial desert crusts. *Journal of Phycology* 32: 774–782.

Goudie, A. S. 1978. Dust storms and their geomorphological implications. *Journal of Arid Environments* 1: 291–310.

Grondin, A. E., and J. R. Johansen. 1993. Microbial spatial heterogeneity in microbiotic crusts in Colorado National Monument: I. Algae. *Great Basin Naturalist* 53(1): 24–30.

Harper, K. T., and J. R. Marble. 1988. A role for nonvascular plants in management of arid and semiarid rangeland. In P. T. Tueller (ed.), *Vegetation science applications for rangeland analysis and management,* pp. 135–169. Kluwer Academic Publishers, Dordrecht.

Harper, K. T., and R. L. Pendleton. 1993. Cyanobacteria and cyanolichens: Can they enhance availability of essential minerals for higher plants? *Great Basin Naturalist* 53: 59–72.

Johansen, J. R. 1993. Cryptogamic crusts of semiarid and arid lands of North America. *Journal of Phycology* 29: 140–147.

Johansen, J. R., J. Ashley, and W. R. Rayburn. 1993. Effects of rangefire on soil algal crusts in semiarid shrub-steppe of the lower Columbia Basin and their subsequent recovery. *Great Basin Naturalist* 53: 73–88.

Ladyman, J.A.R., and E. Muldavin. 1996. *Terrestrial cryptogams of pinyon-juniper woodlands in the southwestern United States: A review.* USDA Forest Service, Rocky Mountain Forest and Range Experiment Station, General Technical Report, RM-GTR-280.

Lange, W. 1974. Chelating agents and blue-green algae. *Canadian Journal of Microbiology* 20: 1311–1321.

Larmuth, J. 1978. Temperatures beneath stones used as daytime retreats by desert animals. *Journal of Arid Environments* 1: 35–40.

Lesica, P., and J. S. Shelley. 1992. Effects of cryptogamic soil crust on the population dynamics of *Arabis fecunda* (Brassicaceae). *American Midland Naturalist* 128: 53–60.

Leys, J. N. 1990. *Soil crusts: Their effect on wind erosion.* Research Note 1/90. Soil Conservation Service of New South Wales, Canberra, Australia.

Loope, W. L., and G. F. Gifford. 1972. Influence of a soil microfloral crust on select properties of soils under pinyon-juniper in southeastern Utah. *Journal of Soil Water and Conservation* 27: 164–167.

Mack, R. N., and J. N. Thompson. 1982. Evolution in steppe with few large, hooved mammals. *American Naturalist* 119: 757–773.

McKenna-Neuman, C., C. D. Maxwell, and J. W. Boulton. 1996. Wind transport of sand surfaces crusted with photoautotrophic microorganisms. *Catena* 27: 229–247.

Martin, P. S. 1975. Vanishings, and future of the prairie. *Geoscience and Man* 10: 39–49.

Mayland, H. F., and T. H. McIntosh. 1966. Availability of biologically fixed atmospheric nitrogen-15 to higher plants. *Nature* 209: 421–422.

Nobel, P. S., E. Quero, and H. Linares. 1988. Differential growth response of agaves to nitrogen, phosphorus, potassium, and boron applications. *Journal of Plant Nutrition* 11: 1683–1700.

Parmenter, R. R., and T. R. Van Devender. 1995. Diversity, spatial variation, and functional roles of vertebrates in the desert grassland. In M. P. McClaran and T. R. Van Devender (eds.), *Desert grasslands*, pp. 196–229. University of Arizona Press, Tucson.

Reynolds, R., J. Belnap, M. Reheis, and N. Mazza. 1999. Eolian dust on the Colorado Plateau—Magnetic and geochemical evidence from sediment in potholes and biologic soil crust. In A. J. Busacca (ed.), *Dust, aerosols, loess soils and global change,* pp. 231–234. Washington State University College of Agriculture and Home Economics Miscellaneous Publication No. MISC0190. Pullman, Wash.

Sagan, C., O. B. Toon, and J. B. Pollack. 1979. Anthropogenic albedo changes and the earth's climate. *Science* 206: 1363–1368.

Schopf, J. W. (ed.). 1983. *Earth's earliest biosphere—Its origin and evolution.* Princeton University Press, Princeton, N.J.

Shields, L. M., and L. W. Durrell. 1964. Algae in relation to soil fertility. *Botanical Review* 30: 92–128.

St. Clair, L. L., J. R. Johansen, and B. L. Webb. 1986. Rapid stabilization of fire-disturbed sites using a soil crust slurry: Inoculation studies. *Reclamation and Revegetation Research* 4: 261–269.

Stebbins, G. L. 1981. Coevolution of grasses and herbivores. *Annals of the Missouri Botanical Garden* 68: 75–86.

Stewart, W.D.P. 1967. Transfer of biologically fixed nitrogen in a sand dune slack region. *Nature* 214: 603–604.

Stolzy, G., and A. G. Norman. 1961. Factors limiting microbial activities in soil. *Archiv für Mikrobiologie* 40: 341–350.

Tiedemann, A. R., W. Lopushinsky, and H. J. Larsen, Jr. 1980. Plant and soil responses to a commercial blue-green algae inoculant. *Soil Biology and Biochemistry* 12: 471–475.

Wallwork, J. A. 1982. *Desert soil fauna.* Praeger Scientific Publishers, London.

Webb, R. H. 1997. *Grand Canyon, a century of change.* University of Arizona Press, Tucson.

Williams, J. D., J. P. Dobrowolski, N. E. West, and D. A. Gillette. 1995. Microphytic crust influences on wind erosion. *Transactions of the American Society of Agricultural Engineers* 38: 131–137.

Wullstein, L. H. 1989. Evaluation and significance of associative dinitrogen fixation for arid soil rehabilitation. *Arid Soil Research Rehabilitation* 3: 259–265.

MESA VERDE COUNTRY'S WOODLAND AVIAN COMMUNITY

George L. San Miguel and Marilyn Colyer

Old-growth piñon-juniper woodlands in and around Mesa Verde National Park support a clearly identifiable bird community. This avifauna consists of a high diversity of species: large and small birds, swift flyers and slow walkers, hunters and hunted, brightly plumed singers and the drably colored with hushed voices, long-distance migrants and anchored residents. Complex woody structure and diverse understory, as described in the preceding four chapters, offer numerous resources and opportunities for birds. Abundant broken limbs, tree cavities, piñon seeds, juniper cones, a layered woodland structure, and a healthy understory create home and food for birds. Structural diversity in old-growth directly correlates with high avian species richness (Chapter 2; Miller 2000). Within the sanctuary of densely wooded mesas and canyons of Mesa Verde National Park today we find every bird species that has been known historically in this region.

Although the piñon-juniper avifauna is characteristic of the piñon-juniper community at Mesa Verde, none of these species resides exclusively in the piñon-juniper woodlands. Even the aptly named pinyon jay and the juniper titmouse sometimes occupy environments lacking either tree species. Nonetheless, the members of the piñon-juniper avifauna occupy important positions within the community. For the purposes of this chapter, our commentary on the different birds on the mesas' dense, old woodlands is limited primarily to the most representative species. Although the lifeway of each species is distinct from all others in at least one significant way, it is convenient to arrange birds into categories based on how they live; and the section headings of this chapter reflect this simple classification (for example, raptors, generalists, nut birds, omnivores). Avian species that are known from the piñon-juniper of Colorado are summarized in Appendix 6.1. Most of this information was gleaned from the 1998 *Colorado Breeding Bird Atlas* (Kingery 1998) and applied to our observations in the Mesa Verde Country. Observations are on record with the Natural Resource Office at Mesa Verde National Park, so they are not cited individually in the following text.

RAPTORS

Among the diurnal birds of prey, the sharp-shinned hawk utilizes the old-growth piñon-juniper most frequently. With a wingspan of only 50 to 72 cm and a long, narrow tail for quick steering, this hunter of small birds is ideally adapted to the tight confines of the dense pygmy woodlands. Sharp-shinned hawks may build their stick nests in tall old junipers, usually at the base of a slope. This species tenaciously holds its territories, even returning to nest following devastating wildfires. Two larger accipiter hawks, Cooper's hawk and northern goshawk, also use the piñon-juniper to hunt for prey, moving in from breeding territories in adjoining canyon forests.

Due to the dense canopy cover, old-growth piñon-juniper gets little use from soaring raptors such as the red-tailed hawk and golden eagle. Turkey vultures, however, can effectively scavenge for carrion concealed from view with the aid of their remarkably well-developed olfactory sense (Poole 1983). Both the golden eagle and red-tailed hawk hunt for small and medium-sized mammals along piñon-juniper margins and in woodland openings. The more open the stands, the more time these species spend in the piñon-juniper. All three soaring raptors nest in ledges and cavities in the sandstone cliffs.

Mesa Verde National Park supports at least six nesting territories for the peregrine falcon. These quick aerial hunters pursue swifts, swallows, jays, and other birds flying over the old woodlands and adjoining canyons and for several kilometers beyond their aeries. Like the eagles and vultures, peregrines nest on cliff ledges and cavities under the piñon/juniper-mantled mesa tops. Another falcon, the American kestrel, occupies more open piñon-juniper country and woodland margins. This small falcon nests in larger tree cavities and hunts rodents, lizards, small birds, and large insects from tree perches.

THE NIGHT SHIFT

When the sun goes down, the hawks, eagles, and falcons give way to a complete cadre of piñon-juniper woodland owls. The most common of these nocturnal hunters is the northern saw-whet owl. Only 20 cm long, with compact wings, this diminutive owl silently negotiates the interlocking branches of the old-growth in the dark to ambush prey ranging from moths to small rodents. The northern pygmy-owl and the flammulated owl share this habitat. Like the saw-whet, these small owls occupy cavities in larger trees for nesting, usually an old woodpecker hole. Although these small raptors may range as much as a few kilometers from their nest to hunt, they remain closely tied to the old-growth piñon-juniper for nesting and safe daytime roosting. Oddly enough, pygmy-owls primarily hunt small birds during daylight hours (Pearson 1936).

Some larger owls use the woodlands as well. Both the long-eared owl and great horned owl forage and nest in most habitats throughout the Mesa Verde Country. A third species, the Mexican spotted owl, is far more closely associated with the old-growth piñon-juniper woodlands of the southern mesas and canyons. Even in preferred habitat within the national park, spotted owls are uncommon. Their breeding territories center around groves of tall canyon-bottom Douglas-fir trees, but the actual nesting and roosting sites are located within natural rocky recesses on the canyon walls among the piñon and juniper (Reynolds and Johnson 1996). Spotted owls favor woodrats and other small nocturnal animals as prey.

GENERALISTS

The Colorado Plateau certainly has its share of generalist, omnivorous, year-round resident bird species, including the highly successful and ubiquitous members of the crow family (Corvidae) along with the tit family (Paridae). Many of the corvids change their behavior when opportunities arise. The common raven, American crow, and black-billed magpie may scavenge the same deer carcass left over from a mountain lion kill, raid the nests of other birds, or consume grains, fruits, nuts, insects, or any number of alternative food sources in their environment. Like ravens, the western scrub-jay and Steller's jay are opportunist omnivores but more often settle for smaller fare such as insects, acorns, and piñon seeds. Whereas the corvids build well-concealed stick nests, the juniper titmouse and mountain chickadee nest in tree cavities. Both of these species are adept at combing the woodland from canopy foliage, to tree bark, to understory shrubs, and down to the forest floor searching for insects, fruits, nuts, or any other edibles. Mountain chickadees even collect tent caterpillar cocoons from the branches of the antelope bitterbrush, cache them in tree bark, and consume them during the leaner winter months. Unlike the titmouse, the chickadee is equally common in forested communities of higher elevations. The juniper titmouse rarely inhabits areas far from piñon-juniper and is quite at home in the densest old-growth stands. It spends the winter on southern exposures and mesa rims and begins defending nesting territories as early as February.

NUT BIRDS

Anyone who has fed birds in the backyard knows how much corvids, parids, woodpeckers, and white-breasted nuthatches (Figure 6.1) enjoy eating "nuts" (seeds or fruits that easily separate from their seed coats). Among the corvids, however, are two highly evolved nut specialists, the pinyon jay and the Clark's nutcracker.

Figure 6.1. The white-breasted nuthatch is a common resident of the piñon-juniper woodland. Illustration by Sue Harper. Courtesy, Mesa Verde Museum Association.

These two species, among the many corvids that feed on piñon seeds, can open green cones and disperse seeds across long distances (Christensen and Whitham 1991). Several members of this family of highly intelligent bird species are known to cache food and remember where they were hidden long afterward. The nutcracker makes an extraordinary effort to harvest and hide vast reserves of pine nuts for the winter larder. Each nutcracker may collect several thousand seeds each fall from a variety of pine species (Tomback 1982), but the large seeds of the Colorado piñon pine are prized for their concentrated nutritional store (Vander Wall 1988). Even with the remarkable memory of corvids, acorns and pine nuts cached in the ground are not always retrieved but are sometimes forgotten, only to sprout into new seedlings. For example, it has been estimated that a flock of Clark's nutcrackers, during a mast year, cached between 2.2 and 3.3 times their caloric needs, leaving seeds in microsites that support germination (Vander Wall and Balda 1977).

Mesa Verde's nutcrackers, Steller's jays, western scrub-jays, and other birds rely heavily on the piñon seed crop. In most years good seed production areas are only scattered lightly throughout the woodland. Stands of piñon trees that perhaps escaped frost, received extra moisture, or were exposed to low temperature (Forcella 1981) will produce the best crop two growing seasons later because megastrobili (immature cones) take two years to mature. In some productive years up to half of the seeds are lost to insects or other causes. Within the mesa top old-growth stands, "mast" years occur every five to eight years (Christensen and Whitham 1991; Ligon 1978) and as frequently as every two to three years on the mesa rims, canyons, escarpments, and deep loamy soils near the park entrance

(Colyer, unpublished). Masting is understood as a defensive mechanism allowing trees to pass a threshold of cone production that attracts dispersers (Christensen and Whitham 1991) and satisfies numerous seed predators. The long interval between mast crops may starve predators in the intervening years (Silvertown 1980). The fluctuation may allow photosynthates to replenish depleted stores after masting (Floyd 1987). Piñon cone crops also vary considerably within a stand and among stands due to herbivore pressures (Whitham and Mopper 1985). For the highly territorial and less nomadic western scrub-jays, the boon from a crop is highly undependable; but for more mobile species, including the Steller's jay but particularly the pinyon jay and Clark's nutcracker, the annual search for the producing populations is a critical part of their survival strategy.

Among the piñon-juniper avifauna, the pinyon jay rates as a true specialist (Balda 1972; Kingery 1998). The high level of co-evolved adaptation between this small corvid and piñon pines is unequaled in the woodlands (Ligon 1971, 1978). Like the highly mobile Clark's nutcracker, pinyon jays cache an enormous number of piñon seeds when a good crop is located, working in flocks from mid-August to mid-October (Balda and Bateman 1971). In Mesa Verde Country it is not unusual to observe a squawking flock of fifteen to a hundred pinyon jays leaving a harvest site, their throats bulging with up to thirty piñon nuts each, or to see them caching the nuts on south-facing steep and sparsely vegetated shale hills. This activity will go on from dawn to dusk every day for weeks at a time until all the available nuts have been concealed at the winter storage site. Then these sites become equally tumultuous when the noisy flocks begin retrieving the nuts a few months later. Good cone crops increase the chances of successful breeding, as the presence of green strobili stimulates pinyon jay breeding (Ligon 1974). This highly gregarious species sometimes nests in colonies, although little is known of its breeding tendencies in the Mesa Verde Country.

INSECT EATERS

Probably the most diverse grouping of piñon-juniper birds is the insectivores, whose feeding activities are critical for keeping insect populations in check. The seasonal abundance of invertebrates has selected for a variety of methods for capturing these strands in the food web. Even birds specially adapted to eating seeds, nectar, berries, or nuts depend a great deal on insects for protein, particularly when feeding their young. As this section shows, often there is a fair amount of adaptive overlap from one group of insect eaters to the next. Because most insects cannot tolerate cold temperatures, insectivorous birds are far less abundant in number and less diverse in variety during much of the year.

Sky Sifters

Among the fastest-flying birds at Mesa Verde is the white-throated swift. Although these sickle-winged missiles never alight on a piñon or juniper tree, their foraging habits play a clear role in the piñon-juniper woodland ecology. Basically they fly around at high speed with their mouths open, ready to snap up flying insects over the woodland mesas and in the rugged canyons in between. Before the onset of the summer rains, swifts congregate at canyon heads, swirling in the updrafts where aerial insects concentrate. They nest in the splitting sandstone crevices under the piñon-juniper–covered mesas. Living a life similar to swifts is a songbird that nests in tree cavities, often old woodpecker holes. The violet-green swallow, although not as common as swifts in this community or as fast a flyer, inhabits many of the same canyons and mesas. Excellent insect-catchers, the swallows also fly around much of the day with mouths agape. The swallows reduce their competition for flying insects with the swifts by foraging much lower to the ground than the higher-flying swifts.

Like owls among the raptor clan, the common nighthawk takes over sky-sifting duties from the swifts and swallows before dusk and again around dawn. In a separate family (Caprimulgidae), nighthawks are neither hawks nor owls and hunt much like bats but without sonar. They nest on bare ground and roost on tree branches, stumps, and snags.

Flycatchers

Another aerial means for catching flying insects, but closer to the ground, is "hawking" or flycatching. A nocturnal relative of the nighthawk with similar nesting habits is the common poorwill. These birds perch on bare ground or rock in woodland openings or edges and wait for a moth or other night-flying insect to come by before sallying forth and snatching it on the wing (Pearson 1936). Even most kinds of owls take advantage of night-flying insects. During heavy hatch years of army cutworms, the neighboring mountain shrub community plays host to numerous nesting owls of all sizes that come over from the piñon-juniper for a frantic and ghostly feast on these fat moths.

During the day members of several other bird families are known to fly out from a perch to catch flying insects. These species include the western bluebird, yellow-rumped warbler, and even some woodpeckers. But in the piñon-juniper perhaps the best insect hawkers are members of the tyrant flycatcher family (Tyrannidae).

Several kinds of flycatchers (such as the cordilleran flycatcher) may be seen migrating through the old-growth piñon-juniper or occupying mesic microhabitats adjacent to the piñon-juniper. Other flycatchers (such as the western wood-pewee and dusky flycatcher) may use side canyons or transitional piñon-juniper habitat that intergrades with other communities and may support deciduous, leafy shrubs and forbs. In more open, often more arid conditions we are more likely to find the Say's phoebe.

Only two flycatchers fully utilize the old-growth piñon-juniper woodland proper. Like so many piñon-juniper birds, the gray flycatcher matches the drab hues of the gray-green pygmy forest. These inconspicuous birds would hardly be noticed if it were not for the abrupt, weak songs delivered by the males from the tops of tall piñons and junipers. They construct their tiny cup nest on or near the ground under the spreading cover of a woody shrub such as antelope bitterbrush. Like the cordilleran and dusky, the gray flycatcher belongs to a nondescript genus of neotropical migrants. In Latin the name *Empidonax* means "emperor of the gnats." More numerous, larger, more colorful, and with a much stronger voice, the ash-throated flycatcher gets greater notice. Ash-throated flycatchers feed on moths, flies, caterpillars, and other invertebrates snared among the tree branches in small gaps or within the bitterbrush understory. Nesters in tree cavities, they take advantage of this abundant old-growth resource.

Wood Borers

Without woodpeckers, far fewer cavity-nesting birds and other wildlife would be able to occupy woodlands. Each year woodpeckers chisel out a new nest cavity of their own, thereby leaving previous nest holes for bluebirds, chickadees, small owls, or other secondary cavity nesting birds. Abandoned woodpecker holes are reused by rodents to cache juniper "berries," piñon seeds, and bitterbrush fruit or as nursery wards and hibernation quarters for bats, chipmunks, and many other animals. Another ecological "service" performed by woodpeckers is the surgical removal of insects infesting tree trunks and limbs (Pasquier 1980).

Because trees are available as prey year-round, within the old-growth piñon-juniper there are plenty of resident wood borers. The old-growth piñon-juniper's three principal woodpecker species vary considerably in size, allowing them to coexist without excessive competition for the same resources. The largest and most commonly seen woodpecker is the 30-cm-long red-shafted variety of the northern flicker (Figure 6.2). Its ability to nest within piñon-juniper depends on the presence of large-diameter trees. Smaller and more nimble is the hairy woodpecker, commonly found in old-growth, where it often seeks out patches of trees weakened or recently killed by fire, drought, insect outbreaks, or disease such as black-stain root wilt. The black-stain fungus frequently results in secondary infestations by wood-boring and bark-boring beetles, highly sought after by flickers and hairy woodpeckers. The smaller and less common downy woodpecker looks like a miniature hairy woodpecker with a stubby little beak. Its light weight allows it to cling to the slender terminal branches of trees and shrubs to reach prey unavailable to larger woodpeckers.

Many vigorous trees in the old piñon-juniper show the horizontal parallel rows of sap wells tapped by sapsuckers, a distinct group of woodpeckers. The two local species are the red-naped sapsucker and the less common Williamson's sapsucker. Neither species nests in the piñon-juniper, but the red-naped sapsuckers can be found at Mesa Verde in the spring as early as February and extending well into April on their way north to the aspen forests of the Rocky Mountains. In fall they pass through during September and October. In Mesa Verde red-

Figure 6.2. *The largest and most commonly seen woodpecker is the foot-long red-shafted variety of the northern flicker, whose ability to nest within piñon-juniper depends on the presence of large-diameter trees. Illustration by Walt Anderson.*

naped sapsuckers have been known to return to the same trees, making the same well-recognized circuit from one tapped sap tree to the next and arriving at the starting point in time for freshly flowing sap to be available once again (Colyer, unpublished).

The rapid-fire drumming of male woodpeckers is not an attempt to feed but instead is a territorial pronouncement, like the melody of a songbird (Poole 1983). When woodpeckers search for insect food within a tree, the pecking, carving, and gouging are far more deliberate and concentrated in trees they recognize as sick or infested, identified by the munching sounds they hear under the bark. After opening up an access to the insect, the woodpecker distends its long, barb-tipped tongue to harpoon its prey. The flicker also invests much of its time on the ground hunting and consuming ants. When measured in overall animal biomass, ants make up a large proportion of the available prey in the woodland; but ants bite and are mostly exoskeleton, and some contain formic acid. Flicker adaptations have overcome these defenses and taken advantage of this protein source, with ants making up about 50% of their diets (Kingery 1998). Of course, sapsuckers do consume sweet tree sap but otherwise make their living like most woodpeckers, including eating tree insects. Not all insects consumed by woodpeckers are taken from under the bark. Woodpeckers, particularly the pint-sized downy, also make a good living patiently searching the hiding spaces within the roughened ridges of tree bark.

Bark Gleaners

In addition to woodpeckers, such piñon-juniper birds as chickadees, titmice, wrens, and sometimes bushtits and jays take advantage of insect prey hiding on and in tree bark. Bark-gleaning specialists, the red-breasted nuthatch and brown creeper, occasionally visit the old piñon-juniper from higher-elevation coniferous forests. At Mesa Verde, however, the only common, breeding, bark-gleaning specialist is the white-breasted nuthatch. A resident cavity nester, the white-breasted nuthatch focuses its search for insects on rough-barked trees while climbing head down on the trunks and branches so as not to compete excessively with bark gleaners that search heading upward. As its name implies, it also feeds on nuts and seeds when available.

Foliage Gleaners

Leaves and needles (moist photosynthetic organs) are among the most vulnerable parts of trees and shrubs. So it is no wonder that the thin tissues of active green foliage often are infested with insects and other arthropods seeking to obtain some of these nutritious fluids. Among the generalists discussed previously, the mountain chickadee and juniper titmouse also are adept at foraging for this multitude of tiny organisms. Certain characteristics are needed for a bird to be an effective foliage gleaner. First, small size and light weight allow these birds to reach the leaves and needles. Even downy woodpeckers can do some foliage gleaning. Second, they need to have fine-pointed beaks, giving them the oral dexterity of tweezers. Finally, they need to have the ability to focus their eyes on tiny insects less than 2 cm from their eyes.

The other year-round foliage gleaners of the piñon-juniper woodlands are the bushtit and the Bewick's wren. Bushtits resemble miniature chickadees, but one of their most interesting adaptations is their propensity to live in flocks virtually year-round. A flock of these birds acts like a twittering broom, sweeping through the tree canopy across the woodland. Highly conspicuous, the flocks probably add some additional level of vigilance against predators and the safety-in-numbers strategy employed by fish schools. In breeding, the pendulous woven nest will hold the eggs of two to four females, with nonbreeding adults helping to raise the young (Kingery 1998). So even before hatching, bushtits begin life in a communal society adapted to utilize foliage insects more efficiently.

Bewick's wrens live a much more solitary and conventional life in the piñon-juniper. Quiet and secretive during most of the year, the musical males begin announcing their courtship intentions by March from the piñon and juniper treetops.

Their thin, pointed beaks allow them to capture foliage insects, leafhoppers, flies, and others from branch tips, within the understory, on the forest floor, and on the canyon rimrocks.

Among the neotropical migratory species so prevalent in the piñon-juniper, the most abundant foliage gleaners are the blue-gray gnatcatcher, gray vireo, plumbeous vireo, yellow-rumped warbler, black-throated gray warbler, and western tanager. Of these only the yellow-rumped warblers continue beyond the piñon-juniper to breed in higher-elevation habitats, but not before their large numbers have refueled on the woodland's foliage insects. While the yellow-rumped warblers continue north, or south in the fall, their more drably colored relative, the black-throated gray warbler, sets up breeding territories right in the old-growth piñon-juniper. These and other members of the wood-warbler family (Parulidae) are expert foliage gleaners, but the black-throated gray warbler also searches the leaf litter on the forest floor. In areas where Gambel's oak makes up part of the understory, the Virginia's warbler forms part of the piñon-juniper bird community.

Birds in the vireo family (Vireonidae) resemble wood warblers in many ways, but they are surprisingly more rugged birds with a more powerful beak for catching larger prey and picking berries. The genus name *Vireo* translates from the Latin to "I am green," and this is true for most vireos. In the piñon-juniper, however, the vireos are not green at all; instead they are drab and gray. As their name suggests, the gray vireos blend in with the somber colors of the piñon-juniper foliage. In the Mesa Verde Country they occupy the shorter, more open stands at lower elevations dominated by junipers and seem to prefer areas with rocky slopes. The larger plumbeous vireo replaces the gray vireo in the dense old-growth at higher elevations of the mesa tops. "Plumbeous" means "lead-colored," a darker gray than the gray vireo. Both vireos are effective foliage gleaners in the trees and in the understory shrubs such as Utah serviceberry.

Unlike the drab vireos, the male western tanagers' bright yellow and red coloration and their loud song make them a fairly conspicuous species. Also unlike the vireos, tanagers are not solely piñon-juniper birds. Instead they breed in the denser forests in more mesic sites along the piñon-juniper perimeter but incorporate the drier woodland within their foraging territories. Although they are foliage gleaners, tanagers also hawk flying insects and pick berries from shrubs.

Among the smallest of Mesa Verde's foliage gleaners—even smaller than the wood warblers—are the blue-gray gnatcatchers (family Sylviidae), which winter well to the south. Their remarkably light weight allows them access to the most delicate branchlets. Their equally small relatives include the ruby-crowned kinglet (family Regulidae), which breeds in coniferous forests at higher elevations but passes through the piñon-juniper in spring and fall. Flitting from leaf to leaf, the gnatcatcher and kinglet are like tiny winged pincers incessantly in search of seemingly microscopic packets of insect nourishment to fuel their hot-blooded engines.

Winter Flocks

As happens in other forests, a winter observer in the piñon-juniper woodland may walk a few kilometers or more and not see or hear a single bird, then come across a mixed flock of several birds busily foraging and moving through the treetops with an invisible sense of direction and purpose. Much like a flock of bushtits, variable numbers of mountain chickadees, juniper titmice, and white-breasted nuthatches make up these flocks, sometimes with other species in tow, combing through evergreen foliage, over bark, and onto the forest floor (Pasquier 1980; Poole 1983). It is commonly assumed that these groups travel together for protection, although they surely must attract more attention this way. Some observers believe that the groups search together for concentrated food or food caches left over from the previous summer.

Rock Gleaners

We have described the tendency for Bewick's wrens sometimes to hunt down insects on the sandstone rim rocks at the woodland's edge. In Mesa Verde two other wren species have perfected that hunting strategy on a more full-time basis and only venture occasionally deeper into the piñon-juniper woodland. Thus the canyon wren and rock wren are more accurately described as piñon-juniper edge species. Although they are cryptically colored, their easily audible songs are an inseparable part of the park's mesa and canyon setting.

NECTAR SIPPERS

Like children in a candy shop, hummingbirds have adapted to exploit nectar, a very specialized part of the food web produced only by some kinds of flowers. Hummingbird coevolution with flowers stems from some plant species' need for accurate dispersal of their pollen. Some flowers are red or purple to attract them and tubular to exclude larger insects and other birds that might otherwise get the nectar first. When a hummingbird sucks out the nectar of a flower, it also receives a dab of pollen from specially positioned anthers. Yellow spots on the foreheads of hummingbirds attest to the effectiveness of this unconscious cooperation.

In the old piñon-juniper there are three hummingbird species (Figure 6.3) that occur in significant numbers. The black-chinned and broad-tailed hummingbirds are summer resident breeding species. Their arrival in spring precedes the blooming of nectar-producing flowers. During their first weeks on site they glean tree foliage for insects. Tubular blossoms of the genus *Penstemon* are a big summer favorite among the Mesa Verde hummingbirds, with large stands of Indian paintbrush available later in summer. The highest points of the treetops serve as courtship stations. Females perch at these conspicuous heights while males swoop and dance around them for their favor. The male broad-tailed hummingbird performs several long vertical arcs and then dives nearly to the ground, repeating this five or six times. The male black-chinned hummingbird makes short, zigzag flights above the attentive female. After mating, the female repairs an old nest or constructs a new tiny cup nest out of lichens and grass blades bound together with spiderweb strands. When feeding the nestlings, once again the female hummingbirds rely heavily on insect protein gleaned from the canopy foliage.

By early July rufous hummingbirds begin arriving from their breeding grounds in the Northwest, among the earliest of southbound migrants. Their inland routes follow the higher elevations where summer-blooming nectar flowers offer refueling stops along the way. By the time the broad-tailed and black-chinned hummingbirds have finished raising young, the rufous hummingbirds have arrived in substantial numbers to compete for Mesa Verde's summer-long nectar supplies. Of the three local species, the rufous is the bully on the block, forcing the black-chinned and broad-tailed hummingbirds to work much harder for their nectar.

BERRY PICKERS

Many plants rely on animals to disperse their fruit and enclosed seeds. Seed dispersal by birds that consume and defecate berries is not common in Mesa Verde. There are fewer than expected berry specialists in Mesa Verde Country, including the American robin, mountain bluebird, Townsend's solitaire, cedar waxwing, and evening grosbeak. Several other species, such as black-headed grosbeaks and vireos, can switch their diets to take advantage of summer berries produced by shrubs. Mistletoe and some berry-producing shrubs are found in the old piñon-juniper woodland, but the more attractive berry producers like chokecherry are more abundant in adjacent habitats.

SEED EATERS AND GROUND FORAGERS

Other bird families evolved specialists in seed diets. Some of these birds spend most of their time on the ground, whereas others stay perched up in trees and shrubs. The largest of Mesa Verde's arboreal seed birds is the black-headed grosbeak. As the name implies, it has a massive beak for cracking open large seeds. Unlike many piñon-juniper birds, the male of this species is colorful and conspicuous, with a loud musical song. The finch family (Fringillidae) offers the piñon-juniper woodland several tree-dwelling seed eaters as well. Two other midsized seed-eating finches that use the piñon-juniper at Mesa Verde are the Cassin's finch and evening grosbeak. It appears that these three species do not remain in the piñon-juniper year-round. A few male Cassin's finches may be heard singing in the Mesa Verde piñon-juniper, but they are not known to nest here. Two small finches of the genus *Carduelis* are the pine siskin and lesser goldfinch. These two species arrive in the Mesa Verde piñon-juniper in the summer to breed and specialize in obtaining small seeds from understory forbs. With the proliferation of non-native thistles, the presence of both goldfinches and siskins has increased. These small finches relish thistle seeds.

Figure 6.3. Many species of hummingbirds, including the black-chinned hummingbird, frequent the piñon-juniper woodland, feeding on Penstemon *and other nectar-producing flowers. Illustration by Walt Anderson.*

Eventually everything found in the trees and shrubs falls to the ground. The thick leaf and needle litter readily retains moisture. The forest floor of healthy piñon-juniper woodlands supports a bountiful food resource for several kinds of birds. Various seeds, fruits, nuts, and invertebrates can be obtained by searching on the ground. Of all the seed-eating piñon-juniper birds, the spotted towhees are the only resident breeding species closely tied to the woodland. Abundant and relatively conspicuous, they hop around in the understory and scratch through the ground litter looking for these edibles, with the males rarely climbing up to the treetops except to sing. In summer another member of the New World sparrow family (Emberizidae) makes full use of the piñon-juniper. The chipping sparrow is a generalist among the sparrows and is commonly found in many habitats. In piñon-juniper it scratches for food in the litter more conspicuously than the towhees and nests under the protective cover of antelope bitterbrush or among "doghair" thickets of piñon saplings. Later, adults and juveniles fatten up on the seeds of forbs and grasses growing in woodland openings but stay close to the woodlands for protection. As fall approaches, they are replaced by their winter generalist counterpart, the dark-eyed junco. Likewise, it seems as if the juncos move north at the first sign of chipping sparrows in early April after a strong southwest wind. Like chipping sparrows, juncos forage in small groups or sizable flocks within small weedy clearings or edges throughout the piñon-juniper.

Another ground sparrow in the piñon-juniper is the white-crowned sparrow. Like the chipping sparrow and junco, white-crowned sparrows migrate through the area in large numbers. But unlike the other two, white-crowned sparrows never stay long in the spring or fall; rather, they overwinter nearby in open dryer lowlands, grassy pastures, and piñon-juniper margins.

Of all the Mesa Verde piñon-juniper ground birds, the wild turkey is by far the largest. As long as there is a drinking water source and plenty of the tall (9 to 12 m) piñon-pine night-roosting trees—for which these mesa-top woodlands are well-known—the old-growth piñon-juniper can support breeding flocks of these gallinaceous birds. The Mesa Verde turkey population was wiped out in the 1960s due to an outbreak of an alien poultry disease but was successfully restored in 1990. Now the park's flocks are thriving again. Perhaps turkeys are not a typical member of the piñon-juniper woodlands, but their large size, highly observable nature, and prominent association with the Ancestral Puebloan people of Mesa Verde make them an interesting addition to the old-growth piñon-juniper avifauna.

SUMMARY

We have portrayed the most common avian stories known from old-growth piñon-juniper woodlands in Mesa Verde Country. Despite the breadth of species, the avifauna shows some clear characteristics. To provide a summary of old-growth avifauna, Appendix 6.1 lists the birds in the Mesa Verde piñon-juniper woodlands and includes breeding statistics from the *Colorado Breeding Bird Atlas* in all piñon-juniper types statewide. The data in this list reveal some generalities:

- There is a discernible group of bird species closely associated with piñon-juniper woodlands.
- Some piñon-juniper bird species are common in some piñon-juniper types and rare or absent in others.
- Some piñon-juniper bird species that are common in other areas of Colorado are absent in the Mesa Verde piñon-juniper.
- Some common Mesa Verde piñon-juniper birds are even more common in other habitat types.
- At least one species associated with the Mesa Verde old-growth piñon-juniper, the flammulated owl, has not yet been shown to utilize other piñon-juniper areas in Colorado.

Based on what we see in the Mesa Verde old-growth piñon-juniper woodlands, the avian community has many fascinating members but few truly unique species. It is difficult to make sweeping generalities about these birds as a group. We see many drably colored species that blend into the drably colored woodland. We find many cavity nesters. Many species are neotropical migrants. But each of these characteristics is matched by the opposite strategies of bright plumage, open cup nesting, or permanent residency. So what we have in the old-growth piñon-juniper really is a diverse, well-rounded avian community that does not define what makes this rare habitat exceptional, although the birds certainly do contribute to this habitat's special nature. Old-growth cannot support every piñon-juniper bird species. In fact, mid-seral-stage piñon-juniper may have a higher overall bird species richness than old-growth. Each piñon-juniper growth form, however, including old-growth, has its own distinct bird community. The key to conserving the piñon-juniper avifauna is to support the full range of natural habitat variability on a landscape level.

It is no secret why people have a special affinity for wild birds. We share their diurnal nature, so they are active when we are. Some are pleasing just because they are colorful. All birds may be heard easily, because they vocalize using sound frequencies that we can readily detect and identify. One of the most delightful experiences in the piñon-juniper is the springtime morning chorus of singing titmice, towhees, tanagers, nuthatches, wrens, robins, flycatchers, grosbeaks, vireos, and warblers, punctuated by the drumming of the woodpeckers. It is an exciting contrast to the cold, quiet winter.

The deer, chipmunks, wildflowers, and native old-growth piñon-juniper avifauna make up the charismatic inhabitants of what many people consider a forlorn environment. To safeguard these birds, we must learn how to sustain the whole ecosystem—and that includes the remnant wintering grounds for the neotropical migrants and the summer breeding grounds for northern species that only winter in or pass through the piñon-juniper woodland. One important step toward protecting what is left is to begin valuing the piñon-juniper woodland for what it is, just as it is. The whole community's conservation is within our hands.

APPENDIX 6.1. Birds of the Mesa Verde piñon-juniper woodlands, including breeding statistics from the *Colorado Breeding Bird Atlas* (Kingery 1998) in all piñon-juniper types statewide

Species	Mesa Verde National Park: Status in Piñon-Juniper	Mesa Verde National Park: Residency in Piñon-Juniper	Approximate % of Blocks in Colorado Piñon-Juniper	Breeding in Mesa Verde National Park Piñon-Juniper	State Piñon-Juniper Use Rank
Turkey Vulture *Cathartes aura*	C	Summer	21	Yes	1
Sharp-Shinned Hawk *Accipiter striatus*	U	Summer	13	Yes	3
Cooper's Hawk *Accipiter cooperii*	U	Summer	19	Yes	1
Northern Goshawk *Accipiter gentilis*	R	All Year	4	Nearby	8

continued on next page

APPENDIX 6.1—*continued*

Species	Mesa Verde National Park: Status in Piñon-Juniper	Mesa Verde National Park: Residency in Piñon-Juniper	Approximate % of Blocks in Colorado Piñon-Juniper	Breeding in Mesa Verde National Park Piñon-Juniper	State Piñon-Juniper Use Rank
Red-Tailed Hawk *Buteo jamaicensis*	FC	All Year	13	Yes	1
Ferruginous Hawk *Buteo regalis*	A	Absent	10	No	3
Golden Eagle *Aquila chrysaetos*	FC	All Year	11	Yes	2
American Kestrel *Falco sparverius*	U	Summer	10	Yes*	3
Prairie Falcon *Falco mexicanus*	R	All Year	10	Nearby	3
Peregrine Falcon *Falco peregrinus*	U	Summer	35	Yes	1
Wild Turkey *Meleagris gallopavo*	FC	All Year	17	Yes	2
Scaled Quail *Callipepla squamata*	R	Wandering	12	No	3
Gambel's Quail *Callipepla gambelii*	A	Absent	12	No	3
Band-Tailed Pigeon *Columba fasciata*	A	Migrant	8	No	5
Mourning Dove *Zenaida macroura*	R	Summer	13	Yes	2
Greater Roadrunner *Geococcyx californianus*	A	Absent	45	No	1
Flammulated Owl *Otus flammeolus*	U	Summer	0	Yes*	–
Western Screech-Owl *Otus kennicottii*	A	Summer	11	No*	3
Great Horned Owl *Bubo virginianus*	C	All Year	10	Yes	3
Northern Pygmy-Owl *Glaucidium gnoma*	U	Summer	15	Yes*	4
Mexican Spotted Owl *Strix occidentalis lucida*	U	All Year	21	Nearby	2
Long-Eared Owl *Asio otus*	FC	All Year	16	Yes	2
Northern Saw-Whet Owl *Aegolius acadiscus*	FC	All Year	22	Yes*	2
Common Nighthawk *Chordeiles minor*	U	Summer	22	Nearby	1
Common Poorwill *Phalaenoptilus nuttallii*	FC	Summer	43	Yes	1

continued on next page

APPENDIX 6.1—continued

Species	Mesa Verde National Park: Status in Piñon-Juniper	Mesa Verde National Park: Residency in Piñon-Juniper	Approximate % of Blocks in Colorado Piñon-Juniper	Breeding in Mesa Verde National Park Piñon-Juniper	State Piñon-Juniper Use Rank
White-Throated Swift					
Aeronautes saxatalis	C	Summer	11	Yes	2
Black-Chinned Hummingbird					
Archilochus alexandri	C	Summer	31	Yes	1
Broad-Tailed Hummingbird					
Selasphorus platycercus	C	Summer	9	Yes	5
Rufous Hummingbird					
Selasphorus rufus	C	Migrant	0	No	-
Lewis's Woodpecker					
Melanerpes lewis	R	Migrant	9	No*	5
Red-Naped Sapsucker					
Sphyrapicus nuchalis	U	Migrant	0	No*	-
Williamson's Sapsucker					
Sphyrapicus thyroides	R	Migrant	2	No*	9
Ladder-Backed Woodpecker					
Picoides scalaris	A	Absent	57	No*	1
Hairy Woodpecker					
Picoides villosus	FC	All Year	10	Yes*	4
Downy Woodpecker					
Picoides pubescens	U	All Year	6	Nearby*	6
Northern Flicker					
Colaptes auratus	C	All Year	12	Yes*	4
Olive-Sided Flycatcher					
Contopus cooperi	U	Migrant	2	No	9
Western Wood-Pewee					
Contopus sordidulus	U	Summer	6	Yes	7
Dusky Flycatcher					
Empidonax oberholseri	U	Summer	4	Yes	8
Gray Flycatcher					
Empidonax wrightii	U	Summer	85	Yes	1
Cordilleran Flycatcher					
Empidonax occidentalis	R	Summer	2	Nearby	13
Eastern Phoebe					
Sayornis phoebe	A	Absent	9	No	4
Say's Phoebe					
Sayornis saya	U	Summer	9	Yes	3
Ash-Throated Flycatcher					
Myiarchus cinerascens	FC	Summer	67	Yes*	1
Cassin's Kingbird					
Tyrannus vociferans	A	Absent	47	No	1
Western Kingbird					
Tyrannus verticalis	R	Summer	5	Yes	5

continued on next page

APPENDIX 6.1—continued

Species	Mesa Verde National Park: Status in Piñon-Juniper	Mesa Verde National Park: Residency in Piñon-Juniper	Approximate % of Blocks in Colorado Piñon-Juniper	Breeding in Mesa Verde National Park Piñon-Juniper	State Piñon-Juniper Use Rank
Violet-Green Swallow *Tachycineta thalassina*	FC	Summer	8	Yes*	5
Cliff Swallow *Petrochelidon pyrrhonota*	R	Summer	2	No	7
Steller's Jay *Cyanocitta stelleri*	C	All Year	9	Nearby	6
Western Scrub-Jay *Aphelocoma californica*	C	All Year	57	Yes	1
Pinyon Jay *Gymnorhinus cyanocephalus*	FC	All Year	87	Yes	1
Clark's Nutcracker *Nucifraga columbiana*	U	All Year	16	Nearby	2
Black-Billed Magpie *Pica pica*	U	All Year	12	Nearby	3
American Crow *Corvus brachyrhynchos*	U	All Year	9	Yes	4
Chihuahuan Raven *Corvus cryptoleucus*	A	Absent	16	No	2
Common Raven *Corvus corax*	C	All Year	16	Yes	1
Black-Capped Chickadee *Poecile atricapillus*	R	All Year	5	No*	6
Mountain Chickadee *Poecile gambeli*	C	All Year	17	Yes*	2
Juniper Titmouse *Baeolophus griseus*	C	All Year	92	Yes*	1
Bushtit *Psaltriparus minimus*	C	All Year	65	Yes	1
Red-Breasted Nuthatch *Sitta canadensis*	U	Migrant	7	No*	6
White-Breasted Nuthatch *Sitta carolinensis*	C	All Year	22	Yes*	1
Pygmy Nuthatch *Sitta pygmaea*	R	Migrant	2	No*	7
Rock Wren *Salpinctes obsoletus*	FC	Summer	20	Yes	3
Canyon Wren *Catherpes mexicanus*	FC	All Year	15	Yes	3
Bewick's Wren *Thryomanes bewickii*	C	All Year	67	Yes	1
House Wren *Troglodytes aedon*	FC	Summer	5	Yes*	7

continued on next page

APPENDIX 6.1—*continued*

Species	Mesa Verde National Park: Status in Piñon-Juniper	Mesa Verde National Park: Residency in Piñon-Juniper	Approximate % of Blocks in Colorado Piñon-Juniper	Breeding in Mesa Verde National Park Piñon-Juniper	State Piñon-Juniper Use Rank
Ruby-Crowned Kinglet *Regulus calendula*	FC	Migrant	0	No	–
Blue-Gray Gnatcatcher *Polioptila caerulea*	FC	Summer	51	Yes	1
Western Bluebird *Sialia mexicana*	C	All Year	17	Yes*	2
Mountain Bluebird *Sialia currucoides*	R	Summer	23	Yes*	1
Townsend's Solitaire *Myadestes townsendi*	U	Winter	7	No	6
Hermit Thrush *Catharus guttatus*	R	Summer	4	Yes	7
American Robin *Turdus migratorius*	U	Summer	7	Yes	6
Northern Mockingbird *Mimus polyglottos*	A	Absent	17	No	3
Sage Thrasher *Oreoscoptes montanus*	A	Absent	3	No	6
Curve-Billed Thrasher *Toxostoma rufum*	A	Absent	3	No	3
Loggerhead Shrike *Lanius ludovicianus*	A	Absent	9	No	4
Gray Vireo *Vireo vicinior*	U	Summer	88	Yes	1
Plumbeous Vireo *Vireo plumbeus*	C	Summer	34	Yes	1
Virginia's Warbler *Vermivora virginiae*	FC	Summer	18	Yes	2
Yellow-Rumped Warbler *Dendroica coronata*	C	Migrant	2	No	12
Black-Throated Gray Warbler *Dendroica nigrescens*	C	Summer	83	Yes	1
Hepatic Tanager *Piranga flava*	A	Absent	59	No	1
Western Tanager *Piranga ludoviciana*	FC	Summer	10	Nearby	4
Black-Headed Grosbeak *Pheucticus melanocephalus*	C	Summer	13	Yes	4
Blue Grosbeak *Guiraca caerulea*	R	Summer	8	No	4
Lazuli Bunting *Passerina amoena*	U	Summer	8	Yes	4

continued on next page

APPENDIX 6.1—*continued*

Species	Mesa Verde National Park: Status in Piñon-Juniper	Mesa Verde National Park: Residency in Piñon-Juniper	Approximate % of Blocks in Colorado Piñon-Juniper	Breeding in Mesa Verde National Park Piñon-Juniper	State Piñon-Juniper Use Rank
Indigo Bunting					
Passerina cyanea	A	Absent	9	No	5
Green-Tailed Towhee					
Pipilo chlorurus	U	Summer	6	Nearby	4
Spotted Towhee					
Pipilo maculatus	C	All Year	26	Yes	1
Canyon Towhee					
Pipilo fuscus	A	Absent	52	No	1
Rufous-Crowned Sparrow					
Aimophila ruficeps	A	Absent	43	No	1
Chipping Sparrow					
Spizella passerina	C	Summer	25	Yes	1
Brewer's Sparrow					
Spizella breweri	R	Migrant	2	No	9
Vesper Sparrow					
Pooecetes gramineus	U	Summer	3	Yes	9
Lark Sparrow					
Chondestes grammacus	R	Summer	10	No	4
Black-Throated Sparrow					
Amphispiza bilineata	A	Absent	33	No	1
Sage Sparrow					
Amphispiza belli	A	Absent	5	No	4
Dark-Eyed Junco					
Junco hyemalis	C	Winter	1	No	14
White-Crowned Sparrow					
Zonotrichia leucophrys	FC	Migrant	0	No	–
Western Meadowlark					
Sturnella neglecta	A	Absent	1	No	13
Brewer's Blackbird					
Euphagus cyanocephalus	R	Migrant	2	No	13
Brown-Headed Cowbird					
Molothrus ater	U	Summer	11	Yes	3
Bullock's Oriole					
Icterus bullockii	A	Summer	5	Nearby	4
Scott's Oriole					
Icterus parisorum	A	Absent	64	No	1
Cassin's Finch					
Carpodacus cassinii	U	Summer	7	Nearby	6
House Finch					
Carpodacus mexicanus	R	Summer	23	Nearby	2
Red Crossbill					
Loxia curvirostra	U	Wandering	3	No	8

continued on next page

APPENDIX 6.1—*continued*

Species	Mesa Verde National Park: Status in Piñon-Juniper	Mesa Verde National Park: Residency in Piñon-Juniper	Approximate % of Blocks in Colorado Piñon-Juniper	Breeding in Mesa Verde National Park Piñon-Juniper	State Piñon-Juniper Use Rank
Pine Siskin					
Carduelis pinus	FC	Summer	7	Yes	6
Lesser Goldfinch					
Carduelis psaltria	FC	Summer	20	Yes	2
American Goldfinch					
Carduelis tristis	A	Absent	3	No	6
Evening Grosbeak					
Coccothraustes vespertinus	U	Migrant	0	No	–

Key: C = Common, FC = Fairly Common, U = Uncommon, R = Rare, A = Absent
* = Tree Cavity Nester
Block = atlas sampling area division of approximately 10 square miles
Use Rank = order of habitat preference for piñon-juniper of each species, with a total of 63 habitat types defined in the atlas for all of Colorado (total of 470 piñon-juniper blocks in the atlas)

REFERENCES CITED

Andrews, R., and R. Righter. 1992. *Colorado birds: A reference to their distribution and habitat.* Denver Museum of Natural History, Denver, Colo.

Bailey, A. M., and R. J. Neidrich. 1965. *Birds of Colorado.* Volumes 1 and 2. Denver Museum of Natural History, Denver, Colo.

Balda, R. P. 1972. The breeding biology of the pinyon jay. *Living Bird* 12: 5–42.

Balda, R. P., and G. C. Bateman. 1971. Flocking and annual cycle of the pinyon jay, *Gymnorhinus cyanocephalus. Condor* 73: 287–302.

Christensen, K. M., and T. G. Whitham. 1991. Indirect herbivore mediation of avian seed dispersal in pinyon pine. *Ecology* 72: 534–543.

Floyd, M. E. 1987. The significance of variability in cone production in *Pinus edulis.* In R. L. Everett (ed.), *Proceedings: Pinyon-Juniper Conference,* pp. 58–64. United States Forest Service General Technical Report INT-215.

Forcella, F. 1981. Ovulate cone production in pinyon: Negative exponential relationship with late summer temperature. *Ecology* 62: 488–491.

Kingery, Hugh E. (ed.). 1998. *Colorado breeding bird atlas.* Colorado Breeding Atlas Partnership and the Colorado Division of Wildlife, Denver.

Ligon, J. D. 1971. Late summer-autumnal breeding of the piñon jay in New Mexico. *Condor* 73: 147–153.

———. 1974. Green cones of piñon pine stimulate late summer breeding in piñon jay. *Nature* 250: 80–81.

———. 1978. Reproductive interdependence of pinyon jays and pinyon pines. *Ecological Monographs* 48: 111–126.

Miller, Rick. 2000. Songbirds in the pinyon and juniper ecosystems. Oral presentation at the Pinyon-Juniper Workshop, November 2000, Zion National Park, Utah.

Monsen, Stephen B., and Richard Stevens. 1999. *Proceedings: Ecology and management of pinyon-juniper communities within the intermountain West.* Rocky Mountain Research Station. U.S. Forest Service, Fort Collins, Colo.

Pasquier, Robert F. 1980. *Watching birds: An introduction to ornithology.* Houghton Mifflin Company, Boston, Mass.

Pearson, T. Gilbert. 1936. *Birds of America.* Doubleday and Company, Garden City, N.J.

Poole, Robert M. 1983. *The wonder of birds.* National Geographic Society, Washington, D.C.

Reynolds, Richard T., and Charles L. Johnson. 1996. Research objectives for Mexican spotted owl in Mesa Verde National Park. Unpublished paper. Rocky Mountain Forest and Range Experiment Station. U.S. Forest Service, Fort Collins, Colo.

Silvertown, J. W. 1980. The evolutionary ecology of mast seeding in trees. *Biological Journal of the Linnean Society* 14: 235–255.

Tomback, D. F. 1982. Dispersal of whitebark pine seeds by Clark's nutcracker: A mutualism hypothesis. *Journal of Animal Ecology* 51: 1–46.

Vander Wall, S. B. 1988. Foraging of Clark's nutcracker (*Nucifraga columbiana*) on rapidly changing pine seed resources. *Condor* 90: 621–631.

Vander Wall, S. B., and R. P. Balda. 1977. Coadaptations of the Clark's nutcracker and the piñon pine for efficient seed harvest and dispersal. *Ecological Monographs* 47: 89–111.

Whitham, T. G., and S. Mopper. 1985. Chronic herbivory: Impacts on architecture and sex expression in pinyon pines. *Science* 228: 1089–1091.

MAMMALS OF THE OLD-GROWTH PIÑON-JUNIPER

Preston Somers, David D. Hanna, Marilyn Colyer, and M. Lisa Floyd

If you sit silently at length on the Mesa Verde cuesta, you just might be rewarded with a glimpse of one of the 400 species of vertebrates that spend time in the piñon-juniper woodland. The protection of sandstone-rimmed mesas has allowed piñon-juniper communities to mature, and as they approach old-growth, the complexity of physical structure increases (Chapter 2). Branches of adjacent trees form a dense network, downed limbs persist on the woodland floor, juniper bark is shed, and crevices form in rotting pockets. The intricate array of wood, leaf, cones, and fruit provides a home, or at least a temporary shelter, for many of these species, but no species are confined to the Mesa Verde piñon-juniper woodland or endemic to our region. Roughly 250 of the recorded species are birds, about 17 are bats, 10 are amphibians, and 27 are reptiles (groups treated elsewhere in this volume). Much of what we have gleaned about vertebrate associates of piñons and junipers comes from observations, experiences, and research (Anderson 1961; Colyer, unpublished; Douglas 1963, 1966, 1969; A. Spencer, unpublished) in Mesa Verde National Park and other areas of southwestern Colorado (Somers 1976a, 1976b, 1977, 1979, 1980).

In this chapter we focus first on the mammals that currently or historically range into piñon-juniper woodlands and illustrate a few of their interesting ecological relationships. Then we turn our attention to one of the most common small mammals of our region, *Peromyscus maniculatus* (deer mouse), describing myriad interactions with other mammals, piñon and juniper trees, shrubs and forbs, fungi, and viruses that shape its unique and integrated place in the piñon-juniper woodlands.

MAMMALS OF THE PIÑON-JUNIPER WOODLANDS

Of the 120 native mammals counted among the Colorado fauna (Armstrong 1972), 74 have been seen in Mesa Verde Country (Anderson 1961). Including humans, 8 mammal orders are represented. The Rodentia are most abundant, with 32 species of rodents; also well represented are Insectivora, with 5 species of shrews, and Chiroptera, with 15 species of bats. In addition, 4 lagomorphs (hares and cottontail rabbits), 20 carnivores (3 of which have been extirpated, and 2 are limited to riparian habitats), and 4 native and 5 exotic ungulates are present. This discussion ignores the bats, a very important group of mammals, because they are considered in greater detail in Chapter 8.

To what extent is the life of a mammal enhanced by the physical structure and biological complexity of the old-growth woodland? And, in a reciprocal way, what do mammals contribute to the woodland? We approach these questions by digging into four decades of natural history observations and collected data (Anderson 1961; Colyer, unpublished; Lechleitner 1969; A. Spencer, unpublished). Using this wealth of information, we estimate whether each species spends much of its time in the piñon-juniper woodlands or if it is more widely distributed (resident status). We also estimate the density of each species in Mesa Verde Country. Density is multiplied by the size and weight of an individual, allowing us to estimate the biomass contributed by each species. These life characteristics are summarized in Appendix 7.1.

A large mammalian species, while rare, might influence the biotic interactions in the piñon-juniper woodland in a significant way, while other tiny mammals that are numerous may also have a great impact. Thus the importance of mammals can be considered by their mass or by their number. The desert shrews are the most abundant Insectivores; the most abundant carnivores are striped skunks and long-tailed weasels; and by far the most numerous rodents are the deer mice, with chipmunks also in the running. Cottontails lead the lagomorphs in density, and mule deer are the most abundant ungulates. If one also considers the size of each organism and the resultant total biomass, however, the "winners" would be different in some categories. The black bear, gray fox, and badger would come out on top of the carnivores. Porcupines and Gunnison's prairie dogs would overwhelm even the deer mouse. Lagomorphs and ungulate "winners" remain the same whether one considers density or mass.

By either criterion, the woodland must support these mammals; and they must be able to contend with the environmental challenges imposed by the woodland. We have selected here a few examples to demonstrate how several Mesa Verde mammals may adapt to the challenges imposed by a semiarid woodland home: food availability and aridity.

AVAILABLE FOOD IN THE PIÑON-JUNIPER WOODLAND

Food is one of the main factors limiting populations of most animals in piñon-juniper woodlands, and we find that most successful vertebrates are generalist feeders. They are capable of facultative exploitation of a wide variety of foodstuffs while finding cover in rocks, shrubs, or trees. Their diets vary according to an ever-changing balance of nutritive value, availability, palatability, and abundance.

Food, at least of desirable quality, is particularly limited in the fall and spring of the year. In the piñon-juniper stands on Mesa Verde, the most abundant ground cover consists of *Poa fendleriana* (muttongrass). Muttongrass stays green late into the fall and greens up early in the spring (Colyer, unpublished), and it may be one of the few forbs available at these times under the piñons and junipers. Deer and cottontails avidly exploit the green grass in preference to all other foods during these seasons. Other examples of seasonal shifts in foods abound. White-tailed antelope ground squirrels feed mainly on succulent green herbs in the spring. As the herbs dry in the heat of summer, the squirrels switch to consuming mainly seeds, insects, and occasionally small vertebrates or carrion. Similar shifts characterize the feeding choices of most rodents. Carnivores likewise are opportunistic in their choice of fare. Coyotes feed on small voles when they are abundant but shift to cottontails and jackrabbits when they are numerous. The use of juniper female "berries" by coyote and foxes is currently being investigated (Colyer, unpublished). During the dry autumn and winter of 1999–2000, the scats of these "meat eaters" consisted almost entirely of juniper seeds, the residuum left after the pulp of the "berry" has been digested.

Only a few vertebrates depend upon specific plants for food. One that ranges into the edge of our area is Stephen's woodrat (Vaughan 1982), which feeds almost exclusively on juniper foliage. Juniper is consistently available and little exploited by other animals, but presents a problem in assimilation. Its content of terpenoids, quinones, and resins inhibits bacterial digestion. Thus the Stephen's woodrat's habitual diet provides a dependable source of food but limits the rodent's potential for rapid increase; the rat's litter is usually only a single small, slow-growing, late-weaned young per season. Females lose weight while lactating and are slow to regain it on a diet of juniper (Vaughan and Czaplewski 1985).

Some plants are less desirable as food sources than others for a given consumer. Plants produce a great variety of secondary products, and the function of many is still unknown. Piñons produce a variety of monoterpenes, for example. The specific types of monoterpenes may vary among trees in a stand or within different tissues in an individual. In ponderosa pines, monoterpene composition of the phloem and cones differs from needles, whereas roots and woody stem monoterpenes are similar (Latta et al. 2000). One of the few mammalian herbivores that feeds on piñons, the Abert's squirrel, discriminates among individual conifers on the basis of the terpenes (Farentino et al. 1981; Reynolds 1966; Snyder and Linhart 1998). Rabbits and deer often feed on scattered shoots

Figure 7.1. Odocoileus hemionus *(mule deer) is commonly seen in piñon-juniper woodlands and open sagebrush shrublands in Mesa Verde. Illustration by Walt Anderson.*

---

dropped by the squirrels, possibly taking advantage of the squirrels' discriminating palate.

Other plants may be so distasteful that they are ignored by a specific herbivore. Our piñon-juniper communities include two species of sagebrush, *Artemisia tridentata* (big sagebrush) and *A. nova* (black sagebrush). The latter is preferred by mule deer (Figure 7.1). Such discrimination favors the increase of big sagebrush at the expense of black sagebrush. Herbivores exert many similar effects on the composition of the community. Fortunately, different herbivores choose different forage species at different times. During the dry, snow-free winter of 1999–2000, Dr. Albert Spencer (professor emeritus at Fort Lewis College) observed that cottontails mowed down nearly all *Gutierrezia sarothrae* (snakeweed), a low shrub that ordinarily is spurned by grazing animals. The rabbits were eating only a very short section of the base of the annual shoots, no more than 2 or 3 cm in length, discarding the remaining 15 to 20 cm. Similarly, most vertebrates shun *Chrysothamnus nauseosus* (rabbitbrush); but Spencer also observed during one heavy snow winter that voles stripped the bark of hundreds of rabbitbrush plants on a 3-ha experimental plot southeast of Mesa Verde, killing many of the shrubs.

Finally, we must remember that availability incorporates both physical accessibility and a measure of the risk involved in harvesting. Rodents may forego seeds scattered in open stands of herbs for scantier stores located under the screen of shrubs, less exposed to predators. Mountain sheep and deer may forsake lush fodder and retreat to barren windswept ridges to escape biting midges and flies. In bright moonlight the deer mouse does not forage as extensively in open areas as it does during dark nights (A. Spencer, unpublished).

ADAPTATIONS TO ARIDITY

The piñon-juniper zone is semiarid. Free water is absent most of the year, so obtaining and conserving water is a constant problem for all inhabitants. The problem is exacerbated by the variability of the climate both regionally and locally. For example, annual precipitation totals for water-years (October 1 to September 30) at the weather station in Mesa Verde National Park have ranged from 23 cm to 85 cm over the last decade. During that period there have been many intervals in the late spring and early summer when several weeks have passed without a trace of rain. The longest dry interval in a seven-year sample of spring and summer daily precipitation was forty-eight days.

When experimentally fed dry food at room temperature, the deer mouse (Figure 7.2) drinks 0.16–0.25 g H_2O/g body wt/day (A. Spencer, unpublished). Somehow a deer mouse has to generate or obtain 3 to 6 ml of water per day to replace what is lost to evaporation, urine, and feces. Deer mice consume succulent vegetation when it is available, gain some water from insect and vertebrate prey, and generate some by oxidation of carbohydrates and fats.

Coyotes (Figure 7.3) and kit foxes illustrate different solutions to the problem. Kit foxes have small bodies and correspondingly small ranges and rarely encounter—much less drink—free water. Their food of insects and small mammals supplies their water needs as well as their energy. They hunt at night and retreat to their den during the day; thus they lose very little water cooling themselves even in the warmest weather (Golightly and Ohmart 1983). Coyotes are famous travelers, ranging across many kilometers in their search for food, and can be active at almost any time of day or night (Henke and Bryant 1999). They

Figure 7.2. Peromyscus maniculatus *(the deer mouse) is a common mouse of the piñon-juniper woodland, which acts as a seed disperser for both tree species. Illustration by Walt Anderson.*

are relatively profligate with their water reserves, using evaporation from their lungs and tongues to maintain their body temperature in the heat. They actually lose 155 percent of the water predicted for an animal of their size during warm weather. Compared with kit foxes, coyotes are protected to some extent by their greater thermal mass. The water obtained in their food, however, is inadequate for their needs. They make up the deficit by drinking at springs and streams within their range (Golightly and Ohmart 1983).

Cottontails, the coyote's favorite prey, face the same stresses. Nuttall's cottontails are competent climbers. They use their ability to consume juniper foliage when the grass dries up. This bolsters their water intake. Moisture obtained from the juniper's scaly foliage supplements the metabolic water obtained from respiration of carbohydrates.

Animals at the lowest elevations on Mesa Verde or in the deep canyons face even greater stresses, so that they rely on all the methods outlined above. Antelope ground squirrels are diurnal, exposed to great heat stress. They change their feeding habits through the year in response to the change in quality and water content of food. In the spring they feed largely on succulent herbs. When these dry up, they forage for seeds (a more labor-intensive task); but, by respiring the

Figure 7.3. Coyotes have continuously captured people's imagination. They are famous travelers, ranging across many kilometers in their search for food, and can be active almost any time of day or night. Coyotes are therefore one of the more visible mammals of Mesa Verde Country. Illustration by Walt Anderson.

carbohydrates and fats in the seeds, they obtain increased metabolic water. They take every opportunity to supplement this source with water from any insects, small vertebrates, or carrion that they come across. This species, like many desert dwellers, can sustain its body water by drinking saltwater at concentrations of nearly 150 percent that of seawater (Bartholomew and Hudson 1959). That is rarely an option, but the capacity indicates the concentrating power of their kidneys.

Arid-land mammals lose very little water in their urine. The urine of pocket mice can attain concentrations half again greater than that of antelope ground squirrels. Pocket mice can survive indefinitely on a diet of seeds alone with no supplemental water, deriving sufficient water from metabolism of carbohydrates and fats to maintain their osmotic balance. Like kit foxes, pocket mice are nocturnal. They meet stress in yet another fashion. During periods of food shortage they spend much of their time in torpor. One pocket mouse held in captivity became cool and unresponsive during much of the day. At these times its inspirations dropped to only one or two per minute. Toward evening it would rouse itself, shivering violently for several minutes to raise its body temperature (A. Spencer, unpublished). By suspending activity, both water loss and the need to forage for food are minimized.

Because rodent urine is so concentrated and often contains calcium oxalates derived from plants, some of the salts precipitate, making the urine cloudy and viscous. Deposits of packrat urine coat the rocks of their dens and glue their feces and midden materials into a solid mass (Figure 7.4). This has benefited students of past climates and prehistoric plant communities. The layered remains preserve a sequential record of the composition of the vegetation through time. Much of what we know of the distributions of plants over the past 10,000 years has been learned by the analysis of packrat middens accumulated in rock shelters (Betancourt 1990).

These examples illustrate just a few of the myriad physiological and behavioral adaptations of mammals living in the arid Mesa Verde Country. It is not simply coincidental that in making shifts in diet the mammals are also maximizing their intake of water and other essential nutrients.

A CASE STUDY: DEER MICE AND THEIR INTERACTIONS WITHIN THE PIÑON-JUNIPER WOODLAND

Arid conditions and scarce food present challenges to mammalian species, but they do not contend with these alone. While we can easily appreciate the myriad biotic relationships that mammals enter into, understanding the intricacy of these mutual interactions is a formidable task. Let us consider simply the known relationships surrounding the life of the deer mouse, a rodent we have shown is one of the most abundant in the woodland (Appendix 7.1).

First, what plant material do deer mice utilize? Deer mice diligently collect and transport seeds from a great variety of plants, among them *Purshia tridentata* (bitterbrush) and many other low shrubs, piñons, and junipers (Figure 7.5). The rodents are not collecting and burying the seeds with the tree's interest in mind. The seeds are their food supply during the fall, winter, and early spring, when (in many years) snow lies on the ground for five months. But some of their cached seeds escape being devoured and germinate, a significant factor in recruitment of new plants into the community.

Figure 7.4. Neotoma mexicana *(Mexican woodrat) is the most common woodrat in the piñon-juniper woodland, shown here with materials used to build its stick nests. Illustration by Walt Anderson.*

"Larders" of bitterbrush seed are stashed in hollows of juniper trees in Mesa Verde (Colyer, unpublished). The mice also bury many seeds in the soil under bushes and around trees, in "scatterhoards." Observations at Mesa Verde suggest that bitterbrush seedlings germinate more successfully when several are placed

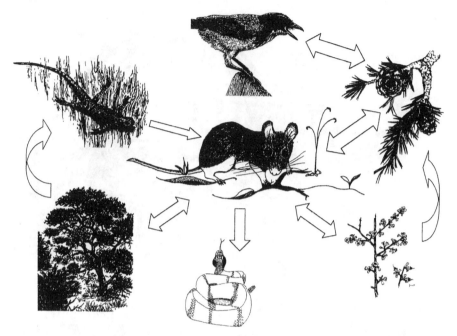

Figure 7.5. The complexity of interactions among woodland organisms is illustrated in this simplified web from the perspective of the deer mice. The direction of the arrow indicates the organism that receives a positive impact from the association. Illustration by Sarah Luecke and Walt Anderson.

together than as single seeds. Rodents tend to plant several seeds together in a cluster. Furthermore, remnants of the flower parts surrounding the fruit inhibit germination. Mice carefully remove this portion before burying the seed in clusters (Mozingo 1987). In the spring little clusters of seedlings pop up, and mice return to the clusters to consume the carotene-rich cotyledons. Yet some seedlings escape as recruits to the bitterbrush population. Bitterbrush clusters were commonly seen poking through the charred soils after the Bircher and Pony fires that occurred in 2000 on Mesa Verde. Caching behavior is not confined to piñon-juniper woodlands; S. B. Vander Wall (1994) found chipmunks similarly involved in bitterbrush recruitment in ponderosa-pine forests.

Piñon pines produce large cone crops every five or seven years and few cones in the intervening years, a phenomenon known as masting. In heavy mast years mice gather a significant fraction of the seeds, which disappear throughout the stand in a few days. In a study of rodents and *Pinus monophylla* (single-leaf piñon), the half-life of piñon seeds was measured in terms of hours (Vander Wall 1994). Recent experimental evidence suggests that on sandy loam soils *Peromyscus* may cache more seeds, whereas on cinder soils more are eaten, a phenomenon with possible impacts on piñon pine dispersal (Covert et al. 2001). Piñon nuts buried by mice in scatter hoards had higher germination rates and often (36%) were located in better sites for establishment than seeds placed at random (Vander Wall and Balda 1977). It is well-known that corvids, especially piñon jays, scrub-jays, and Clark's nutcrackers, also cache piñon seeds in variously sized caches after mast crops are produced (Ligon 1978; Vander Wall and Balda 1977). Many of these caches are forgotten, becoming excellent germination sites. Marilyn Colyer has observed that piñon jays preferentially cache their seeds on south slopes that are more likely to be snow-free, making the seeds more accessible throughout the winter. S. B. Vander Wall and R. Balda (1977) found a piñon seed may be transported by rodents as much as 38 m; this is a short dispersal compared with the distances a nutcracker or jay might carry it, but mice are less likely to put the seed in an unfavorable environment for establishment. In fact, he found that rodents tended to put seeds under shrubs (potential nurse plants) in more favorable sites for germination, an observation confirmed at Mesa Verde (A. Berry and M. L. Floyd, unpublished).

How do birds and deer mice compare as dispersal agents for piñon (and other) seeds? Considering the substantial volume of seeds gathered by mice, the abundance of mice in the community, and their full-time presence, one must recognize the importance of mice as agents of regeneration in the community. Mice are permanent, full-time residents. In contrast, piñon jays and nutcrackers are present only part-time. Although the birds travel in big flocks, they are constantly moving over enormous areas, so that their average density on a particular day at a particular site is comparatively small. The birds are very specialized in their food choice, whereas mice are much more eclectic in their harvest, affecting a wider assortment of plant species. It may also be important that mice are relatively short-lived compared with seed-caching birds and are more vulnerable to predation and stress. Their cached seeds may go unclaimed more often than those of birds, which are renowned for their ability to recall the sites where they buried their seeds (Ligon 1978; Vander Wall and Balda 1977; Whitham 1982).

The effect of the magnitude of piñon reproduction on deer mice populations is significant. Deer mice populations balloon in years when the piñons mast. At these times deer mice continue breeding into the winter, presumably as long as abundant stores of the energy-rich seeds remain. This has become a matter of great ecological interest in recent years. Deer mice harbor the Sin Nombre strain of hantavirus, which has proved fatal to about one-half of humans contracting the illness. The prevalence of mice carrying the virus has been monitored for several years in various localities in western Colorado, and a relationship between seed production and mouse numbers was detected (Root et al. 1999). In some sample periods the proportion of mice carrying the virus or possessing antibodies to it has exceeded 25% (K. Abbot, unpublished).

Piñon pines not only serve the mouse as a food source but also provide foraging substrate, shelter, cache sites, perches, and cover from predators. In addition, they buffer and modify the physical environment.

Junipers also provide physical shelter and food for the mice. *Juniperus* species are less nutritiously important but more consistent as food (Vaughan 1982; Vaughan and Czaplewski 1985). The longevity of the juniper snags and the shaggy characteristics of the bark may make these trees more important as nesting sites than the piñon. Interactions between the deer mouse and the multitude of shrub species (sagebrush, fendlerbush, skunkbush, Gambel oak) in the community are similar to those with the trees. Many are seasonal sources of food, and they provide shelter for the mice.

How is the deer mouse influenced by other herbivores? First, consider how the deer mouse avoids direct competition with other mice in the same genus—primarily by spatial partitioning of resources (Douglas 1969; Wolff et al. 1985). The very abundant and widespread deer mouse, perhaps the most euryoecious (living in widely diverse habitats) mammal in North America, is found in all the habitats that occur in Mesa Verde (Douglas 1969; Vander Wall 1997). The deer mouse may be less abundant at times in the piñon-juniper woodland than *P. truei* (piñon mouse), but it is by far the most abundant over all habitats combined. The piñon mouse is most abundant in the piñon-juniper woodland, although it also occurs in ponderosa pine–Gambel oak woodland, in mountain shrub, in sagebrush, and on rocky slopes in our area (Anderson 1961; King 1968). *Peromyscus boylii* (brush mouse) is most abundant in the mountain-shrub stands and occurs in adjacent piñon-juniper woodlands and sagebrush shrublands in smaller numbers than either the piñon mouse or the deer mouse (Holbrook 1978). *Peromyscus crinitus* (canyon mouse) is limited to the near-vicinity of precipitous rock outcrops. It is the most stenotopic (narrow in habitat choice) of the four species.

Habitat preferences of brush mice, piñon mice, and deer mice are clearly demonstrated in a series of studies conducted in southwestern Colorado (Somers 1976a, 1976b, 1977, 1979, 1980). While deer mice are nearly equally abundant in all habitats, piñon mice are far more abundant in piñon-juniper woodlands than in riparian areas or sagebrush habitats. Although overall abundance is relatively low, brush mice are more prevalent in piñon-juniper habitats, mixes of woodland and shrublands, and riparian areas (Figure 7.6).

Interactions between mice and other herbivore genera revolve around competition for resources. The least chipmunk is commonly encountered in the woodland. The chipmunk's small size, similar food selection, and population density make it the most important sciurid interacting with deer mice. Rock squirrels and golden-mantled ground squirrels are relatively abundant in old-growth piñon-juniper (Appendix 7.1). The rock squirrel consumes substantial volumes of seeds that otherwise would be available to *Peromyscus*. Many rodents, including the deer mouse, make use of the castoff prunings of squirrels for food. All of these species share many predators and infectious agents as well.

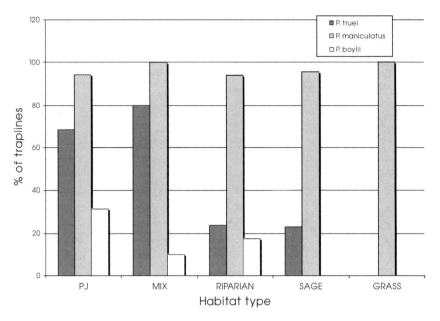

Figure 7.6. The relative abundance of three species of Peromyscus *near Dolores, Colorado (Somers, unpublished).*

Pocket gophers probably compete with deer mice indirectly for resources but in a different milieu. Their most important relationship possibly lies in their roles as alternative prey for predators and as creators of burrows that deer mice and snakes both utilize.

Porcupines are so different in size, foods, and choice of surroundings that any interaction seemingly would be rare and relatively insignificant. But porcupine quills, skin, and bone have been found chewed upon by mice. Porcupine hair is frequently incorporated into *Peromyscus* nests (A. Spencer, unpublished).

The main impact of ungulates on deer mice must be indirect through effects of grazing. They might also trample mice and their nests and exploit shared resources (e.g., seed and fruit production of herbs and shrubs). Mice consume parts of carcasses and skeletons and use hair for nesting materials. Mice may also benefit from waterholes that are dug out by ungulates, bringing water to the surface. The mice, by acting as alternative hosts or reservoir hosts of pathogens, may have significant effects on ungulates.

Do mice interact with lizards, snakes, and other reptiles? Little has been documented about the relationships between deer mice and the various lizard species that inhabit the old-growth piñon-juniper. While they both eat arthropods, there is

Figure 7.7. The spotted skunk is a frequent inhabitant of the old-growth piñon-juniper woodlands of Mesa Verde Country. Illustration by Amy Wendland.

an ecological separation between the diurnal reptiles and the nocturnal mice. More directly, some large lizards like the collared lizard are minor predators of mice, and mice may occasionally prey on young lizards. The predator-prey relationship between mice and snakes is obvious. Rattlesnakes and the larger constrictors (e.g., gopher snakes, kingsnakes, and whipsnakes) eat mice; mice may prey on young snakes and eggs. Mice and snakes are prey for many of the same avian and mammalian predators, thereby constituting links in an important multispecies web.

What can be said about the relationships between mice and the carnivores that roam Mesa Verde Country? All predators have an impact on the population dynamics of prey species, reducing reproductive potential and affecting behavioral patterns; naturally, mice forego some resources when doing so would expose them to predators.

Coyotes and gray, red, and kit foxes feed on mice to some extent (Koehler and Hornocker 1991; Lechleitner 1969). Raccoons and ringtails are determined mousers. Bobcats and mountain lions take mice at least occasionally (Leichleitner

Table 7.1. Parasites identified in *Peromyscus truei* and *Peromyscus maniculatus*

Parasite	Number of Parasitic Species	
	P. truei	P. maniculatus
Trematoda (flukes)		4
Cestoda	1	18
Nematoda	2	16
Acarina (mites)	18	78
Acarina (ticks)	7	17
Acarina (chiggers)	15	58
Insecta (fleas)	31	98
Insecta (lice)	2	6
Insecta (skin-bot)	1	2
Pentatomida	1	
Acanthocephala		1
Total species	78	298

Source: From Whitaker 1968.

1969). Bears have an unusual impact on rodents—they regularly exploit and destroy mouse nesting sites in logs, in stumps, in tree holes, and under large rocks, locations that are inaccessible to most other predators. Thus, despite the scarcity of bears in the woodland, their impact may be greater than that of more abundant predators.

Weasels are potentially among the most influential predators of mice because of their small size and quickness. The weasels' relatively short life history and high fecundity make them the mammal group most able to expand their numbers rapidly to exploit bursts in mouse populations. Badgers are very important predators, but their presence may have a positive effect as well, since abandoned and collapsed burrows are frequently taken over as dens and nest sites by mice, snakes, foxes, weasels, skunks, and even coyotes. Skunks, especially spotted skunks (Figure 7.7), are fairly efficient predators of mice and utilize many of the same resources.

Finally, deer mice interact with bacteria, viruses, and fungi as well. Seed-eating mammals (including deer mice, piñon mice, pocket mice, and kangaroo rats) often hull the seeds before caching them or putting them into their larders. In fact, pocket mice and kangaroo rats prefer slightly moldy seeds, one likely consequence of removing the seed coat (Rebar and Reichman 1983). Perhaps the fungus removes or reduces the toxic secondary compounds in the seeds. It is also possible that the mice and kangaroo rats are just taking advantage of the predigestion accomplished by the fungus.

Animals themselves are communities. Each is host to many endo- and ecto-parasites, and their abundance can limit mammal populations. Deer mice are no exception to this; for example, seventy-seven different species of helminth and arthropod parasites have been described on piñon mice (Whitaker 1968). Deer mice had three times as many parasitic species as did piñon mice (Table 7.1). Many parasites are not host-specific. They could transfer from deer mouse to piñon mouse or vice versa, or possibly to a number of other mammals in the community.

Sometimes another mammal can be passively involved in these host-parasite interactions. Coyotes could easily transport fleas from one rodent to another, possibly across many kilometers of country. Coyotes are constantly putting their snouts into one mouse hole or another. Many fleas are waiting in the holes for a new host to appear. The larval phases of fleas develop there, with the larvae feeding on blood deposited as fecal pellets by the adults chewing away at the mammal. They are attracted to any warm presence and hop on the coyote but, after sampling its blood, may find it unsuitable and disembark at another hole.

Deer mice are apparently resistant to most of the viruses currently identified in their communities (Fairbrother and Yuill 1984). Their relationship with the Sin Nombre hantavirus as described earlier, however, has significant impacts on other species, especially humans. No one has identified the full complement of symbiotic and parasitic bacteria, protozoa, viruses, and other microorganisms associated with any of our native species; nor has anyone worked out the range of relationships between these species and their hosts. Nevertheless, these interactions have important influences on the mammals that live in the piñon-juniper woodland.

Life in the piñon-juniper woodland is a complex network. Although the story of the life of a deer mouse portrays a few of its interactions (Figure 7.5), we have barely scratched the surface. We can only guess how the gnarled and shaggy old trees, full of holes and pockets, provide a home for deer mice and speculate on the interactions among the groups we have discussed above. At best, we have glimpsed the intricate web of interactions that may occur by focusing on the common deer mouse, whose interactions throughout the piñon-juniper community are indeed profound.

APPENDIX 7.1. Mammals of Mesa Verde Country: Estimated densities and biomass

Common Name	Species	Resident Status	Est. Median Density (no./km^2)	Body Length (mm)	Body Weight (g)	Est. Biomass (kg/km^2)
INSECTIVORES						
Desert Shrew	Notiosorex crawfordi	Shrublands	200	55	7	1.4
Dusky Shrew	Sorex monticolus	Widespread	100	65	10	1
Merriam's Shrew	Sorex merriami	Shrublands	100	55	8	0.8
Dwarf Shrew	Sorex nanus	Chapparal	100	50	5	0.5
Water Shrew	Sorex palustris	Aquatic	10	85	14	0.14
Masked Shrew	Sorex cinereus	Marginal	0.001	55	7	0.000007
CARNIVORES						
Coyote	Canis latrans	Ubiquitous	1	880	15,500	15.5
Black Bear	Ursus americanus	Widespread	0.1	1,435	120,000	12
Gray Fox	Urocyon cinereoargenteus	Widespread	2	675	4,500	9
American Badger	Taxidera taxus	Ubiquitous	1	550	8,200	8.2
Striped Skunk	Mephitis mephitis Richardson	Widespread	2	480	3,100	6.2

continued on next page

APPENDIX 7.1—continued

Common Name	Species	Resident Status	Est. Median Density (no./km^2)	Body Length (mm)	Body Weight (g)	Est. Biomass (kg/km^2)
Mountain Lion	*Puma concolor*	Ubiquitous	0.1	1,265	53,000	5.3
Raccoon	*Procyon lotor pallidus*	Widespread	0.2	690	6,300	1.26
Western Spotted Skunk	*Spilogale putorius*	Widespread	1	250	900	0.9
Bobcat	*Lynx rufus*	Ubiquitous	0.1	750	8,200	0.82
Long-Tailed Weasel	*Mustela frenata*	Ubiquitous	2	325	298	0.596
Ringtail	*Bassariscus astutus*	Widespread	0.5	330	1,000	0.5
Red Fox	*Vulpes vulpes*	Marginal	0.1	625	4,700	0.47
Kit Fox	*Vulpes velox*	Rare	0.05	450	2,000	0.1
River Otter	*Lutra canadensis*	Aquatic	0.01	780	9,000	0.09
Short-Tailed Weasel	*Mustela erminea*	Marginal	0.1	200	103	0.0103
American Mink	*Mustela vison*	Riparian	0.001	425	925	0.000925
Grizzly	*Ursus arctos*	Extirpated	0.000001	1,820	200,000	0.0002
Gray Wolf	*Canis lupus*	Extirpated	0.000001	1,125	36,500	0.0000365
Black-Footed Ferret	*Mustela nigripes*	Extirpated	0.00001	550	650	0.0000065
RODENTS						
Porcupine	*Erethizon dorsatum*	Ubiquitous	20	550	5,300	106
Gunnison's Prairie Dog	*Cynomys gunnisoni*	Shrublands	100	300	1,000	100
Rock Squirrel	*Spermophilus variegatus*	Ubiquitous	100	315	850	85
Deer Mouse	*Peromyscus maniculatus*	Ubiquitous	2,000	68	25	50
Mexican Woodrat	*Neotoma mexicana*	Widespread	125	182	225	28.125
White-Throat Woodrat	*Neotoma albigula*	Shrublands	125	206	200	25
Colorado Chipmunk	*Tamias quadrivittatus*	Widespread	300	140	60	18
Piñon Mouse	*Peromyscus truei*	Widespread	400	96	35	14
Valley Pocket Gopher	*Thomomys bottae*	Ubiquitous	100	150	125	12.5
Northern Pocket Gopher	*Thomomys talpoides*	Marginal	100	119	110	11
Least Chipmunk	*Tamias minimus*	Ubiquitous	200	137	50	10
Yellow-Bellied Marmot	*Marmota flaviventris*	Talus	2	525	3,350	6.7
Bushy-Tailed Woodrat	*Neotoma cinerea*	Widespread	20	212	280	5.6
Mexican Vole	*Microtus mexicanus*	Shrublands	160	100	35	5.6
Long-Tailed Vole	*Microtus longicaudus*	Marginal	80	112	40	3.2
Apache Pocket Mouse	*Thomomys bottae*	Widespread	100	75	25	2.5
Montane Vole	*Microtus montanus*	Riparian	45	112	40	1.8
Brush Mouse	*Peromyscus boylii*	Chapparal	50	95	35	1.75
Silky Pocket Mouse	*Perognathus flavus*	Widespread	100	58	12	1.2
Abert's Squirrel	*Sciurus aberti*	Marginal	1	290	800	0.8
Western Harvest Mouse	*Reithrodontomys megalotis*	Widespread	50	62	16	0.8

continued on next page

APPENDIX 7.1—continued

Common Name	Species	Resident Status	Est. Median Density (no./km^2)	Body Length (mm)	Body Weight (g)	Est. Biomass (kg/km^2)
House Mouse	Mus musculus	Human-made	25	82	30	0.75
Northern Grasshopper	Onychomys leucogaster	Shrublands	10	112	55	0.55
Beaver	Castor canadensis	Riparian	0.2	750	2,205	0.441
Canyon Mouse	Peromyscus crinitus	Cliff	10	75	30	0.3
Muskrat	Ondata zibethicus	Aquatic	0.25	290	1,000	0.25
Golden-Mantled Ground Squirrel	Spermophilus lateralis	Marginal	1	175	225	0.225
Kangaroo Rat	Dipodomys ordii	Sand	2	106	45	0.09
Chickaree Squirrel	Tamiasciurus hudsonicus	Marginal	0.1	190	220	0.022
Antelope Ground Squirrel	Ammospermophilus nelsoni	Marginal	0.1	138	125	0.0125
Western Jumping Mouse	Zapus princeps	Riparian	0.01	87	40	0.0004
Spotted Ground Squirrel	Spermophilus spilosoma	Shrublands	0.0001	160	95	0.0000095
LAGOMORPHS						
Nuttall's Cottontail	Sylvilagus nuttalii	Ubiquitous	250	350	1,050	262.5
Audubon's Cottontail	Sylvilagus audubonii	Ubiquitous	250	350	1,000	250
California Jackrabbit	Lepus californicus	Ubiquitous	40	450	2,600	104
Snowshoe Hare	Lepus americanus	Marginal	0.01	435	1,550	0.0155
UNGULATES						
Mule Deer	Odocoileus hemionus	Ubiquitous	4	1,570	90,000	360
Wapiti (Elk)	Cervus canadensis Bailey	Widespread	0.2	2,400	300,510	60.102
Bighorn Sheep	Ovis canadensis	Rare	0.05	1,480	5,250	0.2625
Pronghorn	Antilocapra americana	Marginal	0.001	1,200	45,000	0.045
NON-NATIVE SPECIES						
Domestic Cow	Bos taurus	Widespread	12	2,250	304,000	3,648
Domestic Sheep	Ovis aries	Widespread	25	1,200	45,000	1,125
Humans	Homo sapiens sapiens	Ubiquitous	10	950	60,000	600
Domestic Goat	Capra hircus	Widespread	2	1,200	45,000	90
Feral Horse	Equus caballus	Widespread	0.2	2,400	410,000	82
Burro	Equus asinus	Widespread	0.01	2,100	280,000	2.8

REFERENCES CITED

Anderson, S. 1961. Mammals of Mesa Verde National Park, Colorado. *University of Kansas Publications: Museum of Natural History* 14: 29–67.

Armstrong, D. M. 1972. *Distribution of mammals in Colorado.* University of Kansas Museum of Natural History Monographs 3, Lawrence.

Bartholomew, G. A., and J. W. Hudson. 1959. Effects of NaCl on weight and drinking in the antelope ground squirrel. *Journal of Mammalogy* 40: 3–54.

Betancourt, J. L. 1990. Late Quaternary biogeography of the Colorado Plateau. In J. L. Betancourt, T. R. Van Devender, and P. S. Martin (eds.), *Packrat middens: The last 40,000 years of biotic change,* pp. 259–292. University of Arizona Press, Tucson.

Covert, K., K. Pearson, L. A. Compton, and T. Theimer. Dispersal of pinyon pine (*Pinus edulis*) seeds by *Peromyscus*: Effects of substrate type. Paper presented at the Sixth Biennial Conference on Research on the Colorado Plateau, November 5–8, 2001, Flagstaff, Ariz.

Douglas, C. L. 1963. Apache pocket mouse in Mesa Verde National Park. *Southwestern Naturalist* 8: 113.

———. 1966. New records of mammals from Mesa Verde National Park. *Southwestern Naturalist* 8: 113.

———. 1969. Comparative ecology of pinyon mice and deermice in Mesa Verde National Park. Colorado. *University of Kansas Publications, Museum of Natural History* 18: 421–504.

Fairbrother, T., and T. M. Yuill. 1984. Experimental viral infections of deer mice. *Journal of Mammalogy* 65: 499–503.

Farentino, R. C., P. J. Capretta, R. E. Kepner, and U. M. Littlefield. 1981. Selective herbivory in the tassel-eared squirrels: Role of monoterpenes. *Science* 213: 1273–1275.

Golightly, R. T., Jr., and R. Ohmart. 1983. Metabolism and body temperature of two desert canids, coyotes and kit foxes. *Journal of Mammalogy* 64: 624–635.

Henke, S. E., and F. C. Bryant. 1999. Effects of coyote removal on the faunal community in western Texas. *Journal of Wildlife Management* 63: 1066.

Holbrook, S. J. 1978. Habitat relationships and co-existence of four sympatric species of *Peromyscus* in northwestern New Mexico. *Journal of Mammalogy* 59: 18–26.

King, J. H. (ed.). 1968. Biology of *Peromyscus (Rodentia)*. Special Publication, American Society of Mammalogists 2: 13.

Koehler, G. M., and M. G. Hornocker. 1991. Seasonal resource use among mountain lions, bobcats, and coyotes. *Journal of Mammalogy* 72: 391–396.

Latta, R. G., Y. B. Linhart, L. Lundquist, and M. A. Snyder. 2000. Patterns of monoterpene variation within individual trees in ponderosa pine. *Journal of Chemical Ecology* 26: 1341–1357.

Lechleitner, R. R. 1969. *Wild mammals of Colorado: Their appearance, habits, distribution, and abundance.* Pruett Publishing Co., Boulder, Colo.

Ligon, J. D. 1978. Reproductive interdependence of piñon jays and piñon pines. *Ecology* 48: 111–126.

Mozingo, H. 1987. *Shrubs of the Great Basin.* University of Nevada Press, Reno.

Rebar, C., and O. J. Reichman. 1983. Ingestion of moldy seeds by heteromyid rodents. *Journal of Mammalogy* 64: 713–717.

Reynolds, H. G. 1966. Abert's squirrels feeding on pinyon pine. *Journal of Mammalogy* 47: 550–551.

Root, J. J., C. H. Callisher, and B. J. Beaty. 1999. Relationships of deer mouse movement, vegetation structure, and prevalence of infestation with Sin Nombre virus. *Journal of Wildlife Diseases* 35: 311–318.

Snyder, M. A., and Y. B. Linhart. 1998. Subspecific selectivity by a mammalian herbivore: Geographic differentiation of interactions between two taxa of *Sciurus aberti* and *Pinus ponderosa. Evolutionary Ecology* 12: 755–766.

Somers, P. 1976a. Fauna inventory of the Animas–La Plata Project Area: Final report. Unpublished report to United States Bureau of Reclamation, Western Colorado Area Office, Southern Division, Durango.

———. 1976b. Fauna inventory of the Paradox Valley Unit: Final report. Unpublished report to United States Bureau of Reclamation, Western Colorado Area Office, Southern Division, Durango.

———. 1977. Fauna inventory of the San Miguel Project Area: Unpublished report to United States Bureau of Reclamation, Western Colorado Area Office, Southern Division, Durango.

———. 1979. Inventory of terrestrial non-game animals of the McElmo Creek Unit Area, Colorado River Basin Salinity Control Project, Colorado: Final report. Unpublished report to United States Bureau of Reclamation, Western Colorado Area Office, Southern Division, Durango.

———. 1980. McElmo Creek fauna analysis, non-game inventory number 2: Final report. Unpublished report to United States Bureau of Reclamation, Western Colorado Area Office, Southern Division, Durango.

Vander Wall, S. B. 1994. Seed fate pathways of antelope bitterbrush dispersal by seed caching yellow pine chipmunks. *Ecology* 75: 1911–1926.

———. 1997. Dispersal of singleleaf pinyon pine by seed caching rodents. *Journal of Mammalogy* 78: 181–191.

Vander Wall, S. B., and R. Balda. 1977. Co-adaptation of the Clark's nutcracker and pinyon pine. *Ecological Monographs* 47: 89–111.

Vaughan, T. A. 1982. Stephen's woodrat, a dietary specialist. *Journal of Mammalogy* 63: 118–127.

Vaughan, T. A., and N. J. Czaplewski. 1985. Reproduction in Stephen's woodrat: The wages of herbivory. *Journal of Mammalogy* 66: 429–443.

Whitaker, J. O., Jr. 1968. Parasites. In J. A. King (ed.), *Biology of* Peromyscus *(Rodentia)*, pp. 254–311. Special Publication No. 2. American Society of Mammalogists.

Whitham, T. G. 1982. Indirect herbivore mediation of avian seed dispersion in *Pinus edulis*. *Ecology* 72: 534–541.

Wolff, J. O., R. O. Dueser, and K. S. Berry. 1985. Food habits of sympatric *Peromyscus leucopus* and *P. maniculatus*. *Journal of Mammalogy* 66: 795–798.

BATS OF THE PIÑON-JUNIPER WOODLANDS OF SOUTHWESTERN COLORADO

Alice L. Chung-MacCoubrey and Michael A. Bogan

As one of the most abundant and widespread forest types in the Southwest, piñon-juniper woodlands are used to varying degrees by many bat species (Chung-MacCoubrey 1996; Findley et al. 1975; Hoffmeister 1986; Jones 1965). Because of the uniqueness and rarity of old-growth piñon-juniper, the ancient woodlands of Mesa Verde Country likely provide bats with a combination of roosting and foraging opportunities not found elsewhere. Although no studies on bat habitat use have been conducted in southwestern Colorado, records of bats from western Colorado and Mesa Verde National Park—coupled with habitat use studies in piñon-juniper woodlands of New Mexico—provide insight into how the bat community in Mesa Verde Country uses the landscape. In this chapter we describe the bat species that occur in the Mesa Verde region, some of their life history, availability of bat roosts in piñon-juniper woodlands, and general patterns of habitat use by each species. We also discuss how the bat community probably uses piñon-juniper woodlands in southwestern Colorado.

Over the last twenty years considerable new knowledge has been acquired on the distribution of bats in western Colorado, including Mesa Verde National Park. Studies on the foraging and roosting ecology of these bats, however, have yet to

---

This paper was written and prepared by U.S. government employees on official time and therefore is in the public domain and not subject to copyright.

be conducted. G. E. Freeman (1984) assessed ecomorphological aspects of community structure in bats from northwestern Colorado. R. A. Adams (1993) provided new information on bats at Colorado National Monument. M. A. Bogan and colleagues (e.g., Bogan et al. 1988; Finley and Bogan 1995) studied mammals in northwestern Colorado, in and near Dinosaur National Monument. K. W. Navo et al. (1992) and K. W. Navo and J. A. Gore (2001) provided important new data on uncommon species in western Colorado. Most pertinent to this chapter are the studies by Bogan (1994, unpublished data) of the bat community at Mesa Verde National Park. Most of these studies emphasized the distributional and occurrence aspects of bats in piñon-juniper forests of western Colorado.

With the advent of miniaturized radiotransmitters, bat research has expanded from distributional studies to habitat use studies. These new studies have fueled a growing recognition that forests are used more heavily by bats for foraging and roosting than historically recognized. Most of these studies, however, have been conducted in tall coniferous or deciduous forests of North America (e.g., Barclay and Brigham 1996). Few studies have been conducted in the southwestern United States. M. J. Rabe et al. (1998) and D. J. Lutch (1996) examined roost selection by bats in ponderosa-pine and pine-oak forests of Arizona. Bogan et al. (1998) studied roost use by bats in mixed coniferous forests of New Mexico. Chung-MacCoubrey (1996) and Rabe (1999) are the only studies that have focused on habitat use by bats in piñon-juniper woodlands (in New Mexico and Arizona, respectively). Results from Chung-MacCoubrey (1996, 2003) and Bogan et al. (1998) are relied upon in interpreting landscape use by bats at Mesa Verde National Park.

HISTORIC RECORDS OF BATS FROM WESTERN COLORADO AND MESA VERDE NATIONAL PARK

Eighteen species of bats are known from Colorado (Armstrong et al. 1994), and sixteen of these species are known from western Colorado, although not all species occur at any one place. Of the sixteen species known from the Western Slope of Colorado, fourteen are now known from Montezuma County, the general focus of this chapter. Some of the earliest records of bats from southwestern Colorado come from a visit that Merritt Cary made to Ashbaugh's Ranch in McElmo Canyon, twenty miles west of Cortez, in 1907. There Cary (1911:205) observed "numbers" of pallid bats (*Antrozous pallidus*) appearing about the cliffs near the ranch and shot two that were within range. He contrasted the body size and flight patterns of *Antrozous* with those of smaller species and speculated, probably correctly, that pallid bats were roosting in the "numerous cracks and crevices" in the cliffs. Cary noted that the long-eared myotis (*Myotis evotis*) was "tolerably common" at the ranch and caught two in the ranch house after dark (both were apparently females). He shot two California myotis (*M. californicus*) as they flew about the cliffs and secured a third in the ranch house. The "most numerous species" during Cary's visit to the ranch was the western pipistrelle

(*Pipistrellus hesperus*). He also saw flying bats that he thought were big brown bats (*Eptesicus fuscus*), although he failed to take any specimens. In his description of the landscape, Cary (1911) noted the dense growth of piñon (*Pinus edulis*) on the Mesa Verde and "practically all the broken country from Montezuma County north to Mesa County" and the abundance of one-seeded juniper (*Juniperus monosperma*) on Mesa Verde, although he made no observations on bats from Mesa Verde.

S. Anderson (1961) studied mammals at Mesa Verde and provided more information on bats of this area. He was aware of eight species of bats based on twenty-five specimens. The species and numbers he obtained were *M. californicus* (three), *M. ciliolabrum* (eight), *M. evotis* (four), *M. thysanodes* (one), *M. volans* (one), *E. fuscus* (one), *Corynorhinus townsendii* (Townsend's big-eared bat; five), and *Tadarida brasiliensis* (Brazilian free-tailed bat; two). He captured five of these species in a mist net set near Rock Springs "at the south edge of the [Wildhorse 1934] burn on Wetherill Mesa" (1961:38). Four of the *C. townsendii* were found roosting in a dimly lighted chamber of Square Tower House in March, and both *Tadarida* came from Cliff Palace (taken by A. E. Borell in 1936). D. M. Armstrong (1972), in his monograph on Colorado mammals, recorded the presence of eleven species in Montezuma County, including a record of *Lasiurus cinereus* (hoary bat) from Mesa Verde (Douglas 1967). The latest account of bats in Colorado by Armstrong et al. (1994) added *M. lucifugus* (little brown bat) and *Lasionycteris noctivagans* (silver-haired bat) to the known fauna of Montezuma County.

SPECIES OF BATS AT MESA VERDE NATIONAL PARK

In 1989, personnel from the Midcontinent Ecological Science Center (and its predecessors) initiated studies of mammals at Mesa Verde National Park as a follow-up to the study conducted by Anderson some thirty years earlier. As part of this effort, mist nets were set periodically at sites where bats might be captured, and capture records and voucher specimens were kept. Records of bats from Mesa Verde National Park summarized in this chapter emanate from those surveys between 1989 and 1994 and are presented in order of descending abundance. Voucher specimens are housed in the Biological Surveys collection in the Museum of Southwestern Biology at the University of New Mexico.

Myotis evotis

At Mesa Verde National Park the long-eared myotis was the most frequently captured species ($n = 73$) and was found at ten different localities. It was very common in the vicinity of Rock Springs but was even more abundant in Morfield Canyon. Overall we captured forty-eight females; of six females captured between June 6 and June 9, four were pregnant. In the Henry Mountains of Utah, where there are extensive stands of piñon-juniper woodlands, this was the second most common species of bat captured and ranged from 1,425 m to 2,696 m (Mollhagen and Bogan 1997).

The long-eared myotis was the second most frequently captured *Myotis* species in piñon-juniper woodlands of central New Mexico, although it is frequently characterized as being more common in tall coniferous forests (Findley et al. 1975; Hoffmeister 1986). In New Mexico long-eared myotis used piñon-juniper woodlands as maternity roosting habitat more effectively than did long-legged and fringed myotis by using more commonly available structures. Solitary and small colonies of female long-eared myotis roosted most frequently in junipers and piñon snags and occasionally in the crevices of rocks. Females often changed roosts, but subsequent roosts were only a short distance away ($\bar{x}$ = 180 m). Roosts were closer to the site where they were netted (for radiotransmitter application), indicating that members of this species may forage closer to their roosts than do long-legged or fringed myotis (Chung-MacCoubrey and Delay 2000). In mixed coniferous forest in New Mexico two females and a male roosted in rock crevices on or near the ground and moved among a small number of roosts (Bogan et al. 1998). The high capture rate of this species at Mesa Verde National Park may be explained by the presence of large tracts of old-growth piñon-juniper and the ability of this species to use this habitat type effectively. The subspecies in western Colorado is *M. e. chrysonotus* (Manning 1993).

Myotis volans

Long-legged myotis was the second most abundant species netted at Mesa Verde National Park (fifty-seven individuals). Of forty females, thirteen were pregnant during the period from June 9 to July 27 and three were lactating on August 2 and 3. This species was captured at ten different localities and was abundant near Rock Springs and Limey Draw on Wetherill Mesa as well as in Morfield Canyon. Long-legged myotis are known to day roost in houses, buildings, trees, and rock "hoodoos" but rarely in caves (Bogan et al. 1998; Vonhof and Barclay 1996). Like long-eared myotis, this species is also believed to be most common in ponderosa-pine and higher-elevation forests in the Southwest (Findley et al. 1975; Hoffmeister 1986). This species was the third most abundant species in the Henry Mountains of Utah, however, where it ranged from 1,430 m to 3,060 m (Mollhagen and Bogan 1997). It was the most frequently captured *Myotis* in piñon-juniper woodlands of New Mexico (Chung-MacCoubrey, unpublished data). Thus this species may use piñon-juniper woodlands more than historically recognized, although its presence may rely to some degree on the existence of tall conifers or other types of preferred roosts within or near piñon-juniper habitat.

Within ponderosa-pine and mixed-coniferous forest, this species prefers to roost in tall, large-diameter conifers (Chung-MacCoubrey, unpublished data; Rabe et al. 1998; Vonhof and Barclay 1996). In piñon-juniper woodlands of New Mexico, maternity colonies of long-legged myotis also preferred to roost in large ponderosa pine trees with vertical cracks that were sparsely scattered throughout the woodland or at the interface of the two habitat types (Chung-MacCoubrey 1996). A medium-sized colony was found roosting in a very large, damaged, but live piñon (diameter at breast height [dbh] = 55 cm). Although this species seemed to prefer large trees for colony roosts, individuals often roosted in piñon snags (most frequently) and junipers. Many females roosted alone in a series of piñon or juniper roosts for the entire tracking period. Other females periodically left the colony to roost alone in nearby piñon or juniper. Of the three *Myotis* species studied, long-legged myotis moved the largest distances between roosts ($\bar{x}$ = 619 m) and likely foraged over greater areas than long-eared myotis (Chung-MacCoubrey and DeLay 2000). At Mesa Verde National Park, Douglas-fir forest and localized stands of ponderosa pine provide tall conifers for colonies of long-legged myotis preferring this roost type. Old-growth piñon-juniper woodlands offer large-diameter piñons for colonies, a wealth of piñon snags and junipers for solitary roosts, and abundant foraging habitat. Thus the combination of forest and roost types available at Mesa Verde National Park is a likely explanation for the abundance of this species there. The race occurring in southwestern Colorado is *M. v. interior*.

Corynorhinus townsendii

Townsend's big-eared bats were the third most abundant species at the park. We captured twenty, of which all but one were females. We have records of this species from six different localities, but they appeared to be most common (n = 13) at lower-elevation localities such as Rock Springs and Limey Draw on Wetherill Mesa. Females may prefer warmer roosting sites at lower elevations at Mesa Verde (e.g., Cryan et al. 2000). We captured four individuals in Morfield Canyon (August 2–6), and all were lactating females.

In the Southwest this species commonly roosts on exposed ceiling surfaces of caves and mines and within abandoned buildings (Findley et al. 1975; Hoffmeister 1986). These bats occasionally have been found roosting in undisturbed buildings during the summer. At Mesa Verde National Park abundant cliff and cave roosting opportunities exist for Townsend's big-eared bats. Elsewhere in its range this species is extremely intolerant of disturbance in the roosts and often abandons a site after such disturbance. Due to the exposed nature of its preferred roost locations, this species is the subject of increasing conservation concern among scientists and managers. The subspecies in the Rocky Mountain West is *C. t. pallescens*.

Myotis ciliolabrum

The western small-footed myotis has a fairly wide elevational range and is most common above piñon-juniper forest. At Mesa Verde National Park we captured thirteen individuals (ten males, three females) from ten different localities. Six individuals came from sites in Morfield Canyon, two from Chapin Mesa, two from Wetherill Mesa, and three from other sites. Our records include animals from April through August. No reproductive information is available from this species at Mesa Verde National Park. In the Gallinas Mountains of New Mexico, a relatively isolated set of mountains vegetated primarily by piñon-juniper, this species was one of the most frequently captured *Myotis* (Chung-MacCoubrey, unpublished data). Half of these individuals were female, many of which were reproductive.

These bats are known to roost in small rock crevices, in trees, and in various human-made structures in the summer. In the Gallinas Mountains two reproductive females were found roosting in junipers, and one was found roosting in a crevice within a rock. Initial netting and roost data from the Gallinas suggest that small-footed myotis readily use piñon-juniper woodlands during the summer, even in isolation from other forest types. The subspecies in southwestern Colorado is *M. c. melanorhinus*.

MYOTIS THYSANODES

At Mesa Verde National Park we captured nine individual fringed myotis (including six adult females) at four different localities. Most captures came from the vicinity of Rock Springs, but some were also captured in Morfield Canyon and Spruce Canyon. In general during the summer months this bat is known to form small colonies in rock crevices, caves, mines, trees, and buildings. Although this species is thought to be most associated with woodland and shrub communities, it appears to prefer roosting in other types of structures (large trees, caves, etc.) within these areas. In mixed conifer forests of New Mexico reproductive females roosted in rock crevices and solution pits high on cliff walls (Bogan et al. 1998). In ponderosa-pine forests of New Mexico and Arizona this species roosted in large ponderosa-pine snags with peeling bark (Rabe et al. 1998).

Individuals caught and radiotracked in piñon-juniper woodlands of New Mexico only occasionally used piñons and junipers as roosts. Most of the roosts located were maternity colonies in live or dead ponderosa pine with vertical cracks. A large colony roost (more than 200 individuals) was located in a mine adit. Occasionally females would roost away from the colony in nearby piñon or junipers. In the Gallinas Mountains and in other areas, reproductive females of this species changed roosts at varying frequencies and remained within a small area on a daily basis (Chung-MacCoubrey and DeLay 2000; Cryan et al. 2001). The average distance between roosts in piñon-juniper woodlands was 200 m. This species may forage more widely than other species of *Myotis*. Roosts of radiotransmittered bats were farther from the site where they were netted than were roosts of long-eared myotis (Chung-MacCoubrey and DeLay 2000). The subspecies in southwestern Colorado is *M. t. thysanodes*.

MYOTIS CALIFORNICUS

We captured six California myotis at three localities. Four of the captures were near Rock Springs; but we also netted one individual in Morfield Canyon, and there is a record from Chapin Mesa. In all cases both piñon and juniper were present at capture sites. We have records for this species from July and August. Our samples included three females and three males; one female contained no embryos (August 2), but one was lactating (August 5), so the species clearly breeds at Mesa Verde National Park.

Typically the California myotis is moderately common at low and intermediate elevations in western Colorado and is perhaps most common below the level of piñon-juniper woodland. Along with pallid bats and western pipistrelles, it is one of three species that are common in arid areas of the Colorado Plateau and throughout the Southwest. Elsewhere during the summer these bats are known to inhabit lowland, rocky canyons and roost in trees, rock crevices, and various human-made structures (e.g., bridges, buildings). During the winter they have been found hibernating in caves, mines, and rock crevices. In some areas of its range, but probably not Colorado, this species may be sporadically active during the winter months. Animals from southwestern Colorado are assigned to the subspecies *M. c. stephensi*.

LASIONYCTERIS NOCTIVAGANS

Six silver-haired bats, all males, were captured at three different localities at Mesa Verde in June, July, and August. Five of six captures were in Morfield Canyon. This species probably prefers higher elevations in the vicinity, but lack of adequate netting sites may play a role in low numbers of captures too. In piñon-juniper woodlands of New Mexico mist-netting only yielded silver-haired bats in areas with permanent sources of water (e.g., spring-fed ponds or intermittent streams) or very large stock tanks (Chung-MacCoubrey 1996). Virtually all animals captured were male.

Silver-haired bats are considered "tree bats" and are known to roost beneath the bark and in cavities of trees. Maternity colonies are usually found roosting in areas of high snag densities within ponderosa-pine forest (Mattson et al. 1996). Sexes segregate geographically during the summer; females are typically found in northern and eastern North America, while males are in the western mountains. These bats are assumed to be migratory, but very little is known about their movements or winter habits. They have been found hibernating in dead trees, buildings, ships, and rock crevices. Members of this species rarely enter mines and caves, although they have been found torpid in cave crevices.

EPTESICUS FUSCUS

We captured only two big brown bats at Mesa Verde National Park, at two different localities in July and August; both animals were males. In much of the Southwest this species is more common at upper-elevation forests, where it is known to roost in ponderosa-pine snags and buildings (Bogan et al. 1998). In piñon-juniper woodlands of the Gallinas Mountains, however, big brown bats were the most commonly captured bat species, constituting 40% of the total captures; 75% of these bats were females.

Big brown bats are known to roost in trees, rock crevices, caves, mines, and buildings during the summer and to hibernate in caves, mines, and buildings. In the Gallinas Mountains, maternity colonies of big brown bats roosted in live or dead ponderosa pine with vertical cracks through the trunk. Individuals were occasionally found roosting under the bark of piñon snags. Like long-legged and fringed myotis, big brown bats may only use piñon-juniper if other suitable types of roosts (e.g., ponderosa pine) also exist within the area. Thus at Mesa Verde National Park it is possible that the few stands of ponderosa pine and mixed conifers do not provide sufficient roosting opportunities for this species to be common there. Individuals in western Colorado belong to the race *E. f. pallidus*.

Myotis lucifugus

We captured one little brown bat over the large sewage disposal ponds in Morfield Canyon, the largest expanse of open water on the park. We know of only one other locality for this species in Montezuma County, near the Dolores River five miles north and two miles west of Dolores (Armstrong et al. 1994). Little brown bats are known to forage over open water. This species is not usually associated with piñon-juniper forests; most records in the Southwest are from higher elevations in mixed coniferous forests or lowland riparian areas, where they often roost in buildings. It is possible that individuals in southwestern Colorado actually represent the related *M. occultus* (Piaggio et al., in press).

Lasiurus cinereus

We captured one hoary bat in Morfield Canyon. This low number of captures is similar to the situation in New Mexico, where mist-netting in piñon-juniper woodlands only yielded this species when netting occurred at large stock tanks or permanent water sources within the woodland (Chung-MacCoubrey 1996). Although mostly males were captured, a few nonreproductive females were captured in late May and early June in New Mexico.

Members of this species roost primarily in the foliage of trees and shrubs. They roost alone or in small family groups, usually not exceeding three or four individuals. Although details are lacking, this species migrates northward in the spring; during the summer months adult males are mostly in the western United States, while females spend the summer in the East giving birth and rearing young (Findley and Jones 1964). Hoary bats may be more common at Mesa Verde than our records indicate, but many waterholes there may be too small for the species to use.

Antrozous pallidus

We captured one pallid bat (a female) in Limey Draw on the south end of Wetherill Mesa. We suspect that much of Mesa Verde is too high in elevation to be preferred habitat for these arid-adapted, ground-feeding bats. This species is known to roost in shallow caves and crevices on cliff faces during the summer. Maternity colonies typically range in size from approximately 12 to 200 individuals and frequently change roost locations, possibly in order to maximize thermal efficiency. Winter habits are unknown, though some researchers have suggested that pallid bats remain within the general summer range but may make short migrations between summer and winter roosts. In some desert areas they may forage on warm winter nights. Individuals in western Colorado belong to the nominate race.

EUDERMA MACULATUM

The spotted bat is known from a variety of life zones in the Southwest and is known to roost at low- to mid-elevation areas (e.g., canyons and cliffs; Bogan et al. 1998) and forage in mixed coniferous forests. At Mesa Verde National Park we captured no spotted bats, although one of us (Bogan) heard the audible cries of one or more spotted bats one evening in Morfield Canyon. These bats are known to fly for long periods well above the ground (Wai-Ping and Fenton 1989) and thus are not often captured in mist nets set over water. One of us (Bogan) examined remains of bats found in pellets collected beneath a spotted owl roost in Navajo Canyon. One pellet consisted entirely of the remains of three individual spotted bats. In New Mexico two lactating females, one male, and two juvenile females were followed for an average of 9.2 days (range = 5–14 days) to an average of 1.6 roosts (range = 1–2 days) (Bogan et al. 1998). All of these bats roosted in rock crevices high on cliff walls. Average height of known roosts was 16 m (range = 7–21 m), and their orientation was southeast. Radiotagged individuals traveled farther from point of capture to first roost (13–17 km) than any other species except *Nyctinomops macrotis*. One female engaged in nightly foraging bouts that covered at least 50 km round-trip. Conservative emergence counts averaged six bats (range = 1–30), but some colonies may have had greater numbers of bats. During four months of intensive netting in piñon-juniper woodlands in New Mexico only a single spotted bat was captured. This lactating female was captured in mid-June in open, rugged juniper grassland. T. R. Mollhagen and Bogan (1997) captured one pregnant female in piñon-juniper woodland in the Henry Mountains of Utah.

PIPISTRELLUS HESPERUS

We captured no western pipistrelles, although there are two records for the park (Armstrong et al. 1994). As Cary (1911) noted, the species is common in arid, lower-elevation habitats in southwestern Colorado. Pipistrelles are commonly found roosting solitarily or in very small groups in rock crevices in cliffs and canyons. They do not roost in large numbers, and maternity colonies rarely contain more than twelve individuals. They have also occasionally been found during the day in dense, low-lying vegetation and beneath rocks on the ground. Individuals of *P. hesperus* are known to hibernate in caves, mines, and rock crevices, but they also may be active during the winter in parts of the Southwest.

TADARIDA BRASILIENSIS

We did not capture Brazilian free-tailed bats at Mesa Verde, perhaps because the water sources there are too small for the species to use. Nonetheless, Anderson (1961) reported the occurrence of two individuals of this species in Cliff Palace, and Armstrong et al. (1994:47) reported one individual from "probably in or near Cortez." This species forms large colonies at several sites in the Southwest; the nearest such colony in Colorado is in Saguache County. One of us (Bogan) is aware of several diurnal roosts of this species in rock crevices in cliffs in New Mexico. In late August 1996 one of these housed 540 bats, presumably migrants in transit; the same roost was used briefly in late spring 1997. It is not likely that this species might use such sites in and near Mesa Verde. During the winter months some individuals remain in the summer range, while most migrate long distances to warmer latitudes. The subspecies in the western United States is *T. b. mexicana*.

MYOTIS YUMANENSIS

The Yuma myotis is not known from Montezuma County and is uncommon in much of southwestern Colorado (Armstrong et al. 1994). During the summer these bats are known to roost in caves, mines, buildings, and bridges. Like *M. lucifugus*, this species is frequently associated with bodies of water, over which it forages for insects. The winter retreats of this species in the Southwest are unknown. The subspecies in southwestern Colorado is *M. y. yumanensis*.

NYCTINOMOPS MACROTIS

The big free-tailed bat is not known from Montezuma County, although both solitary individuals and maternity roosts are known from adjacent Utah. These bats typically roost in crevices high on sheer cliffs. Individuals of this species travel moderately long distances on a nightly basis. In New Mexico the average distance from point of capture to first roost was 18.2 km, and the longest recorded flight distance on a single night was 30 km. The winter habits of this species are unknown, though it probably migrates to the south for winter.

IDIONYCTERIS PHYLLOTIS

Allen's big-eared bat is not known from southwestern Colorado but is to be expected (Armstrong et al. 1994). In much of the Southwest it is a part of the general community described herein. The nearest record for this species is from 12 km north of Blanding, San Juan County, Utah (Black 1970). This species was radiotracked in the piñon-juniper woodlands at Natural Bridges National Monument, where it roosted in one of the rock bridges.

ROOST AVAILABILITY AND THE BAT COMMUNITY AT MESA VERDE NATIONAL PARK

The species composition of any bat community is a function of the abundance and diversity of resources available. Individual species requirements, roost types and availability, insect abundance and diversity, and other factors determine species composition and patterns of habitat use by a bat community. Additionally, bats can access resources in dimensions and scales not available to terrestrial animals and frequently travel across different landscapes to fulfill their resource requirements. Bats at Mesa Verde National Park likely feed and roost within more than one habitat type, and our discussion is not meant to imply that piñon-juniper woodlands alone are sufficient to meet the needs of a diverse bat community. Varied habitat types in close proximity offer a diverse set of foraging and roosting opportunities.

The diversity and relative abundance of roosts can greatly influence bat species composition in an area. The varied landscape of Mesa Verde National Park provides a multitude of roost opportunities for bats. Studies in tall coniferous forests indicate that bats commonly roost in taller snags in early stages of decay (Barclay and Brigham 1996). In the park few Douglas-fir forests and localized stands of ponderosa pine are likely to provide this roost type for bats that prefer them (e.g., *Myotis volans, M. thysanodes, Eptesicus fuscus*) (Figure 8.1). Canyons descending from the mesa tops provide plenty of roost opportunities for species that use high-cliff and cave roosts (e.g., *Corynorhinus townsendii, Tadarida brasiliensis, Euderma maculatum*). Other bat species such as *Pipistrellus hesperus, Antrozous pallidus, M. volans,* and *M. thysanodes* may use these cliff and cave roosts as well as scattered rocks and rock outcrops throughout the forests and woodlands.

For other tree-roosting bats, particularly *Myotis* species, the unique structure and composition of old-growth piñon-juniper create foraging and roosting habitat not found in other forest types. Species that roost primarily in other habitat types (e.g., rocks, tall conifers) may also use old-growth piñon-juniper woodlands for foraging or for alternate roost locations away from the colony. In New Mexico individual female *M. volans* and *M. thysanodes* occasionally left their ponderosa-pine colony roost of 100–200 bats to roost solitarily in a nearby piñon or juniper. Solitary-roosting female bats in New Mexico frequently roosted under the peeling bark of piñon snags (Figure 8.2). Occasionally bats were found roosting in rotting internal cavities of piñon snags. A cavity within a large live piñon was used by thirty-six female *M. volans* in New Mexico (Chung-MacCoubrey 1996). Although rare at the New Mexico site, piñon trees of this size are more common at Mesa Verde National Park, which thus may be another valuable roost type for medium-sized to large colonies of *Myotis*.

Junipers present a greater variety of roost opportunities than do piñons. Internal and external roosts in both live and dead junipers were used frequently by

Figure 8.1. Ponderosa pine (Pinus ponderosa) *snag roost in the Gallinas Mountains, New Mexico. A maternity colony of over 200* M. volans *and* M. thysanodes *roosted in the large vertical crack running through the center of the snag. Dead or damaged tall conifers such as this tree provide roost opportunities for colony-forming bat species. Photo by A. Chung-MacCoubrey.*

Figure 8.2a. Piñon (Pinus edulis) *snag roost in the Gallinas Mountains, New Mexico. Peeling bark on piñon snags in early stages of decay may serve as short-term roosts for solitary or small groups of individuals. Photo by A. Chung-MacCoubrey.*

Figure 8.2b. Piñon (Pinus edulis) *snag roost in the Gallinas Mountains, New Mexico. Close-up of the solitary roost of a female* M. volans *under loose bark. Photo by A. Chung-MacCoubrey.*

Figure 8.3a. One-seeded juniper (Juniperus monosperma) *roost tree in the Gallinas Mountains, New Mexico. Crevices and cavities within the bark, wood, limbs, and trunks of live and dead junipers are common and provide numerous roost opportunities for bats. Photo by A. Chung-MacCoubrey.*

several species of *Myotis* in New Mexico (Chung-MacCoubrey 2003). Crevices created by folds of bark or wood, pockets formed at the juncture of limb and trunk, spaces between intertwining limbs, rot pockets, and internal cavities are examples of places where solitary or small groups of bats commonly roosted in junipers (Chung-MacCoubrey 2003) (Figure 8.3). Bats roosted in the exposed faces of piñon and juniper stumps as well as internal cavities accessed at ground level. Scattered rocks, rock outcrops, and Pueblo ruins within piñon-juniper woodlands also provide sites for many species of bats that roost solitarily or in small colonies.

SUMMARY

Of the thirteen species known from Montezuma County, we captured eleven species at Mesa Verde National Park. The two species we did not capture were the western pipistrelle and the Brazilian free-tailed bat. Although spotted bats were not previously documented for Montezuma County, we heard their calls and identified three spotted bat crania in an owl pellet from Mesa Verde National Park. Suitable foraging and roosting habitat seems to exist for *Myotis yumanensis*, *Nyctinomops macrotis*, and *Idionycteris phyllotis*, although these species remain undocumented for the county. The relatively close proximity of old-growth piñon-

Figure 8.3b. One-seeded juniper (Juniperus monosperma) *roost tree in the Gallinas Mountains, New Mexico. Close-up of the solitary roost of a female M.* evotis *in the crevice of a horizontal limb. Photo by A. Chung-MacCoubrey.*

juniper woodland, Douglas-fir forest, desert scrub, and rocky canyons provides a diverse array of roosting and foraging opportunities for bats in southwestern Colorado.

In piñon-juniper woodlands of New Mexico some bat species effectively exploit piñon and juniper as roosts, whereas other species prefer to roost in tall conifers, rocks, or other features. Bats using piñon snags and juniper usually move among several trees in a close perimeter on a daily basis, whereas bats using tall conifers change roosts less frequently. Bats in southwestern Colorado woodlands probably behave the same. These rare and unique old-growth woodlands likely offer roost opportunities not available in younger woodlands. Abundant piñon snags and craggy junipers probably provide maternity roosts for species such as *Myotis evotis* and *M. ciliolabrum*. The larger diameter and height of very old trees also may attract other species such as *M. thysanodes* and *E. fuscus* that prefer large conifers for maternity roosts. Other species, such as those roosting in cliffs and caves, may forage over piñon-juniper woodlands, preying on the associated insect communities. Research on roost preferences and foraging habits of bats at Mesa Verde National Park is needed, however, to confirm exactly how bats use and partition resources in this unique area.

Seven species of bats that occur in piñon-juniper forests of western Colorado are former federal candidate species (U.S. Fish and Wildlife Service 1994), now referred to as "species of concern"; although rigorous documentation of population declines in these species is lacking (O'Shea and Bogan 2000), there is increasing concern for the species *M. ciliolabrum, M. evotis, M. thysanodes, M. volans, E. maculatum, C. townsendii,* and *N. macrotis.* Population trends are unknown for all of these species. Consensus among researchers, however, is that populations of *C. townsendii* in the West have been declining. Retention of diverse bat communities in southwestern forests will require forest management plans that include maintenance of snags (Cryan et al. 2001), protection of water sources (Mollhagen and Bogan 1997), surveys for roosting bats prior to closure of abandoned mines (Altenbach and Pierson 1995), and consideration of impacts of surface mining on foraging and roosting habitat of bats (e.g., Bogan 2001).

REFERENCES CITED

Adams, R. A. 1993. Follow-up study of bat species abundance and distribution at Colorado National Monument. Processed report, Colorado National Monument Association.

Altenbach, J. S., and E. D. Pierson. 1995. The importance of mines to bats: An overview. In B. R. Riddle (ed.), *Inactive mines as bat habitat: Guidelines for research, survey, monitoring and mine management in Nevada,* pp. 7–18. Biological Resources Research Center, University of Nevada, Reno.

Anderson, S. 1961. Mammals of Mesa Verde National Park. *University of Kansas Publications, Museum of Natural History* 14: 29–67.

Armstrong, D. M. 1972. *Distribution of mammals in Colorado.* Monograph, University of Kansas Museum of Natural History, 3. University of Kansas Printing Service, Lawrence.

Armstrong, D. M., R. A. Adams, and J. Freeman. 1994. *Distribution and ecology of bats of Colorado.* Natural History Inventory of Colorado 15. University of Colorado Museum, Boulder.

Barclay, R.M.R., and R. M. Brigham (eds.). 1996. *Bats and forests symposium.* Working Paper 23/1996. British Columbia Ministry of Forests, Victoria, B.C.

Black, H. L. 1970. Occurrence of the Mexican big-eared bat in Utah. *Journal of Mammalogy* 51: 190.

Bogan, M. A. 1994. Research determines status and distribution of bats on the Colorado Plateau. *Colorado Plateau* 4(3): 1–7.

———. 2001. Western bats and mining. In K. Vorhies and D. Throgmorton (eds.), *Bat conservation and mining: A technical interactive forum,* pp. 41–50. Office of Surface Mining, Alton, Ill.

Bogan, M. A., R. B. Finley, Jr., and S. J. Petersburg. 1988. The importance of biological surveys in managing public lands in the western United States. In R. C. Szaro, K. E. Severson, and D. R. Patton (tech. coords.), *Management of amphibians, reptiles, and small mammals in North America,* pp. 254–261. USDA Forest Service, General Technical Report, RM-166.

Bogan, M. A., T. J. O'Shea, P. M. Cryan, A. M. Ditto, W. H. Schaedla, E. W. Valdez, and K. T. Castle. 1998. *A study of bat populations at Los Alamos National Laboratory and*

Bandelier National Monument, Jemez Mountains, New Mexico. LA-UR-98-2418. Los Alamos National Laboratory, Los Alamos.

Cary, M. 1911. *A biological survey of Colorado.* North American Fauna 33. Government Printing Office, Washington, D.C.

Chung-MacCoubrey, A. 1996. Bat species composition and roost use in pinyon-juniper woodlands of New Mexico. In R.M.R. Barclay and R. M. Brigham (eds.), *Bats and forests symposium, October 19–21, 1995, Victoria, British Columbia, Canada,* pp. 118–123. Working Paper 23/1996. Research Branch, British Columbia Ministry of Forests, Victoria, B.C., Canada.

———. 2003. Ecology and management of bats in pinyon-juniper woodlands of west-central New Mexico. Ph.D. diss., University of New Mexico, Albuquerque.

Chung-MacCoubrey, A., and L. DeLay. 2000. Spatial arrangement of long-eared, fringed, and long-legged myotis roosts in pinyon-juniper woodlands of central New Mexico. In J. D. Greer (ed.), *Remote sensing and geospatial technologies for the new millennium,* Proceedings of the Eighth Forest Service Remote Sensing Applications Conference, April 10–14, 2000, Albuquerque, New Mexico. American Society for Photogrammetry and Remote Sensing, Bethesda, Md.

Cryan, P. M., M. A. Bogan, and J. S. Altenbach. 2000. Effect of elevation on female bats in the Black Hills, South Dakota. *Journal of Mammalogy* 81: 719–725.

Cryan, P. M., M. A. Bogan, and G. M. Yanega. 2001. Roosting habits of four bat species in the Black Hills of South Dakota. *Acta Chiropotologica* 3: 43–52.

Douglas, C. L. 1967. New records of mammals from Mesa Verde National Park. *Journal of Mammalogy* 48: 322–323.

Findley, J. S., A. H. Harris, D. E. Wilson, and C. Jones. 1975. *Mammals of New Mexico.* University of New Mexico Press, Albuquerque.

Findley, J. S., and C. Jones. 1964. Seasonal distribution of the hoary bat. *Journal of Mammalogy* 37: 461–470.

Finley, R. B., Jr., and M. A. Bogan. 1996. *New records of terrestrial mammals in northwestern Colorado.* Proceedings Series 3, No. 10. Denver Museum of Natural History, Denver.

Freeman, G. E. 1984. Ecomorphological analysis of an assemblage of bats: Resource partitioning and competition. Unpublished Ph.D. dissertation, University of Colorado, Boulder.

Gottfried, G. J., and R. D. Pieper. 2000. Pinyon-juniper rangelands. In Carol Raish and Roy Jemison (eds.), *Livestock management in the American Southwest: Ecology, society, and economics,* pp. 153–211. Elsevier, New York.

Hoffmeister, D. F. 1986. *Mammals of Arizona.* University of Arizona Press and Arizona Game and Fish Department, Tucson.

Jones, C. 1965. Ecological distribution and activity periods of bats of the Mogollon Mountains area of New Mexico and adjacent Arizona. *Tulane Studies in Zoology* 12: 93–100.

Lutch, D. J. 1996. Maternity roost characteristics and habitat selection of three forest-dwelling bat species in Arizona. M.S. thesis, Arizona State University, Tempe.

Manning, R. W. 1993. *Systematics and evolutionary relationships of the long-eared myotis, Myotis evotis (Chiroptera: Vespertilionidae).* Special Publications of the Museum, No. 37. Texas Tech University Press, Lubbock.

Mattson, T. A., S. W. Buskirk, and N. L. Stanton. 1996. Roost sites of the silver-haired bat (*Lasionycteris noctivagans*) in the Black Hills, South Dakota. *Great Basin Naturalist* 56: 247–253.

Mollhagen, T. R., and M. A. Bogan. 1997. Bats of the Henry Mountains region of southeastern Utah. *Occasional Papers, Museum of Texas Tech University* 170: 1–13.

Navo, K. W., and J. A. Gore. 2001. Distribution of the big free-tailed bat (*Nyctinomops macrotis*) in Colorado. *Southwestern Naturalist* 46: 370–376.

Navo, K. W., J. A. Gore, and G. T. Skiba. 1992. Observations on the spotted bat, *Euderma maculatum*, in northwestern Colorado. *Journal of Mammalogy* 73: 547–551.

O'Shea, T. J., and M. A. Bogan (eds.). 2000. Interim report of the Workshop on Monitoring U.S. Bat Populations: Problems and prospects. U.S. Geological Survey, Midcontinent Ecological Science Center, Fort Collins, Colo. Online interim report available at http://www.mesc.usgs.gov/BPD/ireport.htm.

Piaggio, A. J., E. W. Valdez, M. A. Bogan, and G. S. Spicer. In press. Taxonomy of *Myotis occultus* inferred from sequences of two mitochondrial genes. *Journal of Mammalogy*.

Rabe, M. J. 1999. Bat habitat use in pinyon-juniper woodlands and grassland habitats in northern Arizona. M.S. thesis, Northern Arizona University, Flagstaff.

Rabe, M. J., T. E. Morrell, H. Green, J. D. deVos, Jr., and C. R. Miller. 1998. Characteristics of ponderosa pine snag roosts used by reproductive bats in northern Arizona. *Journal of Wildlife Management* 62: 612–621.

U.S. Fish and Wildlife Service. 1994. Endangered and threatened wildlife and plants: Animal candidate review for listing as endangered or threatened species: Proposed rule (50CFR Part 17). *Federal Register* 59 (219): 58982–59028.

Vonhof, M. J., and R.M.R. Barclay. 1996. Roost-site selection and roosting ecology of forest-dwelling bats in southern British Columbia. *Canadian Journal of Zoology* 74: 1797–1805.

Wai-Ping, V., and M. B. Fenton. 1989. Ecology of spotted bat (*Euderma maculatum*) foraging and roosting behavior. *Journal of Mammalogy* 70: 617–622.

REPTILES AND AMPHIBIANS OF THE PIÑON-JUNIPER WOODLANDS

David D. Hanna and Timothy B. Graham

Walking quietly through the piñon-juniper woodland, a visitor may hear a sudden rustle in the underbrush. A quick flash of movement draws attention as a western whiptail lizard streaks through the bitterbrush. Seconds later a compact short-horned lizard scurries across the path. Farther along—sunning peacefully on the canyon rim with white-throated swifts rocketing over the abyss—a small, sleek sagebrush lizard does push-ups on the sandstone, eyeing the new intruder. These observations, though not uncommon, add much to the enjoyment of a visit to the ancient woodlands. No species of amphibian, turtle, lizard, or other reptile is restricted to piñon-juniper woodlands; but their diurnal nature and their frequency make them an exciting and readily observable part of these ecosystems.

Little information is available documenting amphibians and reptiles in Mesa Verde Country. G. A. Hammerson's (1999) handbook *Amphibians and Reptiles in Colorado* describes what is known at present about all of these species in the state. The Colorado Herpetological Society also keeps current records on all species statewide. Locally C. L. Douglas (1966) published a small volume about the reptiles and amphibians of Mesa Verde National Park. Resource Management personnel at Mesa Verde National Park inventory amphibians and reptiles annually. Yet we still lack a complete picture of amphibian and reptile populations in Mesa Verde Country. Therefore, in this chapter we expand our treatment area west into Utah, where recent research has begun to develop a comprehensive view of these organisms. This information can be interpolated to Mesa Verde Country.

Figure 9.1. Bufo punctatus *(red-spotted toad) is an important amphibian that commonly breeds in ephemeral pools of exposed slickrock within the piñon-juniper zone. Illustration by Amy Wendland.*

AMPHIBIANS

A chorus of red-spotted toads (Figure 9.1) in breeding ecstasy is a characteristic woodland sound. Scattered pools in canyon bottoms teeming with newly hatched tadpoles make one think twice before splashing through. Yet only ten amphibians have been reported from Mesa Verde Country. The amphibians are restricted for breeding to waterpockets, seeps, and springs or require riparian habitats adjacent to the old-growth woodlands. Many amphibians may forage out into the woodlands adjacent to these aquatic environments. Several range out into upland environments as adults to feed, aestivate, or hibernate. Tiger salamanders (*Ambystoma tigrinum*) have been found two miles from the nearest possible breeding site. Toads and chorus frogs forage or seek shelter under logs and rocks in arid upland situations almost as distant. Amphibians may not select habitat based directly on the composition of upland plant communities, but plants can exert indirect influences on amphibians. Plants that provide a thick litter layer will support more invertebrates, the food of amphibians. Amphibians, and reptiles as well, are more likely to select the microhabitats under these plants than the open interplant spaces or the sparse cover under piñon or juniper trees. Accumulated litter may also provide substrates for diurnal burrows that prevent desiccation, because of the insulating and moisture-holding properties of litter. Nonetheless, amphibians are probably indifferent to the vegetative composition of the upland communities. They may be more particular about the character of the substrate than about the vegetation growing on it.

Tiger salamanders breed in stock ponds, weedy lakes, and sewage lagoons. They are most abundant where water has been impounded. Salamanders have a distinct advantage over fish in that they can travel overland to isolated pools or tiny streams and produce larvae that can exploit these aquatic environments as efficiently as do carnivorous fishes. It is likely that during pre-European settlement they were found most commonly at higher elevations or in the major stream valleys of the lowlands. Permanent pools and streams of sufficient volume and duration for their long larval developmental period are rare in most of the piñon-juniper zone. By developing irrigation systems and building impoundments, humans have expanded the distribution of suitable breeding habitat and, possibly, the numbers of salamanders; and it is conceivable that the Ancestral Puebloans who lived on Mesa Verde until the late 1200s developed permanent water storage as well. Adult salamanders may actually spend very little time near/in water, often living entirely underground except when they emerge and migrate to water to breed. It is thought that adults spend most of their lives in networks of mammal burrows, feeding on soil invertebrates and invertebrates associated with the mammals that construct these burrows.

Although the tiger salamander is the only salamander found in the Mesa Verde Country, several species of toads make their home here. Red-spotted toads (*Bufo punctatus*) and Woodhouse's toads (*Bufo woodhousii*) are the most common toads within the piñon-juniper zone. Red-spotted toads are usually found in the vicinity of large rock outcrops with south-facing exposures, where they find shelter in crevices. Rocks are good heat sinks and maintain temperatures deep within their mass that are much closer to the yearly mean temperature. These are good things for a toad. J. Erdman et al. (1969) demonstrated that the canyon sides on average do not experience the low minimum temperatures that occur on the mesa tops or in the canyon bottoms. The south-facing slopes are more like habitats at lower latitudes and elevations than either mesa top or valley floor. Thus Woodhouse's toads find homes in a broader range of habitats.

Both these species sometimes may breed in surprisingly small pools, often less than 2 m^2 in area with as little as 0.1 m^3 volume (about 25 gallons). The critical requirement is that the pool persists long enough to permit the eggs to hatch and the larvae to metamorphose (A. W. Spencer, personal communication, 2000). Typically the breeding sites of both red-spotted and Woodhouse's toads in the piñon-juniper woodlands are associated with pools in intermittent and temporary streams and small springs in the bottoms of canyons. Tiny streams trickle downcanyon for a short distance until the impermeable bedrock is again buried deep beneath loose sediments.

Records of toad occurrences have been kept annually by Resource Management personnel at Mesa Verde National Park (M. Colyer, personal communication, 2001). Red-spotted toads, considered "park-sensitive," have been recorded in bedrock potholes and seeps within Mesa Verde National Park and Hovenweep National Monument on seven occasions since 1986. All records are on Cliff House Sandstone or Dakota Sandstone in slightly acidic water and in the piñon-juniper woodlands. Woodhouse's toads are found in wet microhabitats, such as perennial streams in Soda Canyon and a plunge pool in Fewkes Canyon. They are also common along the Mancos River, the only sizable perennial water body in Mesa Verde National Park. This common toad can be found at elevations from 1,909 m to 2,151 m. In Mesa Verde National Park it is found in the alkaline water of Menefee Shale. Nearly all of the records are in piñon-juniper woodlands, specifically supported by riparian zones or permanent springs (or irrigation facilities). Woodhouse's toads are prolific, known to lay eggs in late April, in July after monsoon rains, and even in mid-August following flooding. The juveniles have been seen in mid-October having just metamorphosed. The Woodhouse's toad is the dominant amphibian of the piñon-juniper elevations. Both species remain active through the summer, even in dry years, but they may not breed. During the hot, dry summer of 2000 in Canyonlands National Park, both Woodhouse's and red-spotted toads were active, at least at night. Even after rains that caused flows 30–40 cm deep, no males were heard calling, and no eggs or newly hatched tadpoles were observed (A. W. Spencer, personal communication, 2000).

Red-spotted toads are more likely than Woodhouse's toads to breed in tinajas (basins in the bedrock of usually dry watercourses on broad exposures of massive sandstone beds). Such pools depend on rainfall to maintain their volume. The "slickrock" catchment area above them must be large enough so that even a small rain of 0.5 cm will supply enough runoff to replenish the amount lost to evaporation between rains. Heavier rains may bring down fine soil or sand above the slickrock, possibly filling the basin with sediment. Very heavy rains produce flash floods that can scour the basin clean of silt once more but that can also wash the tadpoles or eggs downstream into altogether unfavorable situations. Perhaps the most productive tinajas utilized year after year by red-spotted toads are those in which a delicate balance between siltation and floods is dependably maintained. Flash floods do not always sweep tadpoles willy-nilly downstream, however. Tadpoles have been found in the same shallow pools before and after a strong flood. Water rose from a dry wash to over 70 cm deep in an hour in June 2000 in Mesa Verde. Presumably the same tadpoles were observed, because there were no sources of tadpoles for more than 1 km upstream.

Observations made around Moab, Utah, and in Canyonlands National Park indicate that Woodhouse's toads are more likely to breed in April/May. Red-spotted toads breed more often in summer following rains, and frequently there is only one breeding event or at best two all summer. The first event occurs in washes in late spring when temperatures increase, and a second event follows the first big rain in summer in washes and in potholes. In southeastern Utah red-spotted toad eggs laid on June 28, 2000, had metamorphosed into little toadlets by July 25. Eggs laid in the spring, and at higher elevations, require more time to develop, because temperatures are lower and thus development is slower.

Presumably reproduction commences when female toads have accumulated enough energy to produce eggs; this timing varies considerably among individual toads. Several factors affect this physiological readiness. The time of year of emergence from tadpole to toad, as well as its size at emergence, may be of influence. That is, an individual that emerges in August or September may not acquire enough energy to breed until the following July, so these toads would miss the early spring (April or May) breeding event. A toad emerging in April is more likely to have adequate physiological reserves to breed the following April. This may be a result of genetic differences among individuals of a population in development of reproductive maturity, or it may simply be the result of how much biomass each female can acquire and process into eggs and reproductive energy in a given time frame (which is determined by habitat quality).

Conditions during the breeding season are not reliable or predictable. After the first year of life the quality of the home range is likely to play a significant role in determining how long it takes a female toad to get physiologically ready to breed. There simply may not be enough food to build energy supplies. During droughts there may be no breeding even if it does rain, probably because it is difficult to find enough food to provide the energy required for breeding.

It is not clear whether the same individuals breed more than once each year. Populations of both species probably vary their reproductive effort depending on environmental conditions. Reproduction in toad populations appears to be related to the productivity of the particular habitat. One breeding event (often only a single night) is characteristic of pothole breeding populations, although populations in riparian zones may breed two to four times a year. Late spring sees the most significant breeding effort, with many males calling and new eggs evident in pools each morning. In some areas it appears that most Woodhouse's toad reproduction is concentrated in this spring breeding, whereas red-spotted toads contribute most of the tadpoles to summer rain-filled pools. But even in spring, breeding ceases after only a few days, although water continues to run in the washes. Following heavy rains in summer, toads may converge on wash-bottom pools, and slickrock pothole breeders also move to breeding pools.

Toads face many difficulties as they try to breed. Pools can dry in May and June, torrential rains may sweep the pools clean in July, and early fall weather can delay metamorphosis so long that frost kills the young amphibians. Newly metamorphosed females and males come to the pools with each series of rainy days. Travel during dry weather is hazardous for a toad. The chances of finding a filled pool diminish during dry periods and greatly improve after a rain. In fact male toads cease calling within a day or two of rain during the summer, because most pools will dry out before tadpoles can metamorphose. The population, by hedging its reproductive investments, adapts to the uncertainty. In some years early egg-layers beat the odds, whereas in another year the late layers may provide new recruits.

Females are not the only toads to experience difficulties in breeding. Calling males are exposed to increased predation because they are more apparent as they sit along the edges of pools, calling loudly for females. At one pool more than ten partial carcasses were found, including two females that had come to mate. Males are also subject to the vagaries of environmental conditions in trying to build the physiological reserves to breed, although to a lesser degree.

Toads walk hundreds of meters in a night patrolling their neighborhood for beetles and other prey. But during the day or for weeks in dry weather they bury themselves in soil to avoid desiccation. The Woodhouse's toads are more often found along permanent streams, lakes, and ponds, where they are somewhat buffered against the vagaries of weather. They range higher in elevation than red-spotted toads and are more widely distributed throughout Mesa Verde Country.

Young toads, recently metamorphosed, have a very different activity pattern than larger toads. The metabolism of amphibians (like that of other ectotherms) is regulated by the temperature of the surrounding environment. The more quickly a young toadlet can grow, the less vulnerable it will be to predation, and the sooner it will reach breeding size. Observations in Canyonlands and Arches National Parks show that metamorphs are active all day and are especially evident on warm afternoons in and near pools on intermittent stream channels. Very few, if any, adult toads are seen during summer daytime surveys. After dark the tables are turned. Night surveys inevitably find few if any metamorphs, yet a large number of adult toads can be seen. The risk of predation during the day due to increased visibility is apparently not a strong enough selection pressure to override the benefits gained by metamorphs' maintaining higher body temperatures and thus higher metabolic and growth rates when active during the day. H. B. Lillywhite et al. (1973) found that young toads actually bask in the sun and experimentally documented increased growth rates for basking metamorphs in the laboratory.

Many amphibians undergo seasonal shifts in activity patterns. In the spring and fall (when days may be warm but nights are still cold), amphibians and reptiles are most active in the afternoon, after the air and ground have warmed through the morning. During the heat of summer, adult amphibians and reptiles shift their activity to after dark.

These species of toads are only somewhat more visible than spadefoots. They are better armored against evaporative loss of their body water and so can venture out more often and more widely than do spadefoots. In Mesa Verde Country, *Spea multiplicatus* (New Mexican spadefoot) has potentially the broadest distribution of any of the amphibians. These toads breed during floods produced by summer thunderstorms. Spadefoot toad activity has been documented from late April to late August in southern Utah. Spadefoot toads are often found in broad, shallow impoundments produced by landslides that have blocked abandoned channels as well as stock ponds and barrow pits. Following a storm hundreds of adults might congregate at such places for a few nights. Sometimes calling occurs only for one night. The eggs hatch quickly in the warm, shallow water, and the tadpoles can metamorphose in a few weeks. Both young and adults soon disperse and virtually disappear, spending most of their lives buried in the soil, to reappear when summer storms provide another opportunity to reproduce. The species is listed by the state of Colorado as a species of special concern because its numbers have declined through development and agricultural impact on habitat (Colorado Division of Wildlife 2000). Even when they are abundant in an area, however, spadefoot toads are seldom seen except for brief breeding appearances.

Spadefoots undoubtedly spend considerable time buried in the soil, but we do not know how much time they spend actively foraging to acquire energy for the next breeding effort. M. A. Dimmitt and R. Ruibal (1980) found that New Mexican spadefoots needed at least seven meals to acquire adequate fat reserves to survive a year; *S. couchii* could eat enough in one feeding bout to survive twelve months, consuming up to 55% of its body weight in a single meal of termites!

New Mexican spadefoot tadpoles, like tadpoles of other *Spea* species, may obtain nutrients along two distinct paths, grazing on microorganisms and detritus in the water or being primarily carnivorous in nature. Tadpoles of the second type have modified jaws and musculature and can increase in size at an accelerated rate thanks to the animal protein–rich diet of crustaceans and other spadefoot larvae. Apparently some, but not all, tadpoles of the New Mexico spadefoot can be induced to become carnivorous if fairy shrimp are abundant in the pool (Pfennig 1992; Pfennig et al. 1991). A certain number of fairy shrimp must be eaten before the conversion occurs. Fairy shrimp may be quite abundant in ephemeral pools used for breeding by spadefoot toads and become a staple food item for the carnivorous tadpoles. Carnivorous tadpoles develop and metamorphose more quickly than the grazers, giving them an advantage in very temporary pools. The omnivorous morph develops more slowly but possesses larger fat reserves at metamorphosis, and thus postmetamorphosis survival is greater for this morph (Pfennig 1992).

W. G. Degenhardt et al. (1996) studied *Spea bombifrons* records along the San Juan River in northwestern New Mexico, along the border of Mesa Verde Country. Calls that sound like *S. bombifrons* have been heard in Salt Creek Canyon, Utah, but the callers have not been positively identified. Spadefoot toads that have been caught seem to resemble either *S. multiplicatus* or *S. intermontanus* more than *S. bombifrons*. It is possible that all three species are found in the Four Corners area and might even hybridize.

Chorus frogs (*Pseudacris triseriata*), in the elevation range of piñons, are tied more closely to the vicinity of permanent streams and irrigated lands. Chorus frogs are the smallest but most numerous of our amphibians, only 28 to 35 mm in length as adults. They choose a wide variety of breeding habitats but prefer grassy or cattail-crowded ephemeral pools. These are the frogs that trill from flooded meadows. Their voice seems too loud for such small amphibians. They get their name from their tendency to form trios and larger groups. One male (the "chorus leader") calls, and the others answer. The effect is to confuse the nonamphibian listener about the location of the singer. The chorus frogs have invaded piñon-juniper woodlands wherever irrigation has provided habitat. They probably were limited to the floodplains of permanent streams in earlier times. Their breeding season is often initiated by the spring runoff, but they have adapted to the artificial floods of the irrigator. Chorus frogs in the piñon-juniper zone are most active at dusk and dawn. They forage in moist areas. Their small size enables them to find moist crannies in the heart of shrubs or tussocks of grass. They have the ability to be seemingly frozen solid during the winter period and revive unscathed in the spring.

Chorus frogs are very small and secretive, and thus it is difficult to survey for them except during their breeding season, when their choruses can be quite loud. *P. triseriata* were commonly caught in small mammal traps, however, in wet meadows at Curecanti National Recreation Area, Gunnison, Colorado (Graham, unpublished). Small mammal traps may be an effective way to survey for chorus frogs outside of the breeding season.

Northern leopard frogs (*Rana pipiens*) are absolutely dependent on permanent pools, beaver ponds, irrigation ditches, and stock ponds and are rarely found far from the edge of the water. But they are prodigious travelers in wet weather and at such times can wander far from the pool or stream. Within Mesa Verde National Park sightings of the leopard frog were made at sewage lagoon facilities and other permanently wet sites prior to 1986 but have been absent since that time. Charles Douglas (1966) observed leopard frogs in the Mancos River in 1965. This species has suffered from population decline throughout Colorado, possibly because of widespread habitat alteration and competition from the introduced bullfrog (Corn and Fogelman 1984). In much of the western United States a similar decline in ranid frogs has occurred in the last twenty to thirty years (Sredl and Howland 1995). Anuran amphibian populations in Mesa Verde Country have experienced significant reduction in the past three decades. Many Mancos Valley residents will tell you not only that leopard frogs were once common and have now vanished but that they also no longer hear the choruses of chorus frogs that were once an impediment to sleep. Leopard frogs were the most common amphibian in southwestern Colorado until the late 1960s. The mosquito abatement program was initiated at that time, which may have played a role in their demise; however, the direct cause is not clear. Pesticides sprayed for adult mosquitos, larvicides that affect tadpoles, and draining of mosquito breeding sites (which are also leopard-frog breeding sites) all may have contributed to the decline in the northern leopard frog in Mesa Verde Country.

Bullfrogs (*Rana catesbeiana*) have been introduced repeatedly into this area. They were never native to the region and are the most aquatic anuran among our species. Bullfrogs seem not to persist long or multiply very successfully in the region, for which we may be thankful. In New Mexico, bullfrogs appear to be limited to habitat below 2,100 m (Degenhardt et al. 1996), probably because it is too cold at higher elevations. Hammerson (1999) reported a population in Gunnison

County, Colorado, at 2,743 m in pools fed by warm springs. In some parts of California, bullfrogs may take two years to develop from egg to metamorph, spending one winter as tadpoles. They have been implicated in the decline of numerous amphibians in the western United States (Blaustein and Wake 1990; Jennings and Hayes 1994; Moyle 1973).

Other species present in the zone include *Spea bombifrons* (plains spadefoot), *Hyla arenicolor* (canyon tree frog), and, close to the northern boundary of Mesa Verde Country, *Spea intermontanus* (Great Basin spadefoot). The first two of these amphibians have been found in southeastern Utah. The third species has not been found in Mesa Verde Country, although it may occur there.

The canyon tree frog is known from the Dolores River below the McPhee Reservoir (Hammerson 1999). This frog likely lives in many of the canyons of Mesa Verde Country, although its distribution appears to be very patchy. A single large adult was found on top of a slickrock expanse near Moab, Utah, in July 1999 under a boulder by a pothole. This is the only known sighting in southeastern Utah, although there is a report of some heard calling on a tributary of the Green River in Labyrinth Canyon. These frogs can be distinguished by the large sticky pads at the ends of their toes. They are generally found in canyons with bedrock pools, spending their days tucked into crevices and holes in the rock walls and coming out to forage at night on invertebrates. C. W. Painter (1985) recorded beetles, true bugs, caddisflies, and annelid worms as foods of the canyon tree frog. Males calling in narrow canyons can generate quite a cacophony. The call sounds somewhat like a jackhammer: with a large number of frogs, whose calls are all echoing back and forth between the walls, the sound is deafening. In Zion National Park, frogs on the canyon bottom could be heard 1,500 m above at the top of the east rim trail.

REPTILES

Lizards and snakes are prevalent symbols of the American Southwest. Their likenesses are found chiseled into cliff faces by Ancient Puebloans. Today paintings and sculptures of these reptiles are commonly seen as objets d'art. The reptiles of Mesa Verde Country are no exception.

Reptiles are ideally suited to life in this semiarid region. Lack of water is obviously a limiting resource for organisms, and reptiles are exemplary at adapting to aridity. All animals produce wastes containing nitrogen that must be excreted from their bodies. Mammals excrete a compound called urea, which must be dissolved in water for excretion. Thus mammals in arid zones must give up hard-earned water to eliminate these wastes. Reptiles and birds, however, produce an insoluble compound called uric acid. This can be excreted as a dry, white powdery substance, with no water loss at all. Reptiles, like other ectotherms, conserve a lot of energy by not having to keep their body temperature elevated at all times. This greatly reduces the amount of food they need to find and eat. Therefore, they can

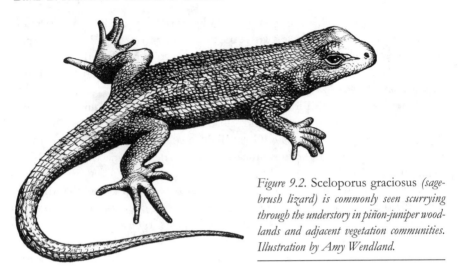

Figure 9.2. Sceloporus graciosus *(sagebrush lizard) is commonly seen scurrying through the understory in piñon-juniper woodlands and adjacent vegetation communities. Illustration by Amy Wendland.*

spend less time exposed to predators and the hot sun while foraging. The amount of water lost by reptiles through respiration and from their skin (which does occur, despite their tough, waxy scales) is greatly reduced by their activity patterns. These characteristics enable reptiles to exploit arid environments on a superior "footing" (even snakes) to mammals; of all the vertebrates, reptiles are the most abundant and diverse in arid environments.

Lizards

About thirteen species of lizards can be found, at least occasionally, in Mesa Verde Country in the piñon-juniper woodland. Characteristics of their residency, density, and body traits are summarized in Appendix 9.1. All the lizards that live in Mesa Verde Country are predators, feeding on insects, spiders, other invertebrates, and in some cases other lizards. Most rely on sharp eyesight to detect movements that translate into prey; and thus most lizards are active during the day (diurnal). Many snakes, in contrast, are nocturnal (active at night), using their chemosensory capacity to detect prey in the dark. Rattlesnakes also have sensors that detect heat given off by endotherms (mammals and birds) to help them find their prey.

Many species are more abundant in other plant communities interspersed with piñon-juniper woodland than among the trees. Two species commonly encountered among the trees, however, are *Sceloporus undulatus elongatus* (northern plateau lizard) and *Sceloporus graciosus* (sagebrush lizard) (Figure 9.2). These small lizards (or swifts) utilize the trees on a regular basis and establish true homes there. Both climb readily and take advantage of cavities, the spaces under loose bark, and tangles of branches and fallen tree trunks as refuges from pursuers and weather. Immature sagebrush lizards have been seen jumping vertically four times their total length from the ground to a limb. They often climb 3 to 4 m above the ground. Sagebrush lizards are common species in the piñon-juniper zone. This small swift has been found resting in a knothole in a hollow, dead piñon 2 m above the ground; and it is suspected that sagebrush lizards may climb into the upper branches of the trees in the morning to catch the sunlight before it reaches the ground. They preferentially select trees or logs for basking sites over soil or rocks when the ground is wet, probably because the wood conducts less heat from their bodies.

When the sun is low or weak, the lizards take advantage of charred surfaces that have absorbed more of the sun's heat. Both species select perches on horizontal limbs, fallen trunks, or exposed roots in order to expand their field of vision while foraging. During hot weather these perches will be in the shade of foliage or branches. They dart out into the sun to seize their prey and dart back into the shade again, thereby avoiding overheating. The ideal habitat is one with lots of down wood but otherwise open canopy and relatively sparse ground cover. Northern plateau lizards favor sites near large rock outcrops or tumbled archaeological ruins. At best, lizards can be classified as commensal associates, contributing little to the vegetation but exploiting its characteristics. Aside from consuming insects, mainly ants, lizard activities have virtually no effect on the trees or other vegetation.

Other lizards that may be encountered in the woodland include *Urosaurus ornatus* (tree lizard), *Phrynosoma douglasii* (short-horned lizard), *Uta stansburiana* (side-blotched lizard), *Crotaphytus collaris* (collared lizard), and *Cnemidophorus velox* (striped plateau whiptail). Again, each of these species is more likely to be found in open situations than among the trees. They readily utilize open slickrock areas or other openings in the tree canopy. Despite its name, the tree lizard is much more at home on big blocks of rock; in fact it is partial to blocks or outcrops with intersecting crevices that offer refuge but also provide passage to many faces of the rock and to the soil. Tree lizards are quite abundant in riparian zones as well, where they live up to their name and spend much of their time climbing in trees. Short-horned lizards and whiptails rarely leave the ground surface and seem to prefer more open stands. The sudden rustling of a whiptail in the litter can be startling: they create a big noise for a small animal, making them one of the more readily observed species in the woodland. The side-blotched lizard is quite common in areas with even small rocky exposures. It is visible in many national parks in the desert, because individuals claim as their territory the stacks of rocks (cairns) piled up to mark trails and routes. Cairns provide a diversity of microhabitats for these small lizards, with shady nooks and crevices, sunny warm sites, and a relatively high place from which to survey their small but for them very important domain.

Snakes

There are fourteen species of snakes in Mesa Verde Country (Appendix 9.1). Like the lizards, snakes are quintessential emblems of the wilderness Southwest. Awe of snakes has pervaded prehistoric as well as historic times; they are commonly seen throughout Mesa Verde Country depicted in petroglyphs. Although they are more secretive than the lizards and much less frequently observed, the region has its fair share of these organisms. The most commonly encountered snakes in the woodland are *Crotalus viridis* spp. (western rattlesnakes), *Pituophis melanoleucus* (gopher snake), *Masticophis taeniatus* (striped whipsnake), and *Thamnophis elegans* (western terrestrial garter snake). The few records of the Mesa Verde night snake have come from roads and trails in piñon-juniper habitat; but the habits of this secretive, nocturnal snake are poorly known. The rattler and bullsnake prey on rodents and lagomorphs, whose activities may be both beneficial and detrimental to vegetation; therefore these snakes may indirectly have a significant role in the piñon-juniper woodland.

For obvious reasons, rattlesnakes are perhaps the best-known reptiles that occur in Mesa Verde Country. The western rattlesnake is common in the area. Many people fear encounters with these rather shy reptiles, although they rarely result in danger to the humans. It is likely that the snake, in fact, has more to fear from the human than vice versa. Nevertheless, surprise encounters do occur that result in poisonous snakebites. Fortunately, humans have produced antivenom treatments for rattlesnake bites. There is some indication that rattlesnake species are altering the composition of their venoms (Grenard 2000). Most species have only hemotoxins and myotoxins—poisons that affect blood and muscle, respectively, and are relatively easy to counteract. Recent studies have shown that some species or subspecies have neurotoxins in their venoms as well. These poisons are much harder to neutralize, both by humans developing antivenoms and by potential prey species evolving resistance or neutralizing compounds. One example that is emerging in the snake literature involves the western rattlesnake (*C. viridis*) complex. There are about eleven subspecies of *C. viridis,* but only one, *C. v. concolor* (the midget-faded rattlesnake), is known to contain neurotoxins in its venom. Jim Glenn and his colleagues found (unpublished data) at the Veterans' Administration Hospital Venom Research Lab in Salt Lake City, Utah, that some individuals from Natural Bridges National Monument, near the eastern edge of *C. v. concolor*'s range, had venom resembling most *C. v. concolor.* That is, it contained neurotoxins. Some specimens had only typical *C. v. viridis* hemotoxins and myotoxins, and other individuals had hemotoxins similar to *C. v. viridis* but myotoxins more like *C. v. concolor,* yet no neurotoxins. The interpretation of this pattern was that the Natural Bridges area might be a zone of intergradation between the two subspecies. It also could be a zone of evolutionary adjustment in venom characteristics.

The remaining species (Appendix 9.1) are rarely seen in the woodland proper and are presumed to have no significant impacts there. Many of the species feed on small rodents, lizards, toads, and insects, however; and thus, like every other organism, snakes play a role in how energy is transferred through the old-growth piñon-juniper ecosystem.

SUMMARY

Overall, the ecological roles played by amphibians (Blaustein et al. 1994) and reptiles in the piñon-juniper woodland are not well understood. What we do know comes primarily from the naturalists' observations and resource management records from Mesa Verde National Park and nearby areas in southeastern Utah—very valuable data indeed. While we cannot say definitively that any reptile or amphibian is critical to continued functioning of the old-growth piñon-juniper ecosystem, these species should be considered integral components in a healthy piñon-juniper system.

Potential impacts on some amphibian and perhaps some reptile species may come from regional and global changes in environmental quality (Blaustein and Wake 1990; Pounds and Crump 1994; Wyman 1990). For example, if global climates warm appreciably, cool-adapted species like the chorus frog and leopard frog may no longer be able to exist in Mesa Verde Country. Bullfrogs may be able to expand their range in the area; thus even if other species are adapted to the physical environmental changes, biological interactions may drive them to at least local extinctions. Changes in global climatic patterns are predicted to include changes in quantities and timing of precipitation (Blaustein et al. 1994; Pounds and Crump 1994). If summer monsoons are reduced or eliminated by climatic shifts, significant breeding opportunities may be lost not only for amphibians but also for reptiles, whose reproductive cycles are timed to take advantage of the increased productivity that follows summer thunderstorms.

Declines in amphibian populations around the globe are being recognized; thus, sadly, amphibians have become indicators of ecological dysfunction (Wyman 1990). For Mesa Verde Country and the rest of the Colorado Plateau our understanding of amphibian species distributions and population sizes and fluctuations is limited. Because we lack baseline data, we cannot yet take advantage of the indicator capabilities of amphibians. The U.S. Geological Survey and National Park Service are initiating inventory procedures to document species of reptiles and amphibians and their population dynamics on the Colorado Plateau. Once we know more about these species, we can develop monitoring programs targeted to particularly sensitive species that will assist in assessing the ecological condition of the piñon-juniper and other ecosystems of Mesa Verde Country and the rest of the Colorado Plateau.

Appendix 9.1. Amphibians and reptiles of Mesa Verde Country: Estimated densities and biomass

Common Name	Scientific Name	Density (no./km^2)	Snout-Vent Length (cm)	Body Wt. (g)	Biomass (kg/km^2)
AMPHIBIANS					
Bullfrog	*Rana catesbeiana*	1	150	200	0.2
Canyon Tree Frog	*Hyla arenicolor*	1	50	5	0.005
Great Basin Spadefoot	*Spea intermontanus*		62	50	0
Northern Leopard Frog	*Rana pipiens*	0	90	80	0
New Mexico Spadefoot	*Spea multiplicata*	25	62	50	1.25
Plains Spadefoot	*Spea bombifrons*		55	50	0
Red-Spotted Toad	*Bufo punctatus*	25	70	60	1.5
Tiger Salamander	*Ambystoma tigrinum*	20	125	100	2
Western Chorus Frog	*Pseudacris triseriata*	2	28	1.2	0.0024
Woodhouse's Toad	*Bufo woodhousii*	25	90	80	2
REPTILES					
Black-Headed Snake	*Tantilla hobartsmithi*	1	240	50	0.05
Black-Necked Garter Snake	*Thamnophis cyrtipsis*	1	500	250	0.25
Coachwhip	*Masticophis flagellum*	0.01	1,100	900	0.009
Collared Lizard	*Crotaphytus collaris*	20	90	80	1.6
Common King Snake	*Lampropeltis getula*	1	700	400	0.4
Desert Spiny Lizard	*Sceloporus magister*	1	110	90	0.09
Glossy Snake	*Arizona elegans*	1	500	250	0.25
Gopher Snake	*Pituophis catenifer*	5	1,200	1,100	5.5
Leopard Lizard	*Gambelia wislizenii*	0.01	90	80	0.0008
Lesser Earless Lizard	*Holbrookia maculata*	1	50	20	0.02
Little-Striped Whiptail	*Cnemidophorus inornatus*	2	60	30	0.06
Long-Nosed Snake	*Rhinocheilus lecontei*		600	300	0
Milk Snake	*Lampropeltis trangulum*	10	350	175	1.75
Mormon Racer	*Coluber constrictor mormon*	5	600	300	1.5
Night Snake	*Hypsiglena torquata*	25	400	200	5
Plateau Lizard	*Scleoporus undulatus*	200	65	30	6
Plateau Whiptail	*Cnemidophorus velox*	25	70	40	1
Sagebrush Lizard	*Sceloporus graciosus*	1,000	50	25	25
Short-Horned Lizard	*Phrynosoma hernandesi*	200	80	30	6
Side-Blotched Lizard	*Uta stansburiana*	10	47	5	0.05
Smooth Green Snake	*Liochlorophis vernalis*	5	370	180	0.9
Striped Whipsnake	*Masticophis taeniatus*	10	950	750	7.5
Tree Lizard	*Urosaurus ornatus*	50	50	5	0.25
Western Terrestrial Garter Snake	*Thamnophis elegans vagrans*	25	500	250	6.25
Western Rattlesnake	*Crotalus viridis*	10	500	500	5
Western Whiptail	*Cnemidophorus tigris*	1	80	45	0.045

REFERENCES CITED

Blaustein, A. R., and D. B. Wake. 1990. Declining amphibian populations: A global phenomenon? *TREE* 5: 203–204.

Blaustein, A. R., D. B. Wake, and W. P. Sousa. 1994. Amphibian declines: Judging stability, persistence, and susceptibility of populations to local and global extinctions. *Conservation Biology* 8: 60–71.

Colorado Division of Wildlife. 2000. Colorado listing of endangered, threatened and wildlife species of special concern [May 2000]. World Wide Web site: www.protectwildlife.org/T&E/list.asp.

Corn, P. S., and J. C. Fogelman. 1984. Extinction of montane populations of the northern leopard frog *Rana pipiens* in Colorado. *Journal of Herpetology* 18(2): 147–152.

Degenhardt, W. G., C. W. Painter, and A. H. Price. 1996. *Amphibians and reptiles of New Mexico*. University of New Mexico Press, Albuquerque.

Dimmitt, M. A., and R. Ruibal. 1980. Exploitation of food resources by spadefoot toads (*Scaphiopus*). *Copeia* 1980: 854–862.

Douglas, C. L. 1966. Amphibians and reptiles of Mesa Verde National Park. *University of Kansas Publications, Museum of Natural History* 15: 744–771.

Erdman, J. A., C. L. Douglas, and J. W. Marr. 1969. *The environment of Mesa Verde, Colorado*. Archeological Research Series No. 7-B. National Park Service, Washington, D.C.

Freed, A. N. 1980. An adaptive advantage of basking behavior in an anuran amphibian. *Physiological Zoology* 54: 433–444.

Grenard, Steve. 2000. Is rattlesnake venom evolving? *Natural History* 109(6): 44–49.

Hammerson, G. A. 1999. *Amphibians and reptiles in Colorado*. University Press of Colorado, Niwot.

Jennings, M. R., and M. P. Hayes. 1994. Decline of native ranid frogs in the desert Southwest. *Southwestern Herpetological Society Special Publication* 5: 183–211.

Lillywhite, H. B., P. Licht, and P. Chelgren. 1973. The role of behavioral thermoregulation in the growth energetics of the toad *Bufo boreas*. *Ecology* 54: 375–383.

Moyle, P. B. 1973. Effects of introduced bullfrogs, *Rana catesbiana*, on the native frogs of the San Joaquin Valley, California. *Copeia* 1973: 18–22.

Painter, C. W. 1985. *Herpetology of the Gila and San Francisco river drainages of southwestern New Mexico*. New Mexico Department of Game and Fish, Santa Fe.

Pfennig, D. W. 1990. The adaptive significance of an environmentally-cued developmental switch. *Oecologia* 85: 101–107.

———. 1992. Proximate and functional causes of polyphenism in an anuran tadpole. *Functional Ecology* 6: 167–174.

Pfennig, D. W., A. Mabry, and D. Orange. 1991. Environmental causes of correlations between age and size at metamorphosis in *Scaphiopus multiplicatus*. *Ecology* 72: 2240–2248.

Pounds, J. A., and M. L. Crump. 1994. Amphibian declines and climate disturbance: The case of the golden toad and the harlequin frog. *Conservation Biology* 8: 72–85.

Sredl, M. J., and J. M. Howland. 1995. Conservation and management of Madrean populations of the Chiricauhuan leopard frog. In *Biodiversity and management of the Madrean archipelago symposium*, pp. 379–385. September 19–23, 1994, Tucson, Ariz. USDA Forest Service General Technical Report RM-GTR-264.

Wyman, R. C. 1990. What's happening to the amphibians? *Conservation Biology* 4: 350–352.

INSECTS ASSOCIATED WITH THE PIÑON-JUNIPER WOODLANDS OF MESA VERDE COUNTRY

David A. Leatherman and Boris C. Kondratieff

In the preceding chapters the microbes, birds, bats, and animals of the piñon-juniper woodland have been described. Here we turn to the insects, a diverse fauna that plays a critical role in determining the health of the piñon-juniper woodland in Mesa Verde Country. It has been estimated that approximately 14,000 to 26,000 species of insects can occur in the old-growth piñon-juniper woodland region of the Colorado Plateau and the Great Basin (Nelson 1994). Many of these species dominate food chains and food webs in biomass and species richness throughout the piñon-juniper woodlands. The feeding specializations by different insect groups range from detritivory, saprophagy, xylophagy, fungivory, and phytophagy (including sap feeding) to predation and parasitism. Their lifestyles encompass solitary, gregarious, subsocial, and highly social modes such as those of termites, ants, and the introduced honeybee. This large group of insect species may be either conspicuous or concealed and active by either day or night. Insect life cycles are adapted to a variety of abiotic conditions, including seasonal extremes of heat/cold, wet/dry, and notably to unpredictable climate, so common in the piñon-juniper woodlands.

It is impossible even to summarize all the relationships of insects within the piñon-juniper woodlands. In the following review we list insect species that have been shown to be specific to either piñon pine or Utah juniper and review the biology of the major associated species. The ecological relationships of the myriad insects that occur within Mesa Verde National Park (Kondratieff et al. 2000) and

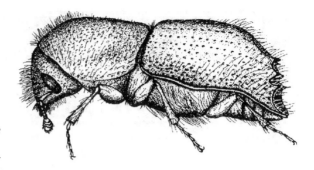

Figure 10.1. Details of mature Ips confusus *(piñon ips).*

the piñon-juniper woodlands are either poorly known or completely unknown and are outside the scope of this review. The reader can consult other sources (such as Essig 1929) for some of this information. Numerous insect species found on these trees have secondary ecological relationships. For example, several species of ants, especially in the genus *Formica,* can be observed tending the various scale insects and aphids (*Cinara* spp.) for honeydew or foraging for other foods. Another example is the cicadas, whose nymphs are fluid feeding below ground, with their beaks in the roots. The adults (Appendix 10.1 and Appendix 10.2) are often abundantly chorusing on the trees during the summer.

Some insects are key components of piñon-juniper woodlands. Beetles, scale insects, moths, and flies use these trees as hosts. Additionally, climatic events can trigger outbreaks that cover large areas. T. J. Rogers (1993) provides a brief review of some of the common insects associated with piñon-juniper woodlands. He indicates that bark beetles of the genus *Ips* and other genera such as *Pityophthorus* are the most important insects associated with piñon-juniper woodlands. Additionally, several species including the piñon needle scale (*Matsococcus acalyptus* Herbert) can be important, especially during droughts. R. L. Furniss and V. M. Carolin (1977) and R. Baskin et al. (1994) also list species, mostly those considered pests of these trees. Little work is available that details the occurrence and biology of insects on piñon-juniper. R. L. James and C. K. Lester (1978) provide information about insect relationships with piñon-juniper woodlands of the Mesa Verde area.

The piñon ips, *Ips confusus* (LeConte), is the dominant scolytid bark beetle found in mature piñon pines over much of the southwestern United States, including the Mesa Verde area (Figures 10.1, 10.2). Like others in the genus, it attacks unthrifty trees via mass onslaughts on the trunk and larger branches. Factors thought to contribute to tree stress and predisposition to ips attack include the presence of black-stain root disease (*Leptographium wageneri* Engelmann), piñon dwarf mistletoe (*Arceuthobium divaricatum* Engelmann) infestation (James and Lister 1978), piñon needle scale infestation (see the following discussion), drought, and root damage. Like most bark beetles, female members of the genus *Ips* vector

Figure 10.2. Mature piñon ips beetles may reach epidemic levels, such as those experienced in Mesa Verde Country in 2002.

fungal pathogens. Successful attacks inoculate various symbiotic fungi, known collectively as "bluestain fungi," which are thought to aid the beetles by helping to negate normal tree defense mechanisms. In most instances piñon ips attacks and subsequent inoculation of bluestain fungi kill the host trees, which then are characterized by foliage that turns reddish-brown throughout the crown over a period of months. Attacks are initiated by males, which bore through the bark, construct "nuptial chambers" just under it for the purposes of mating (Figure 10.3), and initiate the egg galleries, later elongated and finished by females. This passage through the bark results in small accumulations of boring dust spaced at intervals of several centimeters over the trunk and other attacked surfaces. A typical attack rate would be 10 per 0.1 m^2 of bark surface. At least until the tree dies and the foliage discolors, this dust and small masses of coagulated resin at the attack points called "pitch tubes" are the most obvious external symptoms of piñon ips infestation.

The males of this polygamous species emit an aggregating pheromone, which typically elicits a positive response from both females and other males. Stridulation by females while on the bark surface is thought to play a role in the male's allowing entry to the nuptial chamber (Mitton and Sturgeon 1982). Mating occurs within the nuptial chamber, and each of the three to five mated females then elongates one of the male-initiated egg galleries in the cambium tissue between the bark and outer rings of xylem wood. The name "engraver beetle" frequently given to members of the genus *Ips* comes from the fact that these egg galleries often lightly score or "engrave" the outermost layers of wood. The egg galleries of *Ips* can be distinguished from those of related bark beetle genera like *Dendroctonus* found in piñon by their gallery pattern and how free they are of boring dust and frass. Eggs are deposited in egg niches along the egg galleries. Larvae resulting from eggs consume their exuviae and tunnel away from the egg-niche areas, feeding on phloem tissue. After a period of a few months they transform to pupae and then adults, in enlarged areas called pupal cells at the termini of the larval feeding galleries. The piñon ips usually overwinters in the adult stage, occasionally in large accumulations, under the basal trunk bark. In spring, adults emerge through clear holes in the bark and fly to attack live trees. An average flight distance of 4.8 km, with a maximum flight of almost 13 km, has been recorded for *I. confusus* (Kinn 1971).

There are usually three generations per year in southwestern Colorado, with up to five generations in Arizona and other southern portions of its distribution. Other species of *Ips* also colonize piñon on occasion, particularly *I. pini* (Say) and *I. latidens* (LeConte). Although they are much less common in piñon woodlands on a landscape scale, their effect on and development within individual piñons are similar to those of *I. confusus* (Wood 1982).

Piñon ips population surges can result in large numbers of dead trees, which become a concern in park areas, forested subdivisions, and other areas where

Figure 10.3. Galleries, creating by mating of piñon ips, are a common sight in the wood of dead or dying piñon pines in Mesa Verde.

trees have higher than normal values or where heavy fuel loads for potential fires are undesirable. In combination with regional droughts, ips can greatly reduce the piñon component of piñon-juniper woodlands, such as occurred in the early twenty-first century over wide areas of the Southwest, including the Four Corners area.

Common avian predators of piñon ips while under and on the bark include hairy and downy woodpeckers. Clerid beetles are among the most common insect predators, with adults capturing adult ips beetles on the tree surface, and larvae seeking out ips larvae within their gallery systems. Braconid wasps, long-legged flies, cerambycid wood-boring beetles, and others also play a part in natural regulation of populations.

"Twig beetles" in the diverse bark beetle genus *Pityophthorus* are also quite common in the branches and trunks of piñon. No fewer than seven species are listed from piñon in Colorado (Wood 1982). Normally they are found in smaller-diameter portions of the tree than ips beetles. Their habits are basically similar to those described for *I. confusus*, in that they favor stressed trees for colonization, vector bluestain fungi, mass attack, and are polygamous. When the various species of this genus attack small branches or branch tips, the resulting foliage discoloration is confined to that particular branch, giving affected trees a "powder puff" look. That is, the overall crown is green but interspersed with orange patches of foliage here and there at the branch ends. On occasion these tiny bark beetles attack the main trunks of small-diameter piñons, killing entire trees. This would be more common in situations of severe environmental stress such as prolonged drought, nonfatal exposure to wildfire, or root disturbance during the warmer months. Newly planted piñons may also be attacked in this manner.

In both Utah and Rocky Mountain junipers, bark beetles in the genus *Phloeosinus* are common. *Phloeosinus scopulorum neomexicanus* Blackman and *P. furnissi* Blackman are confirmed for Mesa Verde, and *P. hoferi* Blackman probably also occurs, based on collections of this species east, south, and west of Mesa Verde (Wood 1982). Functioning ecologically in much the same manner as *Ips* in ponderosa pine and spruce populations at higher elevations, these small chocolate-brown beetles colonize the trunk and main bole of junipers older than the sapling stage. Males initiate attacks and are monogamous, with the females constructing the egg galleries after first mating with males in the nuptial chamber. The egg galleries engrave the outer xylem wood rather deeply. Larval galleries meander away from the egg gallery in no particular direction. While *Ips* beetles are reported to feed en masse on phloem without necessarily raising broods within certain trees, this type of adult "maturation" feeding is commonplace in *Phloeosinus*. This manner of feeding is characterized by notches cut in the bark of very fine twigs by individual beetles that occasionally result in discoloration or "flagging" of foliage at the tips of those branches. The more obvious trunk attacks that result in broods being developed beneath the bark lead to whole-tree mortality. The bark of trees from which beetles have emerged is peppered with small round holes.

The piñon cone beetle (*Conophthorus edulis* Hopkins) attacks the base or stalk of a second-year cone in late spring or early summer, causing it to abort and never attain full size, regardless of whether brood is produced. Ultimately, tunneling beetle larvae and resulting adults riddle the inside of the cone. From five to twenty adults usually overwinter inside dead cones, either on the tree or on those that have fallen to the ground. There is one generation per year. Up to 90% of cone crops in certain piñon stands can be destroyed (Hedlin et al. 1980). A pteromalid wasp has been recorded as a parasitoid of the piñon cone beetle (Keen 1958).

Piñon needle scale (M*atsococcus acalyptus*) can be important on piñon, particularly during droughts. Population increases can affect piñon-juniper woodlands on a regional scale. The first formal description of its life history was based on observations made in part at Mesa Verde National Park (McCambridge and Pierce 1964). The withdrawal of sap from needles by adult females and nymphs can result in discoloration of foliage, reduced needle length, and premature needle drop, particularly of older needles produced in previous years. This can lead to a "tufted" appearance, with only the newer foliage at the tips of the branches being retained, or can kill entire trees if infestations persist over a period of years. Bark beetles, such as piñon ips, can attack trees weakened by scale feeding.

The developmental stage of the piñon needle scale most commonly observed is the second-instar nymph, often described as the "bean stage," found attached to needles in early spring. Those nymphs, which develop into adult females, are black and mobile. They are about 1.5 mm long and migrate to the trunk, where they produce a cottony mass of eggs in spring. Adult males, present in early spring, are winged, flylike, and rarely seen. Mating is required for the production of viable eggs. Nymphs feed on old needles through the summer and early fall and remain on the foliage during winter. One generation is produced annually (Cranshaw et al. 2000). Birds that prey on piñon needle scale nymphs include bushtits and mountain chickadees (Leatherman, personal observation, 2001).

The seeds of piñon are particularly prized by many species of wildlife and historically by humans. Some of the insects affecting the flowers, cones, and seeds deserve mention. At least two unidentified species of cecidomyiid gall midges are reported to attack first-year piñon cones (Little 1950). Such cones are quite small when attacked, shrivel in August, and fall when touched. Such injury can be quite extensive, yet little is known about the specific identity and life cycles of the midges involved.

The piñon tip moth, *Dioryctria albovitella* (Hulst), tunnels in new shoots and cones. A nodule of pitch, which forms a few inches back from the branch tip, characterizes shoot infestations. Infested cones are hollowed out, and 1.4-mm-diameter exit holes, quantities of frass pellets, and one to a few rusty-brown pupal shells characterize cones from which the adult moths have emerged. Adults are present in late summer from July to September. Damage levels attributed to this insect have not been reported (Furniss and Carolin 1977).

Other insects reported in the literature from piñon microstrobili and female cones include a catkin sawfly (Xyelidae) in the genus *Xyela;* the tortricid moth *Eucosma bobana* Kearfott; various other *Dioryctria* species; a gelechiid moth, *Chionodes periculella* (Busck); a weevil (Curculionidae) in the genus *Conotrachelus;* and a chloropid fly in the genus *Hapleginella* (*Oscinella*) (Keen 1958). Much remains to be learned about the identity and habits of the suite of insects inhabiting piñon cones.

Galls are among the more obvious evidence of insect herbivory in both piñon and juniper. On piñon, various needle galls are formed by cecidomyiid midges. The piñon spindle gall midge (*Pinyonia edulicola* Gagne) life cycle is well described (Houseweart and Brewer 1972). Adults are delicate, orange flies that lay eggs in late June or early July. The orange maggots develop within needles and cause the formation of spindle-shaped swellings at the needle bases. These galls can be red or light green. Six to fifteen larvae are found within each gall. Emergence of the adults occurs the following June (Cranshaw et al. 2000). Galled needles often fall prematurely and in rare instances can lead to total defoliation and death of the host tree. The related piñon stunt needle midge (*Janetiella coloradensis* Felt) causes the formation of needles about one-third the normal length with round bases (Brewer 1971).

Galls on juniper tend to involve the berries or developing buds. The so-called artichoke galls caused by the juniper tip midge (*Oligotrophis betheli* Felt) are quite distinctive and common. They are formed from foliage and appear as yellow, light blue, or purplish rosettes of leaves that resemble artichokes (Cranshaw et al. 2000).

General defoliators of piñon and juniper include larvae of moths and sawflies. Two arctiid moths, *Lophocampa ingens* (Edwards) and *L. argentata* (Packard), form large white tents near the tops of various evergreen trees, including piñon and juniper. These insects are somewhat unusual in that the larvae are present and actively feeding during the winter months. The caterpillars are typical "woolly bears" in appearance. The hairs are known to be irritating to human skin. The tents provide at least partial protection from potential predators like birds.

Two sawflies, *Neodiprion edulicolus* Ross and *Zadiprion rohweri* Middleton, are recorded from piñon in southwestern Colorado (Appendix 10.1). Defoliation events can be intense but are generally short-lived. Only the older foliage is usually consumed, and trees typically survive the outbreak (Furniss and Carolin 1977).

The piñon pitch mass borer (*Dioryctria ponderosae* Dyar) and relatives can be common colonizers of piñon. Attacks occur most frequently along the trunk at the base of branch attachments, particularly if there are preexisting wounds. Infestations are evidenced by large, soft masses of coagulated pitch (Figure 10.4), which overtop the active larvae feeding on phloem under the bark. Larvae are yellow or pale pink in color, with light brown heads, and can reach over 13 mm in length. There is one generation per year, with adults moths flying during the summer. Certain infestations can result in considerable branch mortality and occasional breakage (Cranshaw et al. 2000).

Figure 10.4. Piñon pine trunk with indications of infestation by Dioryctria ponderosae *(the piñon pitch mass borer).*

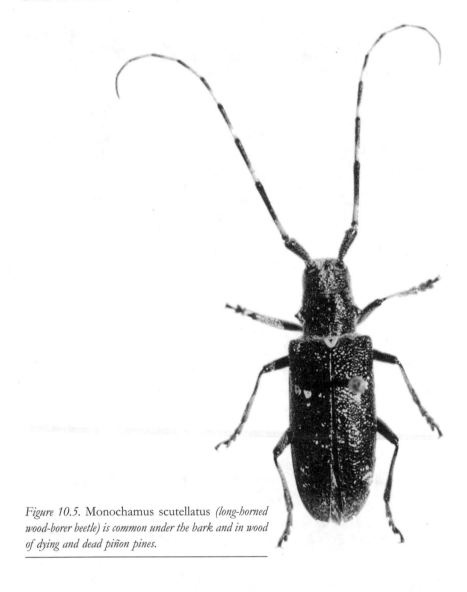

Figure 10.5. Monochamus scutellatus *(long-horned wood-borer beetle) is common under the bark and in wood of dying and dead piñon pines.*

Appendix 10.1 and Appendix 10.2 indicate what species have been observed or are probable in the old-growth woodlands of Mesa Verde Country. It is important to realize that many of these insect species are often sporadic in occurrence (Figure 10.5), common during certain years and almost absent during other years. Their population dynamics are not well-known. The effects of widespread fires on these insect populations are discussed in Chapter 17.

APPENDIX 10.1. Insects reported from or probably on piñon pine (*Pinus edulis*), Mesa Verde National Park

	SPECIES ASSOCIATED WITH FLOWERS AND FOLIAGE	
Homoptera (cicadas, leafhoppers, aphids, scale insects)	Cercopidae (spittle bugs)	*Aphrophora irrorata* Ball
	Diaspididae (armored scale insects)	*Chionaspis pinifoliae* Fitch
	Adelgidae (pine and spruce adelgids)	*Pineus coloradensis* Gillette
	Aphididae (aphids)	*Cinara* spp. (giant conifer aphids)
Thysanoptera (thrips)	Thripidae	*Frankliniella occidentalis* Pergande *Oxythrips pinicola* Hood
	Raphidiidae (snakeflies)	*Raphidia* spp.
Coleoptera (beetles)	Chrysomelidae (leaf beetles)	*Pachybrachus* spp. *Coleothorpa dominicana* F.
	Curculionidae (weevils)	*Magdalis lecontei* Horn *Magdalis gentilis* LeConte
Lepidoptera (butterflies, moths)	Saturniidae (silk moths)	*Coloradia pandora davisi* Barnes and Benjamin
	Pyralidae	*Rhyacionia salmonicolor* Powell
	Tortricidae	*Eucosma bobana* Kearfott
	Arctiidae (tiger moths)	*Lophocampa argentata* (Packard)
	Diprionidae (conifer sawflies)	*Neodiprion edulicolus* Ross *Zadiprion rohweri* Middleton
	SPECIES ASSOCIATED WITH CONES	
Hemiptera (true bugs)	Coreidae (leaf-footed bugs)	*Leptoglossus occidentalis* Heidemann
Coleoptera (beetles)	Scarabaeidae (May and June beetles)	*Dichelonyx* spp.
	Scolytidae (bark beetles)	*Conophthorus edulis* Hopkins
Lepidoptera (butterflies, moths)	Pyralidae (pyralid moths)	*Dioryctria auranticella* (Grote)
	SPECIES ASSOCIATED WITH TWIGS-LIMBS ON/UNDER BARK	
Homoptera (cicadas, leafhoppers, aphids, scale insects)	Cicadidae (cicadas)	*Okanagana magnifica* Davis *Okanagana schaefferi* Davis *Platypedia mohavensis* Davis *Platypedia putnami* Uhler *Tibicen duryi* Davis

continued on next page

APPENDIX 10.1—*continued*

Homoptera (*continued*)	Margarodidae (giant scale insects)	*Matsococcus acalyptus* Herbert
		Matsococcus eduli Morrison
Coleoptera	Buprestidae (metallic wood borers)	*Chrysophana placida* (LeConte)
	Cleridae (checkered beetles)	*Enoclerus sphegeus* (F.)
	Cerambycidae (long-horned beetles)	*Callidium antennatum* Newman
		Monochamus clamator (LeConte)
		Oeme costata LeConte
		Rhagium inquisitor (L.)
	Scolytidae (bark beetles)	*Dendroctonus ponderosae* Hopkins
		Dendroctonus valens LeConte
		Ips confusus LeConte
		Ips pini Say
		Pityogenes spp.
		Pityophthorus conferus Swaine
		Pityophthorus sp.
Lepidoptera (butterflies, moths)	Pyralidae (pyralid moths)	*Dioryctria albovitella* (Hulst)
	Torticidae	*Petrovia arizonensis* (Heinrich)
Diptera (flies)	Asilidae (robberflies)	*Laphria engelhardti* Bromley
		Pogonosoma ridingsi Cresson

SPECIES THAT ARE GALLFORMERS

Diptera (flies)	Cecidomyiidae (gall midges)	*Janetiella coloradensis* Felt
		Pinyonia edulicola Gagne

SPECIES FEEDING ON ROOTS

Coleoptera	Scolytidae (bark beetles)	*Hylastes gracilis* LeConte
		Hylastes fulgidus Blackman
		Hylastes macer LeConte
		Hylurgops porosus LeConte
		Gnathotrichus denticulatus Blackman
		Hylurgops reticulatus Wood
		Orthotomicus caelatus (Eichhoff)
		Orthotomicus latidens (LeConte)
		Pityophthorus confertus Swaine
		Xyleborus intrusus Blandford

APPENDIX 10.2. Insects reported from or probable on Utah juniper (*Juniperus osteosperma*), Mesa Verde National Park

	SPECIES ASSOCIATED WITH FLOWERS AND FOLIAGE		
Homoptera (cicadas, leafhoppers, aphids, scale insects)	Cercopidae (spittle bugs)	*Clastoptera* sp.
	Diaspididae (armored scale insects)	*Carulaspis juniperi* (Bouche)
	Aphididae (aphids)	*Cinara* spp. (giant conifer aphids)
Lepidoptera (butterflies, moths)	Lygcaeidae	*Callophrys gryneus siva* (Edwards)
	Lasiocampidae	*Gloveria arizonensis* Packard
	Tortricidae	*Cudonigera houstonana* (Grote)
	SPECIES ASSOCIATED WITH FRUIT	
Hemiptera (true bugs)	Coreidae (leaf-footed bugs)	*Leptoglossus occidentalis* Heidemann
	Species associated with TWIGS-LIMBS ON/UNDER BARK	
Coleoptera (Beetles)	Cicadidae (cicadas)	*Platypedia mohavensis* Davis *Platypedia putnami* Uhler
	Cerambycidae (long-horned beetles)	*Methia mormona* Linell *Semanotus juniperi* (Fisher)
	Scolytidae (bark beetles)	*Semanotus ligneus* (F.) *Phloeosinus* spp.
	Species that are GALLFORMERS	
	Cecidomyiidae (gall midges)	*Oligotrophis betheli* (Felt) *Rhopalomyia sabinae* Patterson

REFERENCES CITED

Baskin, R., C. Burns, W. Cranshaw, D. Farmer, L. Helburg, D. Leatherman, K. Mullen, J. Potter, L. P. Pottorff, M. Schomaker, V. S. Thompson, and R. Zentz. 1994. *Checklist of tree pests of Colorado*. Colorado Tree Coalition.

Brewer, J. W. 1971. Biology of the piñon stunt needle midge. *Annals of the Entomological Society of America* 64: 1009–1102.

Cranshaw, W., D. Leatherman, B. Kondratieff, P. Opler, C. Sclar, W. Jacobi, and L. Mannix. 2000. *Insects and diseases of woody plants of the central Rockies*. Colorado State University Cooperative Extension, Bulletin 506A. Colorado State University, Fort Collins.

Essig, E. O. 1929. *Insects of western North America*. Macmillan Co., New York.

Furniss, R. L., and V. M. Carolin. 1977. *Western forest insects*. USDA Forest Service Miscellaneous Publication No. 1339.

Hedlin, A. F., H. O. Yates III, D. C. Tovar, B. H. Ebel, T. W. Koerber, and E. P. Merkel. 1980. *Cone and seed insects of North American conifers*. Canadian Forestry Service. Ottawa, Ontario, Canada.

Houseweart, M. W., and J. W. Brewer. 1972. Biology of a piñon spindle gall midge (Diptera: Cecidomyiidae). *Annals of the Entomological Society of America* 65: 331–336.

James, R. L., and C. K. Lester. 1978. *Insect and disease conditions of piñon pine and Utah juniper in Mesa Verde National Park, Colorado*. Biological Evaluation R2-78-4. USDA, Forest Service Rocky Mountain Region Forest Insect Disease Management.

Keen, F. P. 1958. *Cone and seed insects of western forest trees*. USDA Forest Service Technical Bulletin 1169.

Kinn, D. N. 1971. The life cycle and behavior of *Cercoleipus coelonotus* (Acarina: Mesostigmata). *University of California Publications in Entomology* 65: 1–66.

Kondratieff, B. C., D. A. Leatherman, and M. J. Weissmann. 2000. Preliminary survey of the insects of the Park Mesa, Mesa Verde National Park, Colorado. Report to Mesa Verde National Park.

Little, Elbert L., Jr. 1950. Common insects on piñon pine. *Journal of the New York Entomological Society* 51: 239–252.

McCambridge, W. F., and D. A. Pierce. 1964. Observations on the life history of the piñon needle scale *Matsococcus acalyptus* (Homoptera, Coccoidea, Margarodidae). *Annals of the Entomological Society of America* 57: 197–200.

Mitton, J. B., and K. B. Sturgeon. 1982. *Bark beetles in North American conifers*. University of Texas Press, Austin.

Nelson, C. R. 1994. Insects of the Great Basin and Colorado Plateau. In K. T. Harper, L. L. St. Clair, K. H. Thorne, and W. M. Hess (eds.), *Natural history of the Colorado Plateau and Great Basin*, pp. 211–237. University of Colorado Press, Niwot.

Rogers, T. J. 1993. Insects and disease associates of the piñon-juniper woodlands. In E. F. Aldon and D. W. Shaw (tech. coords.), *Managing piñon-juniper ecosystems for sustainability and social needs*. General Technical Report RM-236. USDA Forest Service, Rocky Mountain Forest Range Experiment Station, Fort Collins, Colo.

Wood, S. L. 1982. *The bark and ambrosia beetles of North and Central America (Coleoptera: Scolytidae): A taxonomic monograph*. Great Basin Naturalist Memoirs No. 6. Brigham Young University, Provo, Utah.

PART II

WHAT CHARACTERIZES THE GEOLOGY, WEATHER, AND SOILS OF MESA VERDE COUNTRY?

THE FIVE CHAPTERS IN THIS PART OF THE BOOK examine the physical underpinnings of the biological diversity depicted in the first part. To survive and thrive, wild creatures require basic environmental amenities, which are largely a consequence of local geology, soils, water, and climate.

Chapter 11 provides an overview of the geologic events of the last billion years that have laid the foundation for the landforms and soils of today. The very topography of Mesa Verde as we experience it today is largely a consequence of ancient shifting oceans, rising mountains, and the inexorable wearing away of rock by water and wind over the eons.

Chapter 12 looks more closely at how particular topographic features in Mesa Verde Country have been formed—for example, the canyons that have been eroded down through millions of years of geologic history, the mesa tops with their mantle of red silt carried on the winds from northeastern Arizona, the still-unstable landslides on the north side of the Mesa Verde escarpment, and the broad "sage plains" of the Montezuma valley.

Chapter 13 focuses even more closely on the most immediate component of a plant's environment: the soil. Soils differ greatly throughout Mesa Verde Country in fertility, water-holding capacity, and resistance to erosion. These soil features in turn are a consequence of the rock types, topographic settings, and local microclimatic conditions of the places in which they form.

Water—a critical and limiting resource in a semiarid environment like Mesa Verde—is treated in detail in Chapter 14.

This part closes with a summary of the climate of western Colorado. Anyone who has lived in the Four Corners region for even a few years knows that temperature and precipitation vary enormously from place to place. Major patterns can be identified, however, especially with respect to elevation and topography. The climate also varies greatly over time. Recent research has begun to reveal some of the controls over the alternating wet and dry periods that so characterize this region. Amazingly, atmospheric and oceanic processes in the tropical Pacific have been found to exert a profound influence on the climate of Mesa Verde Country.

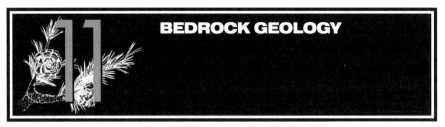

BEDROCK GEOLOGY

Mary Griffitts

THE BEDROCK UNDERLYING ANY AREA is the ultimate cause of the topographic features and soils and therefore directly or indirectly affects plant growth. Geologic processes influence plant and animal communities in various ways. Over very long periods they determine regional elevation and local topography, which in turn affect temperature, precipitation, solar exposure, and rate and degrees of erosion and weathering.

A geologic map of the entire study area (Figure 11.1) was compiled and simplified from state geologic maps. At this scale and for this purpose, the most practical breakdown of geologic units was to the system level, except in the case of the Cretaceous System, where further division is useful because only Cretaceous sedimentary rocks are exposed on Mesa Verde. Within the study area are rocks from Precambrian to Quaternary in age, from deep-seated intrusive crystalline rocks to surficial gravels. Each of the units on the map is composed of a number of smaller lithologic units, which have inherently different chemical compositions, weathering patterns, and physical attributes. Therefore, it is impossible to generalize by a single map unit or even a formation a reasonable correlation between rock formations and specific vegetation types, such as piñon-juniper woodlands. Diverse geologic substrates support piñon-juniper woodlands throughout the study area, and some are correlated in more specific detail in Chapter 2 and elsewhere. In this chapter the various geologic formations and the conditions of their deposition are generally described to set the stage for other chapters that

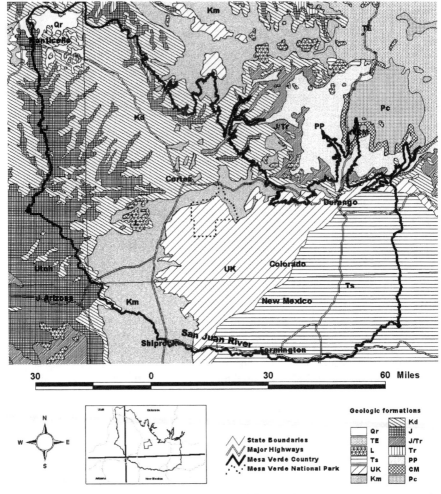

Figure 11.1. A geologic map of Mesa Verde Country.

relate characteristics of piñon-juniper woodlands to specific surficial geologic features (Chapter 12) and soils (Chapter 13).

These woodlands are common across the Colorado Plateau, a physiographic province with predominantly near-horizontal sedimentary rocks of Paleozoic to Tertiary age (Figure 11.2). Local folding and faulting have created uplifts and basins, but overall deformation is less than in adjacent regions. Erosion of alternating soft and hard strata produces a typical stepped landscape of mesas and canyons. Piñon-juniper woodlands also lap onto the foothills of the San Juan

Bedrock Geology

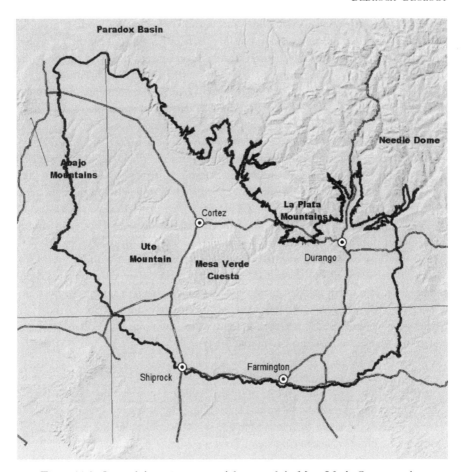

Figure 11.2. Some of the major structural features of the Mesa Verde Country geology.

Mountains, which represent the western margin of the Rocky Mountain physiographic province in this region.

Geologists interpret the origins and ages of rocks by studying features including their composition, texture, bedding, fossils, and relations to other rocks. In this process they define "formations" as bodies of rock that are distinct from adjacent rocks and large enough to be identified and mapped separately. Ages of sedimentary rocks are most often determined from fossils, whereas those of igneous rocks are generally calculated from the decay of radioactive isotopes.

Outcrop patterns of rock units reflect several factors, including thickness and lateral extent of each formation, chemical composition, structural deformation, and topography. In a local sequence of nearly horizontal strata, resistant younger rocks often cap mesas. Older strata may form tongues along canyons or may be exposed where tilted around uplifts such as the Needle Dome in the San Juan Mountains. Conversely, Tertiary strata are preserved mainly in the San Juan Basin, where relatively deep down-folding has left them at low elevations.

Large-scale structural deformation is explained by the concepts of plate tectonics: the earth's brittle crust is composed of large and small plates that fit together like pieces of a spherical jigsaw puzzle. As these plates shift on a deeper layer of hot, ductile rock, their edges collide, under- or override, slide sideways, or spread apart. The resulting compression and tension, with deep-seated melting and movement of magmas, create complex mountain ranges and basins at subcontinental scales. When mountains are raised, sediment eroded from them accumulates in nearby depressed basins either above or below sea level. Rocks in the Four Corners area were formed and partly eroded during several cycles of mountain building and erosion during plate motions in western North America.

In the rest of this chapter, rock units are described briefly under their most common names in those parts of the study area where they are exposed. Lateral variation in lithology and historical usage has given rise to a rather complex nomenclature. By necessity this overview of the rocks cropping out in the study area is very general, but it attempts to summarize the complex geologic environments as a basis for botanical reference.

PRECAMBRIAN ROCKS

The oldest rocks crop out in the core of the Needle Dome within the rugged San Juan Mountains (Figures 11.1 and 11.2). These are intrusive granitic rocks of varying compositions and metamorphic gneiss, schist, and quartzite, which include some of the oldest rocks in Colorado. They were deposited, metamorphosed, and faulted during a very long time interval under varying tectonic conditions. Before the end of Precambrian time, crustal plates in this part of the continent were spreading apart, and existing rocks were fractured into blocks that influenced the formation of all younger tectonic features. These blocks crop out in a compact area surrounded by upturned Paleozoic and early Mesozoic sediments (Blair 1996).

EARLY AND MIDDLE PALEOZOIC:
A STABLE CONTINENTAL MARGIN

A long period of relative stability in the area allowed deep weathering and erosion on the low-lying, almost flat Precambrian surface until the area was covered by a shallow sea in late Cambrian time, probably encroaching from the west. Ignacio Quartzite was deposited in the shallow water and along beaches as sands and occasional pebble beds and is now firmly cemented with silica to form very solid sandstones and quartzites and quartzite conglomerates. The sand probably came from the wearing down of highlands to the east. The sea gradually withdrew. Dating of the Ignacio Formation is difficult, because no diagnostic fossils have been found. A major unconformity (lack of sedimentary record due to erosion or nondeposition) occurs between the Ignacio and overlying Devonian Elbert Formation. Major earth movements broke the preexisting rocks into blocks. Some blocks dropped down, and others were squeezed upward. Now the Ignacio Quartzite is preserved in isolated down-thrown blocks where it was protected from erosion. There is no record of Ordovician and Silurian rocks, a very long interval of about 100 million years. This could mean that no sediments were ever deposited or that they were deposited and later (before Devonian time) eroded.

Another transgression of a sea from the west gave rise to light gray to red shallow-water sandstones of the Elbert Formation on the ancient irregular erosion surface. An upper part of the Elbert Formation consists of green shale and thin-bedded limestones, which have been considered to be tidal flat deposits. Fish remains date the Elbert as late Devonian in age. As the sea continued to encroach, the Ouray Formation of massive, fossiliferous limestones was deposited in clear, quiet water offshore. With an abundant fossil record of corals, brachiopods, snails, crinoids, and microscopic animals, it has been generally dated as late Devonian in age, extending into the early Mississippian period. These conditions continued into the Mississippian with the deposition of the overlying fossiliferous Leadville Limestone. The sea drained off, and once again the area was subjected to a long period of weathering and erosion.

LATE PALEOZOIC: ANCESTRAL ROCKY MOUNTAINS

By late Paleozoic time, tectonic movements to the north and east formed the Uncompaghre Uplift of the Ancestral Rocky Mountains, which shed large amounts of sediment to the west. An ancient red soil developed on the old limestone surface, in some places filling karst surface solution depressions. This paleosol is now the red Molas Formation of siltstone and calcareous shale, and the upper part bears marine fossils that date the age as early Pennsylvanian. A shallow sea gradually entered the area from the south, and the fossiliferous limestone and shale of the Hermosa Formation was deposited. The fossils date this sequence as middle Pennsylvanian in age. To the west these fossiliferous open seabeds grade laterally into the thick evaporite beds, gypsum, and salts of the Paradox Formation, the result of a confining barrier cutting off the western basin from the fresh seawater entering from the south. Toward the end of Pennsylvanian time further strong uplifts in the highlands to the east brought in much more red, clastic material and drove the sea to the south of the study area. Stream-deposited red beds of the Rico Formation are assigned to the late Pennsylvanian.

Red bed deposition continued through the Permian time in the Four Corners area. Few fossils are known, and the deposits are given diverse local names (Cutler Formation, Dolores Formation, etc.) because they vary so much in color and texture and are not easily correlated due to the lack of fossil evidence. From the position in the geologic column above the Pennsylvanian Hermosa Group, this thickness of red beds is generally assigned to the Permian System. There is much interfingering of these late Paleozoic sedimentary units and few fossils, so Pennsylvanian and Permian sediments are shown as one mapped unit (Figure 11.1).

MESOZOIC ERA

A long period of weathering and erosion marked the end of the Paleozoic and the beginning of the Mesozoic era. As a result, there are no early Triassic sediments recognizable locally. In the northeastern part of the study area the late Triassic Dolores Formation, a continental deposit of fluvial, lacustrine, and eolian sandstones, shales, and siltstones is oxidized to a bright red. It lies unconformably (angular in places) on the Permian deposits, which are also red. The color similarity—plus the lithologic interfingering and the lack of marine fossils—often makes the two units difficult to distinguish. As a result, they are sometimes grouped as Permo-Triassic for convenience in mapping. To the southwest the thick, brightly colored, and variegated Chinle Formation is of a similar age and origin. It was laid down in a vast basin, or perhaps a series of separate basins, that extended from northern Wyoming to southwestern Texas and from southeastern Nevada to southwestern Oklahoma. The Chinle Formation has been divided into several formations and members over this wide area.

Jurassic rocks crop out over a large part of the western extent of the study area (O'Sullivan 1997). Unconformably overlying the Triassic Chinle rocks is the Glen Canyon Group of early Jurassic age. The formations of the Glen Canyon Group include, in ascending order, the Wingate, Kayenta, and Navajo. At the time of their deposition the Ancestral Rockies had been worn away, and a widespread, shallow sea transgressed from the north across Wyoming and Idaho in a trough-like depression that lay at the foot of a new mountain belt to the west. Along the southern edge of the sea, varying environments resulted in many interfingered depositional facies. Desert conditions over a very large area produced thick deposits of windblown sands that accumulated in dunes across much of the western United States. The Navajo and Wingate Sandstones are of eolian origin, and the Kayenta is largely a fluvial deposit. The Wingate is dark to light reddish-brown, very fine to fine, cross-bedded sandstone. The Kayenta Formation is yellowish to pinkish-gray, very-fine- to fine-grained sandstone with clay clast conglomerates. The Navajo Sandstone is light gray to white, fine-grained, cross-bedded sandstone. These three formations are widespread. Although locally they vary greatly in thickness, they can be recognized over much of the study area.

Overlying the Glen Canyon Group is another widespread, thick sequence of sandstones and siltstones known as the San Rafael Group (Lucas and Anderson 1997a), which has been dated as middle to late Jurassic age. Six formations are generally recognized within this group. Not all are found throughout the area, and they also vary greatly in thickness and to a lesser extent in lithology from place to place. The lowest formation of the San Rafael Group, lying on an erosion surface above the Navajo Sandstone, is the Carmel Formation. This generally thin formation consists of a wide variety of lithologic units in different parts of Mesa Verde Country. At its base it is a thin limestone interbedded with calcareous shale, overlain by sandstones and interbedded conglomerates, siltstone, and gypsum. Deposited in an arm of a seaway that extended south-southwestward from southwestern Wyoming and southeastern Idaho, these rocks were formed in alternating shallow marine, fluvial, and desert environments. Fossils in the lower, marine limestone beds indicate a middle Jurassic age. Conformably overlying the Carmel Formation is the prominent and extensive Entrada Sandstone. Another significant eolian deposit, it is characteristically a very thick, fine-grained quartzose sandstone. The formation thickens to the west. The texture varies from massive to cross-bedded.

The Curtis Formation overlies the Entrada disconformably (representing an erosional interval) in portions of the study area. This formation represents marine flooding of the central and western part of the San Rafael Basin. The deposits typically consist of greenish-gray, cross-bedded, fine-grained sandstone, with basal beds of siliceous conglomerates with chert pebbles. A sparse fauna of marine mollusks gives an approximate middle Jurassic age. Grading into the Curtis laterally is a thin unit of the Todilto Formation in northern New Mexico and southwestern Colorado. This unit is composed almost totally of carbonates and evaporites and provides nearly all the gypsum mined in New Mexico. It is also a major trap for petroleum reservoirs in the region. Gradationally overlying and in some places intertonguing with the Curtis Formation, the Summerville Formation consists of thin-bedded, very fine-grained silty sandstone, reddish-brown to dark brown in color. In much of southeastern Utah, northeastern Arizona, and southwestern Colorado neither the Curtis nor the Todilto Formation is present, and the Summerville lies disconformably on the Entrada. The Summerville was deposited in quiet, ephemeral, shallow water adjacent to an arid coastal plain of low relief.

At the top of the San Rafael Group, the Bluff Sandstone forms prominent cliffs. It has been given numerous local names, and there has been controversy over whether it should be included within the San Rafael Group or placed as the lowermost unit of the Morrison Formation. No fossils have been reported from this formation, so its late Jurassic age has to be inferred from its stratigraphic position between the Summerville and the lowest Morrison Formation. Whatever its stratigraphic designation, the formation consists of cliff-forming, light-colored, cross-bedded, fine- to medium-grained, well-sorted sandstones that are the result of eolian deposition (Lucas and Anderson 1997a).

During the late Jurassic a major new mountain range began to form far to the west and with the ensuing erosion began to shed sediments into streams flowing eastward. These sediments eventually became the widespread Morrison Formation, which represents a time of climatic change and active fluvial deposition, especially along floodplains. It lies unconformably on the top of the Bluff Formation of the San Rafael Group and crops out all the way from southern Alberta, Canada, to central New Mexico and into Oklahoma. The lower part of the Morrison is referred to as the Salt Wash Member and consists of fine- to coarse-grained, cross-bedded conglomeratic sandstone, interbedded with red-brown mudstone and siltstone. The Salt Wash Member intertongues with the overlying Brushy Basin Member, which is in large part smectitic claystone. The claystone was altered from both primary and reworked volcanic ash. Vertebrate remains are often collected from the Morrison Formation. But, despite the large area of outcrops and detailed study of the many vertebrate fossils it has yielded, it is difficult to determine an age closer than late Jurassic, and there are strong arguments that the uppermost part may even be early Cretaceous (Anderson and Lucas 1997).

The Jurassic rocks in the study area have been combined on the accompanying map (Figure 11.1) because the scale of the map does not permit finer division. In eastern Arizona and part of eastern Utah the Triassic is also combined in part with the Jurassic.

Early in the Cretaceous geologic history in this part of the Western Interior, the very widespread Dakota Sandstone formed. The Dakota or its local lithologic equivalent can be traced almost continuously from Alberta to Texas. The Dakota, consisting of sandstone, shaly sandstone, thin dark shale, and coal deposits with conglomeratic units in the lower parts, forms a major topographic unit. The formation caps many of the mesas and canyon rims across Mesa Verde Country and is very significant in the piñon-juniper ecology. The formation is also a very important aquifer and oil and gas reservoir over much of the West. Because it is such a significant topographic and ecological unit throughout the study area, it is shown as a separate unit (Kd) on the map (Figure 11.1).

A widespread inland sea invaded the central part of the continent both from the north and from the south in Cretaceous time, and its sediments had a major influence on the present topography. In the early Cretaceous few sediments are recorded from the study area. It was a period of erosion and little sedimentation. To the south in Texas and Mexico and along the Gulf Coast, however, thick deposits of marine early Cretaceous are found. In eastern Utah and western Colorado a series of fluvial sandstones, conglomerates, and mudstones was deposited and has been referred to as the Cedar Mountain Formation. Locally a basal conglomerate is present, variously named the Buckhorn, Burro Canyon, Karla Kay, and other designations. The age of this conglomerate is not well documented due to its lack of fossils, but the Cedar Mountain has been dated (by Lucas and Anderson 1997b and others) as late lower Cretaceous or early upper Cretaceous.

Not until the Late Cretaceous did the sea advance to the Colorado–New Mexico area. As the inland sea transgressed over the old, low-relief surface, it reworked the stream sands from the western mountains and left near-shore, shallow-water sand deposits, floodplain sediments, and lagoonal muds. These shallow-

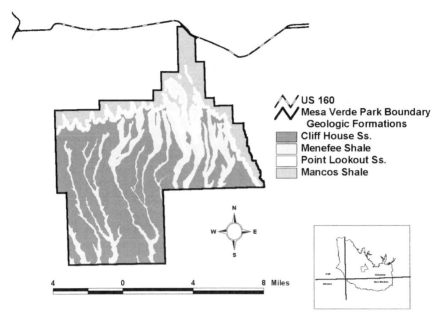

Figure 11.3. The geology of Mesa Verde National Park.

water environments were rich in organic material, later buried and transformed into the coal beds we find today. The sandstones are relatively poorly sorted and show strong cross-bedding in places. The age of the Dakota in this region is difficult to determine (am Ende 1991). There are many fossil mollusks in places, but most species existed across long spans of time. Also, although the formation crops out over a very large area with a consistent lithology, because of its transgressive origin it is not of the same age everywhere.

MESA VERDE GEOLOGY

As the Cretaceous sea continued to transgress to the west, part of the study area became farther and farther from the shore. Hence the sediments received from the western lands became finer and finer and the water quieter, with less disturbance from inflowing streams. Mesa Verde National Park, which occupies the center of Mesa Verde Country and roughly one-half of the Mesa Verde cuesta, can be used to examine in more detail the geologic processes of the Late Cretaceous period in the Four Corners area (Figure 11.3).

Overlying the Dakota Sandstone is the massive Mancos Shale, deposited in the quieter, deep waters of the invading seaway. The Mancos Shale is approximately 670 m thick at its type section along the Mancos River on the northeastern margin of Mesa Verde National Park, thinning to the north and east. It correlates with similar deposits to the north and east. The same conditions of deposition extended over much of the Western Interior at this time, but shallower water and the closer proximity to mountains gave rise to sandier deposits to the west. In Mesa Verde Country the Mancos Formation underlies a large part of the low-relief, cultivated country. Surface expression and ecological environments vary even within this formation. Although it is generally referred to as the Mancos Shale, within its great thickness there are distinctive lithologic and paleontologic units. Most of these have little effect on topography, but two members of the formation, the Bridge Creek (Greenhorn) and Juana Lopez Members, show positive topographic expression. The Bridge Creek (Greenhorn) Limestone Member of the Mancos is recognized throughout Colorado and Kansas. It consists of a series of relatively massive, pure limestone beds that form prominent small cliffs and cap small mesas. These mesas capped by the solid limestone may have little of the piñon-juniper covering but often have prominent large ponderosa pines. This member thickens markedly to the east. The Juana Lopez is a series of quite resistant calcarenites that forms small ridges and caps small mesas in the otherwise soft Mancos Shale topography. Another part of the Mancos, the Smoky Hill oyster bench, is almost a solid layer of limestone made up of oyster shells encrusting the very large bivalve *Inoceramus grandis*. Marilyn Colyer, resource manager at Mesa Verde National Park, has described this environment as occupied by old, very large trees that "never attain great heights" and "tend to 'crawl' over the surface" (Chapter 2).

The Mancos Formation is highly fossiliferous, with an abundant diagnostic ammonite and bivalve fauna spanning a period of nearly ten million years. The shales of the Mancos gradually become sandier in the upper part of the formation as the sea regressed slightly from the region. Much of the shaly part of the Mancos is almost barren of trees, partially because it erodes rapidly and is quite unstable in its outcrops.

As the sea continued to regress, the shaly sandstones gave way to the solid cross-bedded sandstones of the Point Lookout Formation, the lowest formation in the Mesa Verde Group. The Point Lookout is a resistant sandstone unit of marine near-shore deposition, sandbars, deltas, beach sands, and other shallow-water deposits. Few fossils are found, probably because the deposition was in a zone of vigorous wave action. The formation is resistant to erosion, capping many mesas and protecting the soft Mancos Formation along the major escarpments, and supports a healthy piñon-juniper cover.

The sea gradually withdrew a short distance from most of the study area, and the continental sediments of the Menefee Formation were deposited on the old, low-relief seafloor. Floodplain deposits, stream sands, muds in isolated lowlands, and swamp deposits with woody muds and coal form a great thickness (about 200

m) of the middle member of the Mesa Verde Group. Many plant fossils, including palm fronds, conifer tree trunks, and numerous deciduous leaves, are found in the floodplain sandstone layers. This is another soft, nonresistant formation that forms gentle back slopes on top of the cliff-forming Point Lookout outcrops.

Once again the sea transgressed over this western part of the study area, and marine waters covered the land. Shallow-water sandstones with thin breaks of shale were deposited under nearshore conditions and now form the Cliff House Formation, the uppermost member of the Mesa Verde Group. The uppermost layers of the Cliff House bear a mixed vertebrate and invertebrate fauna of reptiles, amphibians, fish, ammonites, and bivalves and are dated by the ammonites as about 78.5 million years old. Many of the sandstones show ripple marks and cross-bedding typical of nearshore deposition. This formation also is resistant to erosion and caps many of the southwestern mesas, including the Mesa Verde. As the name implies, alcoves in the Cliff House Sandstone house the magnificent dwellings of the Ancestral Puebloans of Mesa Verde. In the northern part of the Mesa Verde, the upper part of the Cliff House Formation is composed of interbedded thin sandstones, shaly sandstones, and sandy shales and is only about 21 m thick. This section of the Cliff House thickens and coarsens rapidly to the south, where the cliff dwellings are in massive, solid sandstone layers up to 60 m thick. The northern, softer section was deposited farther from the shoreline and received less coarse sediments. The sandy soil of the Cliff House Formation capping the Mesa Verde is covered in places with significant deposits of loess and supports a large population of piñons and juniper.

As the sea continued to transgress over the Southwest, the waters deepened and gave rise to quiet offshore conditions quite similar to the conditions of deposition of the Mancos Formation, and the Lewis Shale was deposited. This thickness of dark gray shales and thin sandstones has been removed by erosion from Mesa Verde proper, but it crops out to the north, south, and west in the study area. Similar in outcrop and topographic expression to the Mancos, it forms low-relief areas or slopes where protected by more resistant rocks such as the Pictured Cliffs Sandstone. This sandstone formation represents the beginning of the final retreat of the inland sea from the area. The Pictured Cliffs is a fairly resistant sandstone sequence of nearshore and beach sediments, much like the Point Lookout Sandstone. As the sea continued to retreat, stream deposits with channel sands, floodplain muds and sands, and swamp deposits of woody muds were laid down to form the coal-bearing Fruitland Formation. The sandstones of late Cretaceous age are also major oil and gas reservoirs. As the sea continued to retreat, stream deposition of more sandstones and organic-rich shales accumulated.

The geologic age of the late Cretaceous rocks from the Mancos through the Pictured Cliffs is well documented paleontologically and through correlation with bentonites. Throughout the Cretaceous, volcanism to the west intermittently produced large amounts of volcanic ash, which became bentonite through weathering.

These ash deposits can be correlated with ash deposits throughout the Western Interior. The paleontologic zones, especially the ammonite zones, are also well dated and correlated not only with occurrences in North America but also with European zones.

CENOZOIC ROCKS: THE LARAMIDE REVOLUTION

The end of the Mesozoic era was marked by major tectonic changes referred to as the Laramide Revolution. As the inland sea drained off, repeated laccolithic intrusions into the preexisting sediments bowed them up to form the La Plata and Ute Mountains. Rejuvenated erosion of these mountains furnished sands, muds, and gravels to the streams draining these developing highlands and deposited them as the McDermott Formation south of Durango. Regional uplift, folding, and faulting began the formation of the present Rocky Mountains. As the mountains were raised, erosion was at work tearing them down, and streams flowing to the south and southeast carried the sediments away and deposited them as the sandstones, shales, and conglomerates of the upper Animas Formation. The probable, much debated Cretaceous/Tertiary contact is generally placed at the base of the McDermott, though it has not yet been identified in the study area. The Animas Formation has many coarse conglomeratic layers that were probably derived from alluvial fans interbedded with finer floodplain deposits. These deposits grade into the overlying Eocene San Jose Formation, composed of stream sandstones, shales, and tuff beds. A long erosional period left Pliocene and Pleistocene terrace gravels unconformably resting on the Eocene San Jose Formation, thus bringing to an end the region's Tertiary sedimentary history.

Throughout the region Tertiary igneous activity produced numerous relatively small intrusive bodies. In southern Colorado and northern New Mexico, plugs and dikes have been intruded into the sediments with very little deformation of the host rock. These dark, fine-grained, often porphyritic rocks were probably formed in Oligocene time and create prominent topographic features where the surface is otherwise dominated by the soft shales of the Mancos Formation. Shiprock is a prominent and well-known remnant of some of these intrusions. Where the dikes intrude more resistant sediments on the surface, however, such as the Cliff House Sandstone on the Mesa Verde, the dikes erode more rapidly than the sandstone, and the topographic expression may be a trough.

The San Juan Mountain Range is the result of repeated volcanic eruptions that deposited thick flows of lavas and ash over the late Eocene erosional surface. Volcanism and igneous intrusions continued throughout the Tertiary, giving rise to a large variety of rock types and a highly complicated igneous history.

SUMMARY

The geologic column (Figure 11.4) summarizes the depositional history of Mesa Verde Country. The sedimentary history includes a variety of depositional environments throughout Paleozoic, Mesozoic, and early Tertiary time. Jurassic, Cre-

Bedrock Geology

taceous and Tertiary sediments cover the greatest portion of the surface area and have the greatest influence on the ecology and the distribution of the piñon-juniper woodlands. Generally the bedrocks have been weathered, so that (with surficial alteration and soil cover) direct correlation of bedrock with the plant communities, such as the old-growth piñon-juniper woodland, is difficult. Discussions

ERA	Period	Age mya	Epoch or Group	Formation-Dominant Deposits		Map Symbol
Cenozoic	Quaternary	1.5	Pleistocene	glacial deposits		
Cenozoic	Tertiary	-58-	EOCENE	SAN JOSE - sandstone and tuff		
Cenozoic	Tertiary	-58-		ANIMAS - sandstone, shale, conglomerate		
Cenozoic	Tertiary	-58-	Paleocene	McDERMOTT - sandstone		
	LARAMIDE OROGENY including Tertiary igneous			Intrusives and extrusive		
Mesozoic	CRETACEOUS	65		KIRTLAND - sandstone		
Mesozoic	CRETACEOUS			FRUITLAND - shale and coal		
Mesozoic	CRETACEOUS			PICTURED CLIFFS - sandstone		
Mesozoic	CRETACEOUS			LEWIS - shale		
Mesozoic	CRETACEOUS		Mesa Verde Group	CLIFF HOUSE - sandstone		
Mesozoic	CRETACEOUS		Mesa Verde Group	MENEFEE - shale and coal		
Mesozoic	CRETACEOUS		Mesa Verde Group	POINT LOOKOUT - sandstone		
Mesozoic	CRETACEOUS			MANCOS - shale		
Mesozoic	CRETACEOUS			DAKOTA - sandstone		
Mesozoic	CRETACEOUS	-136-		CEDAR MOUNTAIN - sandstone		
Mesozoic	JURASSIC			MORRISON	Brushy Basin Member - claystone	
Mesozoic	JURASSIC			MORRISON	Salt Wash Member - sandstone	
Mesozoic	JURASSIC			BLUFF - sandstone		
Mesozoic	JURASSIC		San Rafael Group	SUMMERVILLE - reddish brown sandstone		
Mesozoic	JURASSIC		San Rafael Group	CURTIS - gray sandstone with chert		
Mesozoic	JURASSIC		San Rafael Group	ENTRADA - massive sandstone		
Mesozoic	JURASSIC			CARMEL - limestone, congl., gypsum		
Mesozoic	JURASSIC		Glen Canyon Group	NAVAJO - white sandstone		
Mesozoic	JURASSIC		Glen Canyon Group	KAYENTA - gray sandstone		
Mesozoic	JURASSIC	-200-	Glen Canyon Group	WINGATE - reddish-brown sandstone		
Mesozoic	TRIASSIC	-295-		CHINLE-DOLORES - red shale, sandstone		
PALEOZOIC	PERMIAN /			HERMOSA - red shale, sandstone, gypsum		
PALEOZOIC	PENNSYLVANIAN	-320-		MOLAS - red shale		
PALEOZOIC	MISSISSIPPIAN	-360-		LEADVILLE - limestone		
PALEOZOIC	DEVONIAN			OURAY - limestone		
PALEOZOIC	DEVONIAN	-400-		ELBERT - sandstone		
PALEOZOIC	SILURIAN			MISSING		
PALEOZOIC	ORDOVICIAN	-500-		MISSING		
PALEOZOIC	CAMBRIAN	-570-		SAWATCH - quartzite		
	PRECAMBRIAN			Intrusive granitic and metamorphic rocks		

Figure 11.4. A geologic column of Mesa Verde Country.

of recent geological processes and soil development in the following chapters complete the story of substrate development of Mesa Verde Country and begin to unravel the plant-substrate relationships.

REFERENCES CITED

am Ende, B. A. 1991. Depositional environments, palynology, and age of the Dakota Formation, south-central Utah. In *Geological Society of America Special Paper 260*, pp. 74–83.

Anderson, O. J., and S. G. Lucas. 1997. The Upper Jurassic Morrison Formation in the Four Corners region. In *New Mexico Geological Society Guidebook, 48th Field Conference, Mesozoic Geology and Paleontology of the Four Corners Region*, pp. 139–155. New Mexico Geological Society, Socorro, N.Mex.

Blair, R. 1996. *The Western San Juan Mountains: Their geology, ecology, and human history.* University Press of Colorado, Niwot.

Delaney, P. T., and D. D. Pollard. 1981. *Deformation of host rocks and flow of magma during growth of minette dikes and breccia-bearing intrusions near Shiprock, New Mexico.* U.S. Geological Survey Professional Paper 1202.

Griffitts, M. O. 1990. *Guide to the geology of Mesa Verde National Park.* Mesa Verde Museum Association, Inc., and Lorraine Press, Salt Lake City, Utah.

Lucas, S. G., and O. J. Anderson. 1997a. The Jurassic San Rafael Group, Four Corners region. In *New Mexico Geological Society Guidebook, 48th Field Conference, Mesozoic Geology and Paleontology of the Four Corners Region*, pp. 115–132. New Mexico Geological Society, Socorro, N.Mex.

———. 1997b. Lower Cretaceous stratigraphy on the Colorado Plateau. In *New Mexico Geologic Society Guidebook, 48th Field Conference, Mesozoic Geology and Paleontology of the Four Corners Region*, pp. 6–7. New Mexico Geological Society, Socorro, N.Mex.

O'Sullivan, R. B. 1997. The Jurassic Section along McElmo Canyon in Southwestern Colorado. In *New Mexico Geological Society Guidebook, 48th Field Conference, Mesozoic Geology and Paleontology of the Four Corners Region*, pp. 109–114. New Mexico Geological Society, Socorro, N.Mex.

Wright, R. 1986. Cycle stratigraphy as a paleogeographic tool: Point Lookout sandstone, southeastern San Juan Basin, New Mexico. *Geological Society of America Bulletin* 96: 661–673.

12 LANDFORMS AND SURFICIAL DEPOSITS

Mary L. Gillam

GEOMORPHOLOGY IS THE STUDY OF INDIVIDUAL LANDFORMS such as mesas and canyons, of larger landscapes, and of related sediments that typically have not yet hardened into rock. As a group, these sediments are often called surficial deposits to distinguish them from the older rocks on which they rest. Geomorphology also includes the study of ancient landscapes, which may either be destroyed by continuing erosion or be preserved where buried by younger rocks. Often the features of ancient landscapes are inferred to explain stages in the development of modern landscapes.

Landforms and surficial deposits affect ecosystems in very basic ways. As mentioned in Chapter 11, elevation and solar exposure strongly influence local microclimates. Also, rock outcrops and surficial deposits are the parent materials for soils (Chapter 13). Finally, relations between landforms and underlying rocks determine how groundwater flows—for example, where seeps occur on the sides of a mesa (Chapter 14). Various factors may combine so that biotic communities favor specific settings, such as the outcrop belt where a certain rock unit reaches the ground surface.

This chapter is divided into three sections. In the first section we consider the general nature of the present landscape and how it developed. In the second we discuss surficial deposits in general, climatic influences on their formation, and the characteristics of common types of deposits. In the last section we describe the landforms and deposits of the piñon-juniper zone in more detail by subregions.

THE PRESENT LANDSCAPE

Overview

The landscape extending west and south from the San Juan Mountains has a variety of landforms that reflect differing rock types, structures, and processes. High peaks in the western San Juan Mountains have been carved from granitic, volcanic, and metamorphic rocks in part by former glacial ice. Seven of these peaks reach elevations above 4,270 m. Surrounding the mountains, at elevations as low as 1,200 m, are mainly sedimentary rocks of varying hardness—mostly sandstone, mudstone, and shale but also limestone, conglomerate, coal, and gypsum. Stream erosion of these rocks has created mesas, canyons, and related landforms. Locally, erosion has also exposed igneous rocks that intruded the sediments and now form low, isolated mountains like the La Plata and Ute Mountains. Most landforms in the region are erosional or cut by partial removal of existing rocks. Some are constructional, however, or built by accumulation of new material like the tops of lava flows or river floodplains. In places older landforms were buried by sediments and later exhumed and modified.

The present topography of this region is very different from the topography that existed in late Cretaceous time before the Laramide orogeny. About 70 million years ago the entire region was slightly above sea level after a long interval of alternating coastal plain and marine sedimentation. How did this striking change occur?

How This Landscape Formed

Generally landscapes form by several basic processes that act over millions to tens of millions of years. The most important are tectonic movement, erosion, rock formation, and isostatic adjustment. Tectonic forces, originating within the earth's crust and upper mantle, move large masses of rock up, down, and sideways; they also deform the rocks by folding and faulting. The most important agent of erosion is water, although ice and wind are effective at some places and times. The amount of water available for erosion depends ultimately on climate, which has varied through geologic time over long to short frequencies. Rock formation includes such diverse processes as intrusion of magmas, which dome overlying rocks, accumulation of lava flows, and deposition of sediments. Rocks of the continental crust are less dense than deeper rocks of earth's mantle and float on them like a cork on water. When large volumes of rock are added to or removed from a region, its surface can be depressed or raised simply by the change in weight. This process, called isostatic adjustment, is influenced by tectonic movements, rock formation, and erosion.

In the Four Corners area existing rocks clearly record some stages of landscape development, but other stages are more obscure. For example, it is possible to estimate the total amounts of uplift that have occurred from late Cretaceous time to the present, but it is very difficult to estimate when and how increments of uplift occurred during this 70-million-year period. Total uplift can be estimated with respect to the pre-Laramide land surface, which was located slightly above sea level. This surface, corresponding roughly to the eroded top of the Kirtland Shale, is now buried in the San Juan Basin at elevations that are locally below 1,070 m (3,500 ft; Kernodle et al. 1990). In contrast, Precambrian rocks in the Needle Mountains are at elevations as high as 4,000 to 4,300 m. This is a minimum estimate of total uplift, because unknown thicknesses of late Cretaceous and older rocks have been eroded from the highest peaks. Unfortunately, post-Laramide rock units in the area did not form at known elevations, and they are not preserved over large enough areas to show later tilting or differential uplift fully. Nevertheless, some educated guesses may be made about the timing of Cenozoic uplift from evidence elsewhere in the Colorado Plateau and Rocky Mountains.

Very briefly, the present landscape seems to have formed roughly as follows. During the Laramide orogeny, uplifts began to erode as soon as they emerged. Much of the eroded sediment was captured in adjacent basins and became formations like the Animas, Nacimiento, and San Jose in the San Juan Basin. After the orogeny ended about forty-five million years ago, continuing erosion reduced peak heights and local relief in the Needle Mountains and other uplifted blocks, while relatively little erosion occurred in lower surrounding areas. From thirty to fifteen million years ago numerous volcanic eruptions built a high plateau of volcanic rocks, related sedimentary rocks, and small intrusions that is now preserved mainly north and east of the Needle Mountains. This activity continued on a much-reduced scale, and the youngest volcanic rock in the area is a small Quaternary basalt flow west of Telluride. Volcanic rocks buried and preserved parts of the prevolcanic erosion surface. Minor local doming may have accompanied the upward movement of subsurface magma bodies that fed the volcanoes.

After most of the volcanic rocks had formed, erosion completely removed their western margins, cutting below the prevolcanic surface into various older rocks. Originally the volcanic rocks may have extended southwestward from the Mt. Wilson area toward the Ute Mountains and southward around the Needle Mountains into the San Juan Basin (Steven 1975). If so, they would have influenced the development of rivers in those areas.

At the eastern margin of the San Juan Mountains, tilted volcanic rocks provide information about uplift. These rocks now descend eastward below the alluvial floor of the San Luis Valley, a part of the Rio Grande rift. The same rocks rise toward the west and can be projected toward an elevation of about 4,200 m in the Mt. Wilson area. Tilting at the edge of the San Luis valley began by early Miocene time and accelerated during middle to late Miocene time, when other evidence indicates that the Colorado Plateau rose with respect to the Rio Grande rift and the Basin and Range province on the west. During part of the time when tilting occurred, from fifteen to five million years ago, moderate erosion produced rolling relief on volcanic rocks near Creede (Steven et al. 1995). Those authors believe that uplift accelerated during the last five million years, when canyon cutting occurred at Creede and in many other areas of the southern Rocky Mountains. Concurrent uplift in the western San Juan Mountains may have contributed to canyon cutting in that area and to removal of the westernmost volcanic rocks.

This story is sketchy and controversial in many respects. Some kinds of evidence are contradictory for unknown reasons. It is not known how much of the post-Laramide uplift is due to tectonic forces and how much is due to isostatic adjustment after erosion. Finally, increased erosion rates during the last two million years may be due in part to the onset of glacial-interglacial climatic cycles, which are discussed in the following section.

SURFICIAL DEPOSITS

We now turn from large-scale features and their development during the last seventy million years to smaller landforms and surficial deposits at the scale of single mesas, slopes, and valleys.

How Surficial Deposits Relate to Landforms and Older Formations

Any landscape includes many small areas with differing bedrock types, slope angles, and water runoff. Therefore, at any moment some places are eroding at varying rates, while others are being covered with sediment from nearby or distant sources, and still others are changing so slowly that they may be considered stable.

Surficial deposits are closely related to the processes that created nearby landforms. Each type of deposit also has characteristic sediments and constructional shapes (as described later in this section). Thus they are sometimes called morphostratigraphic units. For example, stream sediments are found in valleys. They typically consist of bedded sediments that underlie floodplains. Some other common kinds of surficial deposits are glacial till, colluvium, landslide deposits, and windblown silts.

There are several basic differences between the surficial deposits in this region and the sedimentary formations discussed in Chapter 11. Those formations are relatively large bodies of rock that resulted from deposition across very large areas during very long periods. These rocks also become younger toward the top of the sequence. In contrast, surficial deposits are discontinuous, typically thin, found in complex arrangements, and preserved more briefly. Thicknesses up to a few feet or tens of feet are common, whereas hundreds of feet are rare and restricted to special settings. As discussed in the following paragraphs, surficial deposits are more comparable in scale to beds within a formation or to sediments that existed temporarily on erosion surfaces of the past, which are now preserved as unconformities between formations. Formation names have not been given to most surficial deposits because of their small volumes.

Age Patterns and Ranges

The following examples show how surficial deposits can exist in complex and varying space-time arrangements. As rivers erode their valleys, they leave remnants of alluvium along their valley sides in the form of terraces. In a sequence of terraces, the youngest are found at the lowest elevation—the opposite from the age trend in adjacent sedimentary rocks, which become younger toward the top (unless overturned). After a terrace has formed, its alluvium may be buried by younger sediments. Also, while a river forms several terraces, windblown deposits can accumulate more or less continuously on a nearby mesa. These complex spatial patterns in the ages of surficial deposits occur because parts of the landscape evolve differently.

With time the landscape position of a given deposit can also change in a process known as "topographic reversal." For example, alluvial gravels composed of igneous and metamorphic rocks are often more resistant to erosion than the sedimentary rocks beneath them. Alluvial gravels are deposited on floodplains in valley bottoms. As time passes, gradual deepening of valleys and erosion of their sides leave gravel remnants on ridge tops that stand above younger and lower valley floors.

In comparison with sedimentary formations, the age range of existing surficial deposits is also relatively short. Although net erosion has occurred in this region for tens of millions of years, elevations remain so high that erosion will continue to dominate for a comparable time. In this erosional setting most existing deposits have ages from thousands to hundreds of thousands of years. The oldest surficial deposits in the piñon-juniper zone are tiny remnants of alluvial gravels that are estimated to be between two million and five million years old.

In turn, the ages of surficial deposits imply that most of the landscape now seen in this area formed during the last 0.5 million years and that nearly all formed during the last few million years. In the San Juan Mountains, however, the tops of some volcanic rocks have been modified relatively little during the last fifteen million years. Some old landforms have also been buried and later exhumed, such as parts of the Cambrian erosion surface cut on Precambrian rocks.

Effects of Climate Change

During the time when most of the existing surficial deposits formed, alternating glacial and interglacial climates affected erosion and sedimentation. As summarized by Williams et al. (1998), worldwide climate cooled progressively during the Tertiary period, mainly because tectonic processes pushed some continents toward the poles. This altered oceanic and atmospheric circulation patterns that transfer heat from the equator to the poles. A decrease in the atmospheric concentration of carbon dioxide, the principal "greenhouse gas," probably also contributed to the cooling trend. After a gradual buildup of ice in Antarctica and later in Greenland, moderate-sized ice sheets appeared at higher latitudes throughout the Northern Hemisphere between about 2.5 and 2.4 million years ago. Since that time, earth's climate has oscillated between relatively cool glacial periods and relatively warm interglacial periods. Irregularities in earth's orbit, which slightly alter the global distribution of solar energy, control the lengths of these cycles, but many other processes govern their intensities. Early cycles had durations of about 40,000 years. During the last 0.9 to 0.8 million years, the cycles lengthened to about 100,000 years, including roughly 90,000 years of glacial conditions and 10,000 years of interglacial conditions. At the same time, the intensity of cold periods increased.

During the last glaciation, environmental conditions in the Four Corners region were much different than at present. Ice fields and valley glaciers with a total area of about 5,000 km^2 covered nearly all of the upper San Juan Mountains except for the highest peaks. Smaller valley glaciers, totaling about 100 km^2, were present in the La Plata Mountains (Atwood and Mather 1932; Leonard 1984; and unpublished data). Botanical evidence, groundwater chemistry, and glacier volume estimates suggest that mean annual temperatures decreased by about 5–7° C, while precipitation possibly increased by 30 to 100% (based on reviews by Adams and Petersen 1999, Gillam 1998, and Van Devender et al. 1987). Water runoff and infiltration must have increased even more than precipitation, because cooler temperatures would have reduced evaporation and transpiration by plants. Very roughly, the glacial climate at Shiprock might have resembled the modern climate at Durango, while the glacial climate at Durango might have resembled the modern climate at Silverton.

As a result of these climatic changes, vegetation zones were lowered by about 700 to 900 m (Betancourt 1990). Their species composition also changed, with some species migrating into the area and others moving out. At this time piñon was restricted to refugia far south and southwest of the Four Corners in areas that are now deserts (Spaulding 1990; Van Devender et al. 1984). On the Colorado Plateau, Utah and Rocky Mountain junipers shifted to elevations as low as 900 and 1,300 m (Cole 1990; Betancourt 1990). One-seed juniper was absent, although it is common in modern piñon-juniper woodlands of northern New Mexico.

Climatic changes also influenced local rates of erosion and sedimentation, as seen in some kinds of deposits and landforms. In the following paragraphs common types of surficial deposits are described, including glacial till, alluvium, colluvium, landslide, and windblown (or eolian) deposits.

GLACIAL DEPOSITS

The largest glacial deposits in or near the piñon-juniper zone form end moraines of two glaciers. Moraines of the Animas glacier are in Durango at elevations from 2,000 to 2,150 m, and those of the La Plata glacier are north of Hesperus at 2,580 to 2,730 m. Moraine deposits cover small areas, but they are important evidence of climatic cycles.

The Durango moraines formed as arcuate ridges around the tonguelike end of the Animas glacier, which was confined within the Animas River valley. These ridges rise 15 to 60 m above adjacent landforms and are 60 to 350 m wide at their bases. The most recent study attributes these ridges to six glaciations and estimates their ages as 15,000 to 360,000 years (Gillam 1998). The moraine deposits are composed of till (sediment released from ice), debris flow deposits (saturated sediment that slumped off the top of the ice), outwash (alluvium of small meltwater streams that flowed from the ice), and deposits of small lakes (Johnson 1990). The texture, bedding, and internal structures of this sediment vary greatly. The most common materials are poorly bedded, sandy silt and silty sand with varying amounts of pebbles, cobbles, and boulders.

Alluvium

Alluvium is sediment deposited from water flows of any size. In valleys it is associated with stream channels and floodplains, but where valleys open into broader areas it forms alluvial fans. In contrast, sheetwash alluvium forms smooth aprons at the gentle bases of slopes and is deposited by shallow sheetlike runoff after large storms.

Stream alluvium varies in character depending mainly on stream power, available sediment, and channel behavior. Typical features include pods of sediment with varying textures and flat or inclined bedding. If coarse rock particles are available for transport, they are most common in the sediment of rivers or very steep streams. Pebbles, cobbles, and sometimes even boulders become rounded if they are carried far enough, and they often rest in shingled (or imbricated) arrangements. The composition of sediment particles in alluvium reflects the kinds of rock exposed in the source area. Therefore, alluvium of rivers that originate in mountains with igneous and metamorphic rocks can be distinguished easily from alluvium of rivers and streams that originate in lower areas of sedimentary rock. With more care, the specific river that deposited an eroded patch of alluvium can often be identified. The volume of alluvial deposits varies with the size of the stream, but in this area some deposits have areas up to many square kilometers and thicknesses up to 50 m. In comparison, sheetwash alluvium is usually sandy and relatively thin.

Along most streams, sediment deposition and rising floodplain levels alternate with channel incision, which isolates the former floodplain as an alluvial terrace. Causes of incision vary with the stream's history and environment (Schumm 1975). For large rivers with glaciated upper basins, sediment accumulates during glaciations and incision usually occurs early during the following interglacial periods. Along small streams in unglaciated areas, incision can be triggered frequently by minor climatic changes, vegetation changes, or surface compaction that increases runoff from contributing slopes, or gradual steepening of the floodplain by sediment accumulation until it starts to erode. Several cycles of sediment accumulation and incision produce a vertical sequence of terraces along the valley sides.

Colluvium

Colluvium is unsorted sediment of varying texture that forms aprons on slopes. This material is carried downhill mainly by creep, a very slow process that involves repeated, small-scale heaving by frost, roots, and burrowing fauna, followed by settlement. Colluvium often includes some other deposits that are too small to be distinguished, however, such as fallen rocks on slopes below cliffs, minor sheetwash alluvium, and deposits of small debris flows. Sediment in colluvium is derived from the bedrock or other surficial deposits that underlie the slope and thus can be a mixture of different compositions and particle sizes. Most colluvium is unbedded, but weak bedding can be present. In cross sections, most deposits form wedgelike bodies that thicken down slopes and within swales. Thicknesses up to 1 m are common, but several meters can occur. Colluvium in this area appears to form mainly during glacial periods and to be eroded during interglacial periods, especially on south-facing slopes. It is very common on slopes throughout the area.

In some places near the bases of tall escarpments, talus flatirons occur. These are small hills capped by old colluvium, and they resemble hogbacks formed on dipping strata. With time, erosion along escarpments makes them shift toward the cores of adjacent uplands. Talus flatirons mark former locations at the base of a retreating escarpment.

Landslide Deposits

Landslide deposits form by sudden downslope movements of large masses of sediment or rock. Landslides in this area commonly occur where relatively hard caprock allows a very steep slope to form in underlying softer rocks. Through several processes, gravitational stress on rocks beneath parts of the slope comes to exceed their strength. Mainly, erosional steepening of the slope increases stress, while relatively large amounts of seeping groundwater and infiltrating surface water decrease strength. As a result, some of the rock suddenly slides along a fracture plane or disintegrates into a flowing mass. Many kinds of landslides are recognized, but in general they can be identified from the presence of a hollowed source area above a heap of jumbled rock. The deposits are chaotic mixtures of material that can contain angular boulders up to many meters in diameter.

In this area overlapping landslides have formed complex deposits that are as large as several square kilometers and as thick as 50 m or more. Landslides are more common where strata dip in the same direction as the surface slope and where the following combinations of caprock and substrate are present: Dakota Sandstone and Burro Canyon Formation over upper Morrison Formation, Point Lookout Sandstone over Mancos Shale, and Cliff House Sandstone over Menefee Shale. Landslides have also occurred where the sides of glaciated valleys were oversteepened by ice abrasion. Many active landslides are superimposed on older, more extensive landslide deposits that are now stable (e.g., Kirkham et al. 1999). This suggests that slides were more common during moister, glacial periods.

WINDBLOWN DEPOSITS

Windblown (or eolian) deposits in this area are mainly blankets of silt and fine sand that have formed very slowly on nearly flat surfaces where erosion is too slow to remove it. Silt and fine sand can be carried by wind for very long distances. The reddish-brown color of this sediment and downwind fining and thinning west of Cortez suggest that much of it came from fine-grained, reddish rocks that are exposed widely in Arizona (Price et al. 1988). Northward fining from the central San Juan Basin indicates other sources in New Mexico, and some sediment probably came from river floodplains (Gillam 1998). Dunes composed of locally derived sands or shale pellets are rare.

Windblown deposits are found on several kinds of stable surfaces in this area, such as exhumed sandstone beds and alluvial terraces. Their thickness generally increases from centimeters or decimeters on younger surfaces to several meters on older surfaces (for example, on progressively older terraces). This trend suggests that some deposits began to form hundreds of thousands of years ago and perhaps even earlier. Soil features in many thicker deposits show that accumulation rates varied through time, but the climatic phasing of eolian deposition is not well-known. More sediment may be carried from Arizona and New Mexico during interglacials like the present, when vegetation there is limited. More may be carried from river floodplains during and just after glaciations, however, when those areas are mostly bare and covered with loose sediment.

During a one-year period, dust was gathered in a trap at Mesa Verde National Park (Arrhenius and Bonatti 1965). The weight of that dust suggests a short-term accumulation rate of about 2 cm in 1,000 years, much faster than the long-term rates inferred from older deposits (Gillam 1998). Accumulation rates also vary with topography, so that deposits are thicker on the windward edges of bluffs.

A TOUR OF THE PIÑON-JUNIPER ZONE

The piñon-juniper zone (between roughly 1,500 and 2,450 m) sweeps across areas with different rock formations, local structures, and suites of landforms. These areas are described here from west to east, with emphasis on Mesa Verde.

DAKOTA BELT

The western part of the piñon-juniper zone is dominated by broad exposures of Dakota Sandstone (Kd) (Figure 11.1). On a large scale, this outcrop belt is like a long bench that is slightly tilted and warped. At its east end it forms cuestas that slope away from the San Juan Mountains. Along its north and south edges Mancos Shale locally overlaps the sandstone in belts that are transitional to higher elevations. West of the Ute Mountains, where the San Juan River and its tributaries have cut deeply into the underlying Morrison Formation, the trailing end of the bench is dissected into long finger mesas. The average slope of this Dakota surface varies between one and two degrees in most places but is nearly horizontal east of the Abajo Mountains, along the crest of an arch that roughly follows the northwest-trending part of the Dolores River, and near the mouth of Montezuma Creek. On this wide, mainly south-facing slope the axes of larger streams, like those exposing the Morrison Formation, are roughly parallel to local dip (except for the northwesterly Dolores River). The orientations of these streams show that local dip is southerly along the main stem of Montezuma Creek but southwesterly from its eastern tributaries to the San Juan Mountains.

The Dakota Sandstone is very well expressed in landforms wherever it meets the surface because it is one of the hardest sedimentary formations in the region (equally hard strata exist only among early Paleozoic rocks, which crop out in a narrow belt around the Needle Mountains uplift). The underlying Burro Canyon Formation, composed mainly of sandstones and conglomerates, also resists erosion. In contrast, the overlying Mancos Shale is one of the weakest formations, so it is easily eroded from the top of the Dakota. Similarly, the upper Morrison Formation is also very soft and contrasts clearly with the overlying Dakota and Burro Canyon Formations.

Small-scale features on the Dakota surface vary. Although the sandstone is relatively hard, in some places stream erosion has formed a rolling landscape with colluvial and alluvial deposit. Erosion on flatter uplands is extremely slow, so windblown deposits are widespread and characteristic of areas such as the Sage Plain. Landslide deposits are common along canyon sides where the Dolores River and other streams have cut through the Dakota Sandstone into the upper Morrison Formation.

Mancos–Mesa Verde Belt

Farther to the southeast along the piñon-juniper zone are landforms associated with younger Cretaceous rocks, mainly the Mancos Shale (Km) and Mesa Verde Group (Figures 11.1 and 11.3). Their outcrop belt is mostly along a structural feature called the Four Corners platform. Minor deformation within the platform includes the shallow Mesa Verde basin.

The trend of structural contours in this area is partly expressed by the outcrop belt of the Mancos Shale, which curves around the northwestern part of the Mesa Verde basin and then continues east toward Durango. Because the shale is very thick and mostly very soft, broad areas of low topography have formed on its lower part.

In contrast to the Mancos Shale, the Mesa Verde Group of formations contains many strata that are moderately hard. Therefore, the Mesa Verde Group forms a caprock zone that protects the upper Mancos Shale from erosion, and a steep escarpment from 300 to 600 m high separates the lowland from landforms on top of the Mesa Verde Group. The upper shale and lower parts of the Mesa Verde Group (the Point Lookout Sandstone and Menefee Formation) are exposed mainly in a narrow band along the escarpment. Except in a few places where the escarpment rises in several smaller steps, the highest formation of this group (the Cliff House Sandstone) is exposed over most of the area shown on the geologic maps as upper Cretaceous rocks (Figures 11.1 and 11.3).

Many canyons have been cut through the Cliff House Sandstone, but its top is well preserved along the crests of intervening finger mesas, especially in the vicinity of the Mesa Verde basin, where the formation is composed mainly of thick sandstone units and is relatively hard. From Park Point in Mesa Verde National Park, the basin looks like a broad, shallow saucer that is tilted to the south and centered on the middle of Mancos Canyon. Farther east the Cliff House Sandstone consists of thinly interbedded sandstone and shale and is not as hard. Therefore, ridges in that area are more rounded, valleys are wider, and slopes are less steep.

Now that the larger features of this area have been sketched, some smaller features of the Mancos lowland, escarpment, and Mesa Verde upland can noted. The Mancos lowland displays a characteristic suite of landforms. This lowland, the Montezuma Valley, is a curving strike valley that is drained by parts of two intermittent streams, McElmo Creek and Navajo Wash. In the valley bottoms, smaller tributaries flowing toward these axial streams have cut broad, gentle slopes that are covered with alluvium. Remnants of similar but older surfaces form flat-topped ridges (often called pediments) that radiate from uplands like Mesa Verde and the Ute Mountains. Capping sediments are coarse next to the uplands, containing angular boulders in places, and become finer with increasing distance. Such ridges at several heights form a stepped landscape that results from episodic stream incision, not from varying hardness within the shale itself. Although the capping sediments give some protection, these surfaces are formed and eroded very quickly because the shale is soft. Therefore, the lowest and youngest surfaces are most extensive. In other places slightly harder beds within the shale, which are composed of sandstone or fossil-bearing limestone, cap low benches and hills. Piñon-juniper woodland is poorly developed on Mancos bedrock because of its high clay content. Most woodlands within the shale belt are located on surficial deposits that are sandier and more permeable to water. Steep slopes in the shale erode too rapidly to support many plants and are mainly bare. Elevations in this Mancos lowland decrease from about 2,100 m at Mancos to 1,500 m near the mouth of Mancos Canyon, but local relief is generally low.

Mesa Verde is a large, complex cuesta that is separated by the Mancos River canyon from other areas that are also underlain by strata of the Mesa Verde Group. Intermittent streams flowing toward the river have dissected the top of Mesa Verde into long, branching finger mesas. The highest strata on most of these mesas are near the top of the Cliff House Sandstone; however, lower strata within the Menefee Formation and Point Lookout Sandstone also cap smaller mesas at Point Lookout and along the west rim (Figure 11.3). Bedding orientations and rock hardness affect many features of Mesa Verde. Streams are elongated parallel to the slope (or dip) of bedding. Average dips are two to three degrees toward the center of the Mesa Verde basin, so dominant stream trends vary from the western to the eastern part of the mesa. Near the north rim the Cliff House Sandstone is composed of thinly bedded sandstone and shale, so ridge crests are rounded and valleys have V-shaped cross sections. Shale beds decrease quickly toward the south, however, where the Cliff House is composed of thick sandstone units with rare shale partings. As the formation becomes harder, mesa tops become flatter, and near-vertical cliffs appear at their sides. The floors of many canyons are cut below these cliffs into the Menefee Formation, which forms moderately steep slopes. Overall elevations range from about 2,500 to 1,800 m.

Many kinds of surficial deposits are present on Mesa Verde, often occurring in complex associations. Windblown deposits are widespread on parts of the mesa tops where erosion has nearly stopped. Stream alluvium forms floodplains, low terraces, and alluvial fans along the floors of ravines and canyons, especially in the north part of Mesa Verde. Colluvium is widespread on moderate slopes. Landslide deposits occur in some places at canyon sides or heads where streams have cut through harder formations into softer ones. Small deposits of tufa (calcareous spring deposits) are found in a few places where seeps have issued from sandstone beds, mainly during former glaciations when the water table was probably higher. Finally, alluvium of an ancient Mancos River, containing rounded cobbles of igneous and metamorphic rocks, is present on the south ends of some mesas overlooking the canyon and on terraces within the canyon.

Because of its steepness and height, the escarpment rising toward Mesa Verde is a zone of rapid erosion with varied landforms. Along the east rim the scarp is dissected by very steep, closely spaced, shallow gullies of ephemeral streams. Boulder deposits near the bases of these gullies may have accumulated from muddy debris flows after large storms, by rockfall, and by colluviation. In other places landslides are common. Along the north rim, recent landslide activity and older landslide deposits are limited mainly to the heads of very steep valleys that embay the escarpment; in contrast, bedrock is exposed on intervening ridgelines. Along the east and southeast rims, however, boulder landslide and colluvial deposits cover large areas, and bedrock exposures are limited mainly to the upper sandstone cliffs. Truncated (or beheaded) stream valleys along the north rim and talus flatirons in the Mancos lowland show that the north, west, and east rims have retreated toward the interior of Mesa Verde since at least middle Pleistocene time, roughly several hundred thousand years ago.

Landforms of the Mesa Verde area illustrate the kinds of judgments that are needed to interpret stages of landscape development. Because much of the evidence has been eroded, such judgments are partly subjective and must change when new data become available. It is most reliable to work backward and forward in time from the deposits and landforms that are best understood. For example, evidence that the Mesa Verde escarpment has retreated suggests that during an earlier stage the Mancos Shale and Mesa Verde Group extended farther northward onto the present Dakota plains. Homoclinal shifting (or downslope migration of escarpments in bedded rocks) may also have removed some of the missing formations above the Mesa Verde Group.

Other inferences about landscape development can be made from the distribution of alluvial gravels. The absence of pebbles and cobbles of igneous and metamorphic rock on most of Mesa Verde suggests that no river has flowed from the La Plata or San Juan Mountains across the location of the present north rim, at least not for the time that would be needed to remove all traces of gravel, probably millions of years. The highest and oldest remaining deposit of Mancos River alluvium, however, is located well east of the modern river canyon on a ridge that separates the modern drainage basins of the Mancos River and Cherry Creek. Younger alluvium next to and within the modern Mancos canyon suggests that the Mancos River shifted toward the northwest before incising along its present course. Tributary streams north of the Mancos River gradually eroded older rocks from the top of Mesa Verde, exposing the Cliff House Sandstone. At some stage in this process, streams flowing down the top of the sandstone could have had shallow valleys, but the sandstone was probably never exposed as a continuous, planar surface.

Differences between this sequence of events and earlier proposals point out the problems of interpreting landscape development. W. W. Atwood and K. F. Mather (1932) stated that Mesa Verde was once part of a broad erosion surface covered with igneous and metamorphic gravels, which radiated smoothly away from the La Plata and San Juan Mountains. The absence of any such gravels on most of Mesa Verde (and in many other areas) argues strongly against their hypothesis. C. B. Hunt (1956) suggested that the lower Dolores River once continued along a southwesterly course past what is now western Mesa Verde. He proposed that the lower Dolores River was later diverted toward its present northwesterly course in part by surface doming during the emplacement of igneous intrusions in the Ute and La Plata Mountains. Those intrusions are now known to have formed in late Cretaceous time (Semken and McIntosh 1997), however, very early in the Laramide orogeny. During this orogeny, the Dolores River may have been located north of a ridge connecting the Ute and La Plata Mountains. The course of the river may never have passed between the Ute Mountains and present Mesa Verde.

HOGBACK MONOCLINE

Continuing toward the southeast along the piñon-juniper zone, the next area of distinct landforms is along the Hogback monocline. Structurally, the monocline is a moderately sloping ramp that conceals thrust faults at greater depths (Taylor and Huffman 1998). It separates more gently dipping strata on the Four Corners platform and south flank of the La Plata Mountains from broadly downwarped strata in the San Juan Basin (Figure 11.2). At the surface the monocline is expressed by a series of hogback ridges that are supported by late Cretaceous and early Tertiary formations (Figure 11.1). Colluvium is the most common surficial deposit in this area. Valleys of the south-flowing La Plata, Animas, Florida, and Pine Rivers cross this hogback belt. These valleys contain river terraces, as described more fully in the next section. The oldest remaining alluvial gravel of the La Plata River is preserved only within the hogback belt at Bridge Timber Mountain, southwest of Durango. The Bridge Timber Gravel, roughly 340 m above the modern La Plata River, is probably late Miocene or Pliocene in age (Kirkham and Navarre, in press).

San Juan Basin

Exposed within the San Juan Basin are late Cretaceous and mainly early Tertiary strata that consist mostly of mudstones, sandstones, and conglomerates (Figures 11.1 and 11.2). The structure of these rocks has been compared to a stack of saucers with their lips along the Hogback monocline and other, more distant basin margins. Dips within most of the area shown are a few degrees, but the strata are almost horizontal near the basin's structural axis, which passes through Aztec. These rocks are cut by valleys of the south-flowing La Plata, Animas, and Pine Rivers, all of which originate in mountains to the north and join the east-flowing San Juan River at points within the basin.

Landforms in the basin include benches and mesas formed by differential erosion of strata and alluvial terraces along the larger rivers and some of their tributaries. River terraces are underlain by gravelly alluvium from the La Plata and San Juan Mountains, which is covered with younger alluvium of tributary streams and windblown deposits. Among the terraces, those of the Animas River have been studied most carefully (Gillam 1998). Many of the lower and younger Animas River terraces can be traced to moraines at Durango, so their gravelly alluvium appears to be glacial outwash. Older and higher terraces project above existing moraines but probably formed in the same manner. The oldest remaining alluvium is found on a dissected cuesta that forms the present drainage divide between the Animas and Pine Rivers. This alluvium, at heights from 470 to 385 m above the modern river channel, is late Pliocene and may have accumulated about 2.4 million years ago (Gillam 1998).

Other surficial deposits in the basin are mainly colluvium on slopes and windblown deposits on mesa tops. A few landslides have occurred along steep slopes or where more groundwater seepage occurred during past glaciations.

SUMMARY

The present landscape west and south of the San Juan Mountains displays varied landforms of many sizes that strongly influence local environments. Scales of landforms range from the general southwestern slope of the region to intermediate features like Mesa Verde and finally to small features like single knolls, slopes, and gullies. Erosional landforms, which predominate, are influenced strongly by rock hardness. Even in this erosional setting, small-volume surficial deposits have created minor constructional landforms. The overall pattern of this landscape reflects progressive development during the last 70 million years. Mid-scale and smaller features mostly achieved their present shapes at times during the last 2 million to 0.5 million years.

After the last glaciation, as the climate warmed again, piñon and juniper expanded into the Four Corners, and piñon-juniper woodland adjusted to its present range between roughly 1,500 and 2,450 m. On a regional scale postglacial changes in most landforms are minor and have not affected ecosystems as much as short-term climatic variability. In small areas, however, short-term erosion and deposition can greatly alter local ecosystems. Examples of such changes are gully incision that lowers the water table in adjacent sediments, slope erosion after wildfires destroy stabilizing vegetation, and landslides that block streams.

Chapter 13 looks more closely at the soils that have developed on bedrock and surficial units in the piñon-juniper belt.

ACKNOWLEDGMENTS

This chapter benefited from reviews by Robert W. Blair and Emmett Evanoff. My research in the Four Corners area has been partly supported by the U.S. Geological Survey, New Mexico Bureau of Mines and Mineral Resources, Colorado Geological Survey, National Park Service, University of Colorado, and private clients.

REFERENCES CITED

Adams, K. R., and K. L. Petersen. 1999. Paleoenvironmental reconstruction: The last 40,000 years in the northern San Juan River drainage basin. In W. D. Lipe, M. D. Varien, and R. H. Wilshusen (eds.), *Colorado prehistory: A context for the southern Colorado River Basin*, pp. 34–50. Colorado Council of Professional Archaeologists, Denver.

Arrhenius, G., and E. Bonatti. 1965. The Mesa Verde loess. In D. Osborne (ed.), *Contributions of the Wetherill Mesa archaeological project*, pp. 92–100. Memoirs of the Society for American Archaeology, no. 19.

Atwood, W. W., and K. F. Mather. 1932. *Physiography and Quaternary geology of the San Juan Mountains, Colorado.* U.S. Geological Survey Professional Paper 166.

Betancourt, J. L. 1990. Late Quaternary biogeography of the Colorado Plateau. In J. L. Betancourt, T. R. Van Devender, and P. S. Martin (eds.), *Packrat middens: The last 40,000 years of biotic change*, pp. 259–292. University of Arizona Press, Tucson.

Cole, K. L. 1990. Late Quaternary vegetation gradients through the Grand Canyon. In J. L. Betancourt, T. R. Van Devender, and P. S. Martin (eds.), *Packrat middens: The last 40,000 years of biotic change*, pp. 349–366. University of Arizona Press, Tucson.

Gillam, M. L. 1998. Late Cenozoic geology and soils of the lower Animas River valley, Colorado and New Mexico. Ph.D. dissertation, University of Colorado, Boulder.

Hunt, C. B. 1956. *Cenozoic geology of the Colorado Plateau.* U.S. Geological Survey Professional Paper 279.

Johnson, M. D. 1990. Fabric and origins of diamictons in end moraines, Animas River valley, Colorado, U.S.A. *Arctic and Alpine Research* 22: 14–25.

Kernodle, J. M., C. R. Thorn, G. W. Levings, S. D. Craigg, and W. L. Dam. 1990. Hydrogeology of the Kirtland Shale and Fruitland Formation in the San Juan structural basin, New Mexico, Colorado, Utah, and Arizona. In *U.S. Geological Survey, Hydrologic Investigations Atlas* HA-720-C.

Kirkham, R. M., M. L. Gillam, T. D. Losecke, J. C. Ruf, and C. J. Carroll. 1999. *Geologic map of the Durango West quadrangle, La Plata County, Colorado.* Colorado Geological Survey, Open-File Report 99-4.

Kirkham, R. M., and A. K. Navarre. In press. *Geologic map of the Basin Mountain quadrangle, La Plata County, Colorado.* Colorado Geological Survey, Denver. Open-File Report 01-4.

Leonard, E. M. 1984. Late Pleistocene equilibrium-line altitudes and snow accumulation patterns, San Juan Mountains, Colorado. *Arctic and Alpine Research* 16: 65–76.

Price, A. B., W. D. Nettleton, G. A. Bowman, and V. L. Clay. 1988. Selected properties, distribution, source, and age of eolian deposits and soils of southwest Colorado. *Soil Science Society of America Journal* 52: 450–455.

Schumm, S. A. 1975. *The fluvial system.* John Wiley and Sons, New York.

Semken, S. C., and W. C. McIntosh. 1997. $^{40}Ar/^{39}Ar$ age determinations for the Carrizo Mountains laccolith, Navajo Nation, Arizona. In O. J. Anderson, B. S. Kues, and S. G. Lucas (eds.), *Mesozoic geology and paleontology of the Four Corners region,* New Mexico Geological Society, 48th Field Conference Guidebook, pp. 75–80. New Mexico Geological Society, Soccoro.

Spaulding, W. G. 1990. Vegetational and climatic development of the Mojave desert: The last glacial maximum to the present. In J. L. Betancourt, T. R. Van Devender, and P. S. Martin (eds.), *Packrat middens: The last 40,000 years of biotic change,* pp. 166–199. University of Arizona Press, Tucson.

Steven, T. A. 1975. Middle Tertiary volcanic field in the Southern Rocky Mountains. *Geological Society of America Memoir* 144: 75–94.

Steven, T. A., K. Hon, and M. A. Lanphere. 1995. *Neogene geomorphic evolution of the central San Juan Mountains near Creede, Colorado.* U.S. Geological Survey, Miscellaneous Investigations Series, Map I-2504.

Taylor, D. J., and A. C. Huffman, Jr. 1998. *Map showing inferred and mapped basement faults, San Juan Basin and vicinity, New Mexico and Colorado.* U.S. Geological Survey, Miscellaneous Investigations Series, Map I-2641.

Van Devender, T. R., J. L. Betancourt, and M. Wimberly. 1984. Biogeographic implications of a packrat midden sequence from the Sacramento Mountains, south-central New Mexico. *Quaternary Research* 22: 344–360.

Van Devender, T. R., R. S. Thompson, and J. L. Betancourt. 1987. Vegetation history of the deserts of southwestern North America: The nature and timing of the late Wisconsin–Holocene transition. In W. F. Ruddiman and H. E. Wright, Jr. (eds.), *North America and adjacent oceans during the last deglaciation,* pp. 323–352. Geological Society of America: Geology of North America, vol. K-3.

Williams, M., D. Dunkerly, P. De Deckker, P. Kershaw, and J. Chappell. 1998. *Quaternary environments.* 2nd ed. Arnold Publishers, London; and Oxford University Press, New York.

SOILS OF MESA VERDE COUNTRY

Doug Ramsey

IN THE LAST TWO CHAPTERS WE HAVE DISCUSSED THE GEOLOGIC PROCESSES that formed bedrock and surficial substrates in Mesa Verde Country. Our last look at the earth's surface focuses on the soils that have weathered from these rocks. The soil environment is of critical importance in determining plant survival, and it is intricately intertwined with climatic factors, temperature, and precipitation. For example, while our summer thunderstorms may let loose a downpour, the amount of moisture that actually becomes available to the piñon and juniper trees is restricted to the amount that penetrates the soil to depths where their roots reside. In arid regions it is critical that our soils hold and store enough water for plants throughout the year. This "available water capacity" (AWC) varies greatly with the various soil types and soil depths. Soil material acts like a sponge, absorbing water and retaining it in pores between the mineral particles. Moisture may sit in these pores until the time when the plant is in need of it and actively absorbs it into roots. It is from this stored moisture that the productive piñon-juniper woodlands of Mesa Verde have developed; this is especially obvious when we remember that our study areas receive only about 3–4 cm of precipitation most months of the year.

In this chapter I compare soil environments at three piñon-juniper sites—Hovenweep National Monument, Cortez, and Mesa Verde National Park—to illustrate the relationships of precipitation, temperature, soil moisture status, and soil temperature (Table 13.1). Soil scientists typically use a model called the Newhall

Table 13.1. Characteristics of three piñon-juniper habitats, southwestern Colorado

Site	Elevation		Annual Precipitation		Annual Temperature	
	m	ft	cm	inches	C	F
Hovenweep, NM	1,600	5,240	29.2	11.5	10.7	51.3
Cortez	1,890	6,210	33.8	13.3	9.0	48.2
Mesa Verde	2,150	7,070	46.2	18.2	9.6	49.2

Simulation Model (Van Wambeke et al. 1992) developed by Cornell University to demonstrate how the temperature and precipitation affect soil moisture conditions. The model plots the soil moisture balance, allowing a comparison of evapotranspiration for values of precipitation and soil temperature. Using this model it is possible to compare similar soils at three sites, each of which supports piñon-juniper woodlands in our study area. Hovenweep National Monument occurs at the very low elevation and precipitation range of the piñon-juniper ecosystem. Cortez, Colorado, is located in the middle of the range. Mesa Verde National Park is found at the upper extreme of elevation and precipitation in our area. By comparing these three sites, we studied the effects of increased precipitation and decreased temperature on the soil moisture balance, and consequently on woodland characteristics, as we moved to higher elevations. The predictive value of the Newhall Simulation is described in the following discussion for several different periods during a hypothetical growing season.

Consider first this progression at the end of the growing season, about mid-October. At this point soils at all three sites are dry, due to plant growth during the previous summer months, with little soil moisture available to plants. Evapotranspiration (ET) is minimal due to the cool temperatures and the dormancy of many of the plants. From this point through the winter, the majority of the precipitation (PPT) falling as rain and snow is absorbed into the soil and is stored in spaces between soil particles called soil pores.

Hovenweep receives the lowest precipitation of the three sites, and therefore it takes a longer period for the soil profile to begin to fill with moisture. Generally, the soil does not attain a "moist condition" until mid-February. Because of higher precipitation, the soils at Cortez reach a moist state by mid-January. Soils on Mesa Verde reach this point in late December. From this point until spring, the soils continue to absorb moisture and store it for the plants' eventual use.

The amassing of soil moisture continues until a time in the spring when the temperature increases and the plants begin to grow. Temperatures will begin to increase earlier at the lower elevations. The soil temperature will lag behind the increasing air temperature slightly, reaching about 40° F in early April at Hovenweep and mid- to late April at Cortez and Mesa Verde. Plant growth supported by soil moisture begins near this time. From this point the soil moisture state is in flux, as plants begin to remove water from the soil profile and the ever unpredictable precipitation adds to the volume.

The graphical relationship of precipitation, stored moisture, and evapotranspiration (Figure 13.1) suggests that by April the amount of water used by ET will exceed the amount of PPT received at the Hovenweep site. Depending on when plant growth starts in the spring and the PPT/ET balance, the stored soil moisture is eventually depleted and plant growth is limited. At Hovenweep the soil moisture begins to be depleted in early May and is severely depleted by late May. Cortez is slightly cooler and has lower ET and higher PPT. Therefore, Cortez plants begin to deplete soil moisture in late May, and some water is available until early July. Mesa Verde, which sits at a higher elevation and receives greater PPT, does not begin to show evidence of soil moisture depletion until early July, and the soil moisture is not depleted until early August. By comparing these values, we can begin to understand the influence of local climate and soil moisture in determining what plant community is capable of existing on a particular site.

What woodland characteristics of each of these three study sites might be attributed to soil moisture trends? By June the juniper and piñon at Hovenweep begin to suffer from moisture stress. This, along with the lack of sufficient precipitation often experienced during the rest of the summer, drives the number and particular type of plants that survive in the understory. At Hovenweep *Juniperus osteosperma* (Utah juniper) dominates. An open spacing of junipers (with a few piñons in wetter sites) characterizes the Hovenweep area, with few shrubs and forbs occurring in the understory due to the need to harvest moisture from a large area. The predominant grass in this area is *Hilaria jamesii* (galleta grass), a species that is common in the deserts of the Four Corners area and is obviously capable of existence in very low soil moistures. Trees tend to be smaller than similar-aged stands at higher elevations and precipitation. Total annual vegetative production is low, with the majority occurring in woody growth on the trees.

At the middle elevation site—Cortez, Colorado—piñon trees share dominance in the tree layer with juniper. These conifers normally do not begin to show signs of moisture stress until around the first of July. Trees attaining greater heights and relatively high vegetative production characterize this site. The biomass of understory vegetation is greater, and the species composition is unique relative to Hovenweep. For example, galleta grass is replaced by *Poa fendleriana* (muttongrass), a species that requires more moisture.

High elevations on the Mesa Verde cuesta represent the upper range of the piñon-juniper ecosystem in the Four Corners area. Precipitation exceeds 46 cm, allowing for exceptional growth of piñon and juniper as well as a wide variety of understory shrubs and grasses. In old-growth woodlands of the upper Mesa Verde plateau, trees are closely spaced, and there tends to be a clumping of forbs and grasses in the intercanopy spaces. Piñon pines occur in higher density than junipers,

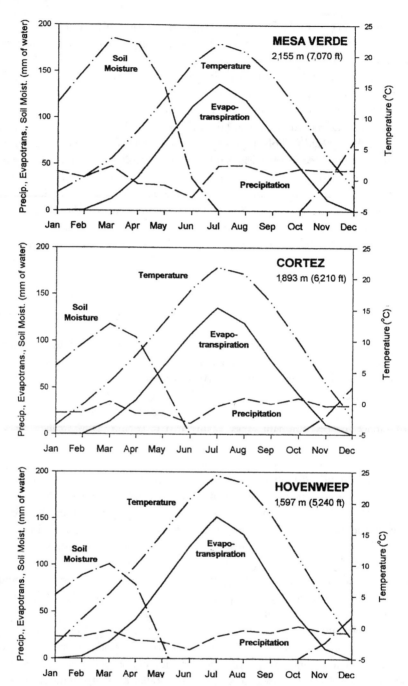

Figure 13.1. Newhall Simulation Models for three sites in Mesa Verde Country: Hovenweep, Cortez, and Mesa Verde. The model depicts the relationship of temperature and precipitation and soil moisture.

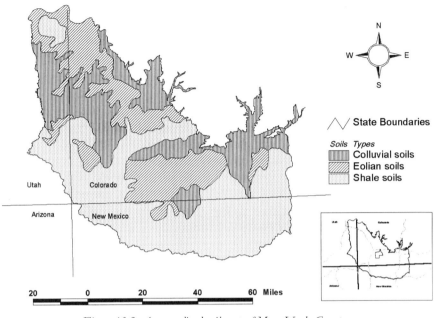

Figure 13.2. A generalized soil map of Mesa Verde Country.

and in wet microsites Utah juniper is replaced with *Juniperus scopulorum* (Rocky Mountain juniper). Muttongrass and *Purshia tridentata* (bitterbrush) are common understory components.

GENERALIZED SOIL TYPES OF THE MESA VERDE COUNTRY

Three major soils groupings are recognized by the Natural Resource Conservation Service (NRCS) in our study area (Figure 13.2). Present on the stable mesa tops and uplands are soils that have developed from eolian (windblown) deposits. Surrounding these stable mesas and uplands are the steep canyons and hills made up of colluvial (deposited by gravity) material fallen from steep slopes. Low rolling hills and valleys cross large expanses, with soils forming from the soft shale bedrock.

EOLIAN SOILS

About 100,000 years ago the environment was a great deal drier and windier in the area we refer to as the Four Corners. Mesas and uplands had been stripped of their soils and soft bedrock, leaving vast areas of bare sandstone. Over time, but at a significant rate, soil material consisting of mainly silt and fine sand was blown in great storms from areas to the southwest of Mesa Verde Country (Price et al. 1988). These particles began to accumulate, building up millimeter by millimeter, on landforms that were level enough for accumulation (where deposition rate exceeds rate of erosion). On the Mesa Verde cuesta, Florida Mesa, and areas northwest of Cortez these reddish eolian deposits accumulated. Thick deposits are the result of several episodes of deposition over approximately the last 100,000 years. The last of these major episodes is believed to have occurred approximately 16,000 years ago. This age is based on C^{14} dating of a buried musk ox (*Symbos cavifrons*) skeleton that was recovered from eolian deposits during an archaeological excavation at the McPhee Reservoir near Dolores, Colorado (McDonald et al. 1987). There is very little evidence of continual deposition of significant amount of eolian deposits over the last few thousand years; for example, Ancestral Puebloan ruins that date to 1,000 years ago remain on the surface and are not buried under recent deposits of eolian material.

In simple terms these deposits are the "red" soils that are the major cropland soils in La Plata, Montezuma, and Dolores Counties. These soils generally have excellent properties for agriculture uses and natural vegetative growth. Local variations in soil depth are demonstrated by a general catina (horizontal sequence). Near the edges of the mesas or on canyon rims, bare sandstone called "slickrock" is normally exposed. Closest to the rim, the soils are generally a mixture of eolian and residual sandstone. They have the texture of sandy loams, with minimal development of soil horizons. Sandstone fragments may be mixed with the soil as small, coarse fragments. Farther from the mesa edge the eolian material has accumulated to greater depths; for example, 50 to 100 cm accumulated at a distance of about 50 to 100 m from the canyon rim on Mesa Verde. These soils are relatively stable eolian deposits with very little influence from the underlying bedrock. Subsoil textures are generally loams to clay loam with the development of subsoil argillic horizons.

Winter moisture has leached very slightly soluble calcium carbonate downward from the surface layers very slowly over thousands of years; calcium carbonate has accumulated in the lower parts of the profile, forming white powdery deposits known as calcic horizons. The depth of carbonate movement can be correlated with the average depth of wetting from winter precipitation.

Still farther away from the canyon edges, near the middle of the mesa, the eolian soils have developed to greater depths—1 to 2 m overlying the geologic sandstone or shale material. Some deposits on the most stable locations exceed depths of 3 m. These deeper deposits show excellent evidence of stability and soil development. Well-developed subsoil argillic horizons along with calcium carbonate deposits in calcic horizons are common. Deep deposits may extend over large areas on level topographies, and depths decline as you move toward the next canyon or steeper, less stable hillslope.

Eolian soils of the Four Corners region exhibit excellent physical properties for the growth of piñon and juniper. Textures are predominately loams and clay loams, with the sand fraction almost exclusively of fine and very fine (0.25 to 0.05 mm) sizes. Remember that eolian deposits were blown for distances of hundreds of miles or more, and the textures we see today are a result of natural sorting processes. Lack of medium and coarser sand particles reduces the size of the soil pores, so eolian soils are able to hold a larger amount of available water per soil volume when compared to clayey and sandy soils. This ability to store soil water complements the winter moisture that is received and stored for use during the spring and summer. The surface textures of these native soils tend to be in the range of fine sandy loams to very fine sandy loams for the first few inches. Sandy textures allow for rapid infiltration of precipitation into the surface; as a result, short-term storage develops, and water moves farther into the soil profile. Nonetheless, exceptionally heavy rainfall events still result in runoff from eolian soils, while sandy soils reduce surface runoff.

The chemical properties of these soils also result in an environment that is suitable for plant growth. The pH tends to be near neutral at the surface and increases in alkalinity with depth. The developed argillic horizons have pHs in the 7.2 to 8.0 range, with the calcium carbonate–enriched horizons ranging from pH 8.0 to 8.4. In response to high precipitation, pH values tend to decrease as the carbonates are leached to lower depths. Low sodium characterizes soils in the elevation ranges at which piñon and juniper are found (Ramsey 2002).

All the plant's nutrition is derived from the soil; fortunately, the natural fertility levels of eolian soils are high. Cation exchange capacity ranges from 10 to 25 milli-equivalents (meq)/100 grams, due to both the presence of smectitic clays and the organic matter component. Base saturation is normally 100%. Detailed soils information can be obtained from the NRCS (Natural Resource Conservation Service, National Soil Survey Laboratory) web site at www.nrcs.usda.gov (see Morefield series, sample number S93CO-083-006; Witt series, sample number S80CO-83-003; and Barx series, sample number S93CO-083-003).

Colluvial Soils of Canyons and Steep Hills

Soils of the steep canyons and hills exhibit the greatest variation of the three major soil types found in the piñon and juniper woodlands of southwest Colorado. Soils of these landforms are strongly influenced by the geology located above them. In most cases, steep canyon and mesa slopes are capped by a band of hard sandstone, with softer and more erosive shale or interbedded materials beneath. The Mesa Verde cuesta is an excellent example of this feature. In the deep canyons of the mesa the upper third of the slopes is composed of the massive Cliff House Sandstone. Sandstones are relatively hard and therefore control erosion and down-cutting of the mesa top. Colluvial soils develop below these near-vertical cliffs, incorporating sandstone pieces from above mixed with the interbedded sandstone and shale layers of the underlying shales in the Menefee Formation.

The general properties of soils on these steep slopes, while variable, are predictable in most locations. On the convex positions of the landscape and nearer the top of the slopes the soils tend to be shallow and have minimal development. This is because the exposed steeper slopes have increased erosion rates, approaching the point where erosion is equal to deposition and weathering of the bedrock. It is not surprising that steep rocky slopes with little soil development tend to be poorly vegetated. The available water capacity is very low due to this shallow depth, and little water infiltrates into the soil. Runoff rates are consequently high. Sparse vegetation cover, allowing abundant sunlight and little protection from wind, increases evaporation rates.

The soils at the lower portions of the slopes accumulate to greater depths as the material erodes from above and is deposited in the fans and toe slopes near the bottom, exceeding 6 m in the larger canyons. Deep colluvial soils are made up of soil and geologic material that has fallen from above. If there are major layers of sandstone, the soil may contain greater than 50% sandstone fragments. While these soils are the result of depositional factors, the rate of addition is extremely slow. Lower slope soils have generally well-developed soil features and horizonation. Weakly to moderately developed argillic horizons and calcium carbonate deposits are common. Soil textures tend to be in the range of loams and light clay loams but have a great deal of variability, depending on the upslope geologic formations. Soil moisture status varies with location on the landscape. Many areas near the lower parts of the slope benefit from increased moisture due to runoff from areas above. The aspect also plays a major role in determining the development and moisture status of the site.

The Mancos River canyon south of Mesa Verde National Park provides an excellent example of this relationship. Although soils are forming from the same parent material (Point Lookout Sandstone and Mancos Shale), the north-facing aspect supports a full stand of piñon and juniper trees with a wide variety of shrubs and grasses, while the south-facing slopes are very sparsely vegetated with only scattered trees and shrubs. The soils are more fully developed on the north-facing slopes due to improved moisture status. The pH, organic material, and leaching of carbonates all are tangible evidence of this increase in the available moisture of the north-facing slopes (Ramsey 2002). Representative samples are the Durango series, sample number S93CO-083-002 (NRCS web site: www.nrcs.usda.gov).

Rolling Shale Hills

In Mesa Verde Country unique soils form on the rolling shale hills and valleys on several geologic formations. Mancos shales occur in the region immediately surrounding Mesa Verde, and in the areas east of Durango the Animas Formation predominates. Soils forming from shale formations generally exhibit a catina sequence that is very predictable. On the higher and steeper hills and knobs, the high erosional rates tend to allow only shallow, poorly developed soils to exist. Clay is abundant and organic material is scant in shale soils; thus infiltration rates are very slow. The result is a sparse vegetation when compared with better soils in the same area. Piñon and juniper, once established, grow on these sites normally with almost no understory plants. Chemical properties vary widely, depending on the geologic formation from which the soil is derived. Mancos Shale contains large amounts of calcium carbonates and soluble salts as well as very high amounts of gypsum in localized areas. The Animas Formation also contains calcium carbonates and soluble salts but in lower quantities than in the Mancos Shale. Growth of piñon and juniper trees is generally not restricted by the levels of salts and carbonates found in these soils, except in some localized strata.

Shallow soils of the summits and shoulders of the hills change rapidly from an erosional surface to a depositional area as you reach the footslopes and valleys. The valleys lack coarse fragments, and resulting soil textures have a direct relationship to the shale material from which they are being eroded. While most soils are high in clay and silt, the shale may contain considerable sand, and the resultant texture may be in the clay-loam range. Abundance of smectitic-type clay causes a "shrink and swell" phenomenon to occur as the soils dry out and then remoisten, leaving deep and wide cracks (Ramsey 2002). A representative sample for the Arboles series is sample number S80CO-067-001 (NRCS web site: www.nrcs.usda.gov).

While piñon and juniper may be the predominate species on the shale hills and knobs, they are conspicuously absent from the lower slopes and valleys of these areas. This phenomenon, while not totally understood, is thought to be caused by the presence of abundant grasses such as *Pascopyrum smithii* (= *Agropyron smithii*, western wheat) and *Artemisia tridentata* (sagebrush) that outcompete the conifer seedlings for limited water and thus restrict their establishment. These plants establish healthy stands of vegetation and utilize the available moisture quickly, causing piñon and juniper seedlings difficulty in establishment. When overgrazing or other disturbances have occurred on these soils, grasses decline, and piñon and juniper slowly move into these areas from the surrounding hills.

SUMMARY

Piñon pines and two species of junipers in Mesa Verde Country interact with the world through their roots. Soil moisture status and soil temperature have much to do with nutrient and water availability, factors critically needed by trees. Each of the soil types discussed—eolian, colluvial, and rolling shale hills—has its own unique set of textural properties that retains water in various ways. Thus adequate conditions for growth are met at a range of sites in our area, as illustrated by the three chosen for this chapter: Hovenweep, Cortez, and the Mesa Verde cuesta. This spectrum represents quite a range of site conditions; hence the growth and density of trees and associated understory are highly variable. Hovenweep can never support the woodland structure that develops on Mesa Verde. Yet a unique type of old-growth condition can be attained on either site, given adequate time free from disturbances. This chapter has laid the groundwork for interpreting the role of local climate and soil on woodland structures. The following chapters describe some aspects of that range of natural variation in the old-growth piñon-juniper woodlands.

REFERENCES CITED

McDonald, J. N., S. W. Neusis, and V. L. Clay. 1987. An associated partial skeleton *of Symbos cavifrons* (Artiodactyla:Bovidae) from Montezuma County, Colorado. *Journal of Paleontology* 61: 831–843.

Price, A. B., W. D. Nettleton, G. A. Bowman, and V. L. Clay. 1988. Selected properties, distribution, and age of eolian deposits and soils of southwest Colorado. *Soil Science Society of America Journal* 52: 450–455.

Ramsey, D. K. 2002. *Soil survey of Cortez, Colorado, parts of Dolores and Montezuma Counties.* United States Department of Agriculture, Natural Resources Conservation Service.

Van Wambeke, A., P. Hastings, and M. Tolomeo. 1992. Newhall Simulation Model, a basic program for the IBM PC (DOS 2.0 or later). Department of Agronomy, Cornell University, Ithaca, N.Y.

WATER RESOURCES IN MESA VERDE COUNTRY

Marilyn Colyer

Free water is a most precious resource in the piñon-juniper of southwestern Colorado. The few permanent water sites that exist are sought out by many bird species, the larger mammals, bats, amphibians, some reptiles, and certain invertebrates. Riparian plant communities are restricted to small wet drainages or seeps. Although small springs and seeps are minor in terms of area, they are vital to the health and diversity of the ecosystem. It has been said that 80% of all animal species depend on the riparian conditions at one time during their life cycle. It is no surprise that on the Mesa Verde cuesta and on the mesas surrounding Hovenweep there are numerous, and sometimes very large, settlements built by the Ancestral Puebloans around reliable cliff springs. Water-holding outcrops and impoundments are often easy to locate, simply by following trails of larger mammals, by seeking bright green patches on slopes or drainages, by checking bedrock areas for groups of potholes, or even by paying attention to certain butterflies, dance flies, and dragonflies that cluster around seeps. Eventually the human visitor to the woodland mesas and canyons is able to determine a pattern in the landscape suggestive of water sites. Check the confluence of two canyons; this is where you will most often find a seep in the drainages (Figure 14.1). Look at the lower end of the geologic dip, where the water-bearing aquifer is cut across by a side canyon or fault; this is where you will most often find cliff-base springs. Look in small, mesa-top drainages in bedrock areas where potholes have developed in Dakota and Cliff House Sandstones. Look also for the bentonite (clay formed by the

Figure 14.1. Water sites, such as this one near Spruce Tree House, Mesa Verde National Park, often form in potholes and small dams in sandstones. Drawing by Mary Vozar.

decomposition of volcanic ash) exposure in drainages or on slopes where the water forms pools. And, of course, permanent streams and rivers flow through the woodland country, supporting more obvious riparian plant formations.

Sources of water in this relatively dry country are frequently cryptic. While snowmelt from the nearby San Juan Mountain peaks of the southern Rocky Mountains feeds larger, often perennial rivers and streams, numerous water-holding geologic structures trap more ephemeral sources. The aquifers and their associated aqui-barriers provide numerous scattered sources of much needed water that are often ephemeral in nature. Here they are broken out by their geologic origin: sandstone aquifers, shale aquifers, alluvial aquifers, "clay slope" aquifers, and potholes that form in various bedrock types. Next I discuss the recharging of our water sources that comes intermittently and often violently from monsoon rains. Finally, I give examples of the water sources scattered throughout Mesa Verde Country.

EPHEMERAL WATER SOURCES

In Mesa Verde Country it is common to find that permanent springs are fed from sandstone aquifers that store and carry the water and are underlain by a clay aqui-barrier that prevents further vertical flow of the water (Colyer 2000). In these situations the water flows with the dip of the geological beds until there is a break across the sandstone layer. It is here that the water "springs" to the surface. On the Mesa Verde cuesta and on the mesas around Hovenweep the dip runs southward. The Cliff House Sandstone, Dakota Sandstone, and portions of the Menefee Formation (a mixed shale and sandstone layer) all serve as aquifers. A thin bentonitic layer at the base of the Cliff House Sandstone, the Morrison Shale below the Dakota Sandstone, and a bentonitic layer in the Menefee Shale serve as aqui-barriers. Point Lookout Sandstone is tightly cemented with silicone and rarely carries water.

Water quality is variable. Water from the Cliff House Sandstone and Dakota Sandstone is relatively pure: the pH is 7.0 to 7.4, and salts are not particularly high. Water from the Menefee Formation is alkaline, however, with pH values from 7.4 to 8.0, and contains calcium, sodium, magnesium, chloride, and selenium (Water Site Records, on file). Interestingly, it is often the Menefee Formation springs that are dug out by elk and wild horses, despite the presence of other water sources. Apparently these animals and others that follow are drinking the brown waters to obtain the salts they contain. Two Mesa Verde springs occurring in Menefee Shale are known as "Old Ugly," referring to the brown water pocketing in gray, barren bentonite, and "Brown Spring," referring to distinctly brown water. Both are heavily used by deer, elk, and wild horses. Signs or sightings of bear, mountain lions, gray foxes, skunks, coyotes, and rock squirrels are frequent at these and other Menefee Shale water sites. Oak-shaded habitats supported at Menefee Shale seeps attract robins, flickers, Steller's jays, several warblers, and several flycatchers. The close association of great horned owls with Menefee sites may be a reflection of high rodent populations near these waters.

While the salty water in Menefee shales does not support an abundance of riparian growth, the diverse riparian plant communities at the Cliff House Sandstone and Dakota Sandstone springs reflect the relative purity of the water. Fremont cottonwood trees, acuminate cottonwood, common reed, poison ivy, helleborine orchid, narrow leaf cattail, two willow trees, and several sedges flourish around these springs, creating a vibrant green patch in the piñon-juniper woodland. Menefee Shale springs are commonly vegetated by Gambel oak, baltic rush, foxtail barley, alkali muhly grass, herb willow, coyote willow, and (rarely) boxelder.

Equally important but ephemeral water sites are fed by water in alluvial sandy silt aquifers in canyon bottoms. The aqui-barriers are Mancos Shale or bentonite layers in Menefee Shale, and these may produce weak flows for about 500 m along a canyon bottom. They produce free water for one to three months a year or rarely throughout a year. While pH is relatively high (7.2 to 7.8) at these sites, they are nonetheless used often by wildlife. Coyote willows, alkaline buttercup, alkali muhly grass, foxtail, and baltic rush grow here. Over the past fifty years it has not been uncommon to find the non-native tree tamarisk at these water sites.

There are only a few instances of water springing from ancient river gravels. On the east side of Ute Mountain, in the Yucca House area, we find a series of permanent springs that arise from gravel layers underlain by Mancos Shale aqui-barriers. Surrounding these springs, in an otherwise semidesert, the Ancestral Puebloans constructed large settlements now called Yucca House, Grinnell Site, and Mud Springs. Imagine how important these permanent water sites were in a land that receives 25–30 cm of precipitation annually (in a good year). The source of the water is Ute Peak, a dioritic laccolith located at 2,727 m whose talus slopes absorb snowmelt and rain and feed the water downslope to below 1,818 m in south Montezuma Valley. The diorite rock and even the overlying formations are all basic; therefore they produce alkaline water. Spring water is high in calcium, sodium, and magnesium. Coyote willows, baltic rush, bulrush, broad-leafed cattail, and alkali muhly grass are the common riparian plants, while Fremont cottonwood trees are infrequently found in drainages. The water is generally used by wildlife, including muskrats, Woodhouse's toads, Hammond's spadefoot toad, western chorus frogs, garter snakes, and the non-native crayfish. At one time these springs supported leopard frogs. The riparian areas are inhabited by voles, shrews, the non-native raccoons, Virginia rails, northern harriers, Scott's orioles, and Bullock's orioles; and of course many other birds visit the water. As many as eighteen species of bats feed on mosquitoes, midges, and various flies that breed in the water and other water-associated insects (Chapter 8).

Clay slopes lack true aquifers that store and transport water; yet in a world where water impoundments of any type are rare, these serve an important water source for wildlife. The 600-m-high Mancos Shale escarpment around the Mesa Verde cuesta and the associated foothills and valleys are scattered with structures that impound water or seep water for several months at a time. A few are even active year-round. These water sites are fed from snowmelt and rain runoff from the immediate vicinity. For example, small plunge pools are formed below sandstone boulders (pieces that invariably fall from sandstone layers that cap the mesas above), and some of these pools hold several hundred gallons of water. Most, however, are much smaller and retain water for a week or two after each storm event. Because the slope clay below this cap is impervious, there is considerable runoff. Gullies may run over 5 cubic feet per second (cfs) during March or April snowmelt or during July or August thunderstorms. Much of this water is subsequently trapped in sandstone pools. During the wet 1980s these pools had water nearly year-round. This type of impoundment is not only found in the region's Mancos Shale slopes. The Morrison Shale forms similar water sites but less frequently, because the slopes are characteristically less than a fourth as long as the prominent Mancos Shale slopes.

The slump pool is another type of water site, developing on clay slopes and enduring for three to ten years. During observations from 1980 to 2000, none of these pools lasted more than ten years on the ever-moving slopes of clay. Yet while they persist, their importance should not be overlooked. One pool perched 500 m above the Montezuma Valley was 30 m long by 6 m wide, vegetated with tamarisk, home to nest mallards. A series of five- to ten-gallon pools formed along the clay escarpment, persisted for a couple of summers, watered western tanagers and yellow rumped warblers, bred mosquitoes, and then disappeared!

Elsewhere on the mesa a point of clay slumped at the confluence of two canyons. Two slump pools were formed, and the clay ran across one canyon, forming a small dam that held water for five years before all three pools slumped and silted in with clay moving from above. But while they did hold water, wildlife came to drink. Skunk, rock squirrels, gray fox, coyote, mountain lion, deer, and feral horse tracks were frequently seen here. A few acuminate cottonwood seedlings moved in and became established. These sites commonly form in Mancos Shale and Menefee Shale but rarely in Morrison Shale, where (in a little-understood process) the water works its way down into and through the kaolinitic clay, forming caverns and tunnels.

A third type of clay-based water site is those that are associated with the red loess. On the upper-elevation mesa top, years of weathering have broken the silt-sized particles of the loess down to finer, water-tight, clay-sized particles. Rarely we find this loess clay collecting in the heads of small drainages, allowing water to pool after rains. Another type of loess clay pool forms where humans dig barrow pits, grade road cuts, or otherwise remove the top soil and expose a clay layer. Whether on the mesa top or in a drainage head, these water sites are relatively shallow and provide short-term water for wildlife. They retain water in the spring, in late summer after thunderstorms, and perhaps during the winter. Because the water level fluctuates, sometimes dropping as much as 2 cm per day in the plunge pools, riparian vegetation and amphibians are not supported. Ducks visit, however, and even lay eggs at some pools. The larger mammals dig road puddles deeper to squeeze out the last water. The water boatmen (a type of aquatic insect) breed in these waters. Gopher snakes have also been seen at the water's edge, where they come to drink during periods of food shortages or to soften their old hide during molting.

Recently ecologists throughout the Southwest have realized that "potholes" support entire ecosystems themselves (Tim Graham, personal communication, 1998). The bedrock potholes are perhaps the most intriguing type of water site found in the piñon-juniper woodland. They are often formed in a series of stair-stepped bowls along a bedrock swale. The method of formation is quite interesting. Some sandstone bedrock areas are "turtle-backed" with a checkerboard of crevices. These crevices collect sand and pebbles, which in turn scour out cavities. Cobbles and boulders later find their way into the cavities, and the scouring action is increased. With each subsequent precipitation event, the sand and rocks are swirled around, gradually scouring out the crevices to form pools. Invariably the pools higher in the bedrock drainage are the shallow pools, and the pools become deeper as you proceed down the drainage. The pools on the cliff rims are sometimes a meter deep and up to a meter in diameter. They are so deep and shaded that they never go dry. A small rain of less than 8 mm will drain toward the swale and into the pools and revive the whole system. In fact, a 0.3 cm rain filled a bedrock pool that held over a hundred gallons from a bedrock area about 0.4 hectare in size (Colyer, unpublished data). These pools grow narrow-leafed cattail and sometimes coyote willow and Fremont cottonwoods. Woodhouse's toads, Hammond's toads, and red-spotted toads breed successfully in the bedrock potholes, even though these water sites are sparse and isolated from each other. Sometimes the amphibian breeding potholes are separated by several kilometers of dry canyon-rimmed country. Birds and mammals seek these mesa-top sources of water. Mountain lion scratches attest to their use of potholes, an especially poignant record when the pools are dry.

Cliff House Sandstone and Dakota Sandstone at the lower ends of the mesas are frequent locations of pothole series in Mesa Verde Country. Two Point Lookout pothole water sites are known. One is a strange little pothole in a crevice on a narrow neck of sandstone. The second is a large fallen boulder where a cavity holds up to 150 liters of water (Colyer, unpublished data). Both sites hold water except during extremely dry years, and both have prehistoric hand- and toe-hold trails to reach the water. The boulder pothole is a site where daphnia and fairy shrimp are consistently found. Both of these Point Lookout Sandstone potholes are fed from direct precipitation and benefit from runoff from a large bedrock area. In contrast, potholes in the Dakota Sandstone are in locations with

25–28 cm of annual precipitation. In these cases the potholes are not scoured out to deep cavities, and they rely on runoff from about a hectare of bedrock. These potholes usually have water only in the spring and after summer thundershowers. Migrating waterfowl, many mammals, amphibians, fairy shrimp, seed shrimp, tadpole shrimp, and aquatic beetles utilize these ephemeral, pure water sites.

PERENNIAL WATER SOURCES

When most people think of water, what comes to mind is a river winding through the landscape, carrying huge flows and supporting galleries of cottonwood trees and dense stands of willows on its banks. We do have such water sources in Mesa Verde Country; although they are fed by snowmelt and summer thundershowers, their flows tend to be erratic. The rivers that flow through much of the piñon-juniper woodland in the Mesa Verde area have headwaters in the San Juan Range of the southern Rocky Mountains. From east to west, these are the Pine, the Animas, the Florida, the La Plata, the Mancos, and the Dolores Rivers, whose peak flows vary considerably (Figures 14.2 and 14.3). All but the Dolores River flow essentially south into the San Juan River. Smaller tributaries include Yellow Jacket Creek, McElmo Creek, and Montezuma Creek. These perennial water sources provide needed hydration for reptiles, amphibians, mammals, and birds, especially when ephemeral sources are dry.

There is little that can be said with certainty about the natural state of the rivers, because not a single one of these permanently flowing streams has remained unaltered by water diversion or water addition. The Pine River feeds the San Juan River, which is dammed at Navajo Lake, where the Arizona and Colorado borders meet. The Animas River is diverted to Vallecito Lake. The La Plata River feeds Mormon Reservoir. The Dolores River is dammed at McPhee Lake. In addition, each river is fringed with numerous irrigation ditches, drawing off water to agricultural fields. In the process of diversion, water chemistry is altered by evaporation, and salts become concentrated. Several of the streams are recharged by water from irrigation that returns to the channel via buried gravels, picking up additional salts, agricultural pesticides, herbicides, and selenium from shale soils.

We know very little about the original floral community of the river zones. We do know that the proportion of large old cottonwoods has been reduced. Cottonwood trees are fast growing, reseed and resprout easily, and die easily. The fallen snags have served as small dams that spread the water table. Beavers, cottonwood trees, and the associated vegetation are all part of the same ecosystem. Beavers cut a few trees during the late summer when dam construction takes place, and in return the beaver pond spreads the water table and results in a wider cottonwood zone. Beavers feed on the phloem of fallen tree limbs, willows, and forbs and grasses. Studies have shown a 50% increase of species diversity around beaver dams (Yucca House water records). In the past few decades, however,

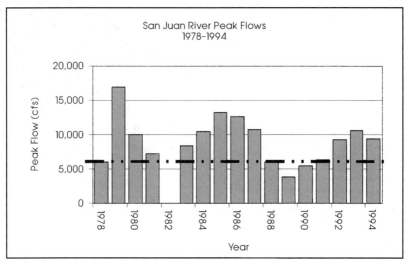

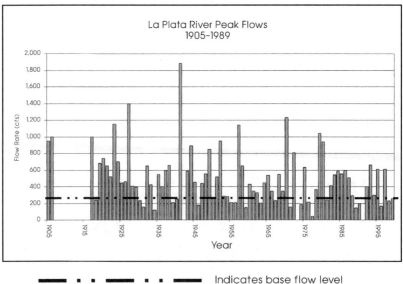

Figure 14.2. Peak flows on the San Juan and La Plata Rivers. The periods for which data are available vary.

---

beavers have been removed; together with cattle grazing in the area, this has left a legacy of raw soil banks that easily erode.

This is but one of many factors that have changed our region's rivers. Because they are excellent avenues for seeds and rootstock, the rivers are now lined

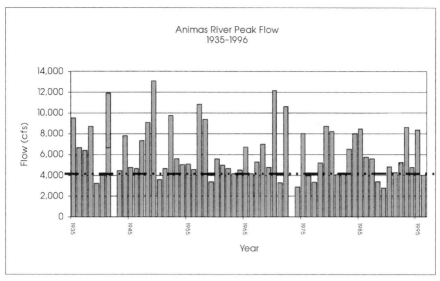

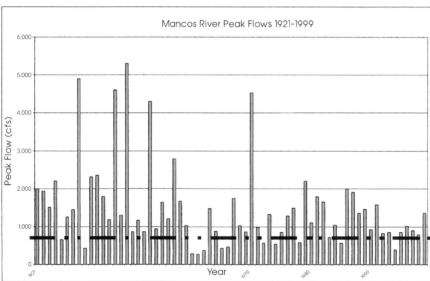

Figure 14.2 (continued). *Peak flows on the Animas and Mancos Rivers. The periods for which data are available vary.*

with almost every non-native species that is found in the region. The wetlands of side channels, oxbows, and pools that once supported an array of native species are virtually missing. Structural changes have resulted from lack of spring flooding that once scoured large pools along the river channel and left constantly

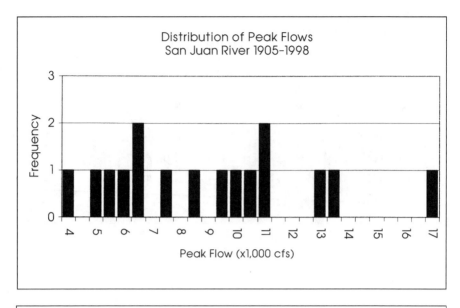

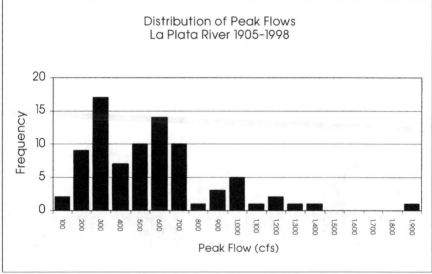

Figure 14.3. The frequency of values (magnitude) for peak flows on the San Juan and La Plata Rivers.

changing deposits of sand and gravel. The deeper pools of cooler water were important for the native fish fisheries and this, in turn, provided food for bears, great blue herons, and other fish-eating birds. According to oral history records

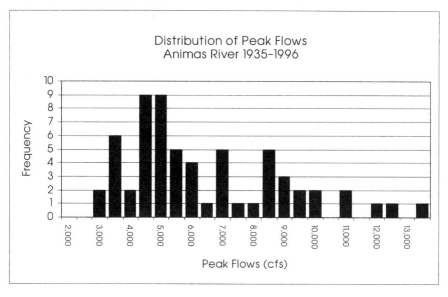

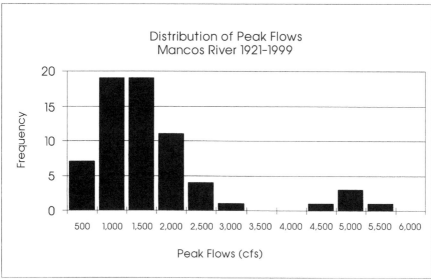

Figure 14.3 (continued). *The frequency of values (magnitude) for peak flows on the Animas and Mancos Rivers.*

the Mancos River once had pools 3.5 m deep by 7.5 m in diameter (Maness 2000). A recent pool survey along 9 km of river found only twelve pools up to 1 m deep.

The importance of rivers in a dry southwestern landscape cannot be overstated. Rivers cut down into the underlying sediments and form canyons or at least deepen drainages. These drainages are extremely important habitats for the riparian floral and faunal communities. The drainages also serve as avenues for seasonal migration of deer, mountain lions, bears seeking wild fruit and fish, migrating birds that use the cottonwood treetops for cover and feed on the ground, and a multitude of other kinds of wildlife that periodically use the river corridors. Of all the disruption that modern humans have caused in the piñon-juniper woodland, the compounded and continuous manipulation of the riparian corridors is perhaps the most damaging. Removal of cottonwood trees and willows from bottomlands, plowing and grazing the understory and replacing it with agricultural crops and non-native species, removal of beavers and their dams, and introducing cattle all reduce habitat and biodiversity.

PRECIPITATION INPUT

Mesa Verde Country, like much of the Southwest, is characterized by two major periods of precipitation. One period occurs during the winter months, when about half of the precipitation falls as snow. The other occurs during the July–August thunderstorm season that provides most of the remaining precipitation (Chapter 15). The summer thunderstorm season can produce considerable flows of water, especially where vegetation is lacking. Small, low-volume flows tend to drop their sediment in basins along the channels, and they gradually fill. Large floods scour out basins behind and beneath large blocks of stone and create plunge pools where they pour off a ledge or over an obstacle. Large flows probably transport tadpoles (which are weak swimmers out of their habitat) and, since most of the channels are typically dry, deposit them in unfavorable situations. Small flows bring deposits of organic material into the pools, providing perhaps the most significant part of the energy influx and thus of the tadpoles' food supply. Large floods speed up the escape of water from the drainage system, decrease influent recharge of aquifers and groundwater relative to small flows of the same total volume, and reduce the dependability of a watercourse as breeding habitat.

The ancient piñon-juniper woodland that occupies mesa tops of the Mesa Verde Country provides a considerable amount of stability to the soils, preventing erosion. This stabilizing effect is clearly demonstrated when the woodland is destroyed by large, stand-replacing wildfires. Nearly all of the piñon and juniper trees exposed to fires were killed in 1989, 1996, 2000, and 2002 wildfires on Mesa Verde. Huge flows ran off the catchment areas, flooding watercourses during rains that followed the fires. For example, postfire floods in Long Canyon scoured out several basins in sandy soil along a reach with low gradient. In Soda Canyon, however, water flowed through the channel, overtopping the bank, and in some instances flowed 2 m deep. It scoured an already deeply entrenched channel, converting it to little more than a uniform gutter with few pools.

The high angle of Mesa Verde canyon walls and the low permeability of the immature soils and rocky substrates make it possible for even a short, moderately intense rain to generate a large flow of water. The over-ground flow on such slopes can easily produce a flow of 100 cfs or more within a few minutes of the start of a storm. For example, on May 22, 1997, nearly one year after the Chapin 5 fire, a brief squall left perhaps 11 mm of total rain in twenty minutes. Within thirty minutes a group of astounded ecologists and resource managers watched as a torrent spilled off a 25-degree slope from about 4 hectares of fire-denuded hillside. The water carried an enormous load of sand, ash, and vegetative materials. It attained a width of 6 m and a depth of 0.15 m and a maximum volume discharge estimated at 0.2 m^3/sec. The flow continued from onset to half its maximum in less than twenty minutes. The total volume discharged in that time was approximately 125 m^3. Projecting this to the catchment area of 200 ha, this is a yield of 4 m^3/sec. It is also reasonable to estimate in Rock and Soda Canyons that flows exceeded 11 m^3/sec based on stranded large logs above the channel floor (Al Spencer, personal communication, 1997).

WATER SOURCES ON THE MESA VERDE CUESTA

Morefield Spring is one of the more reliable springs on the Mesa Verde cuesta. During twenty or more years of observation, this spring has never stopped flowing. The aquifer and aqui-barrier are of the Menefee Sandstone and Shale, and it appears that the point of emergence is at a fault line. The water is highly mineralized, yet it is consistently used by birds, mammals, and nonaquatic insects such as yellowjackets and flies. No aquatic insects or freshwater crustaceans live in this water, and there is no developed riparian plant community. A legacy of human dependence of this water source exists: a very large Ancestral Puebloan settlement sits nearby across the canyon floor, dating to A.D. 1100. Early cattle ranchers inserted a pipe into the canyon slope and installed a trough in the 1870s, and the National Park horse concessionaire used this water until the 1960s. A small reservoir was built about 30 m down-canyon, but during the past twenty years this has filled only from snowmelt or summer rains.

WATERS CANYON SPRING

Depending on the winter and late summer precipitation, the water table varies from the surface of the canyon floor to 1.5 m deep (as seen in a hand-dug well). A bentonitic layer within the Menefee Shale acts as the aqui-barrier. Wild horses and elk have dug 0.6 m to reach the water table, even on the steep slope at the canyon head. In the canyon bottom a line of baltic rush, buttercup, buffalo berry, and helianthella is associated with the moist soil. Moisture-loving butterflies such as the blues and coppers are often abundant at Waters Canyon Spring. One location along the drainage serves as a natural salt lick for large mammals. During extreme drought, all but the well dries up. A medium-sized Ancestral Puebloan site is nearby.

Prater Canyon Turkey Springs

This site is at the lower end of the Mancos Shale aqui-barrier, which occurs in Mesa Verde National Park only in upper Morfield and Prater Canyons. Deep alluvial soils serve as the aquifer. Snow drifts into side canyons at the head of Prater Canyon, and during the spring snowmelt the flow can be considerable. Turkey Springs is the most persistent portion of the 5 km stretch of alluvium that stores water. Elk have been known to dig down 30 cm or more to reach the water table up-canyon from Turkey Springs and produce free water. These shallow elk pools are temporary, but Turkey Springs generally has a trickle of water between small pools along a hundred-meter stretch. Bear and elk tracks are common sights. Riparian associated insects (such as blues and copper butterflies, tiger swallowtail butterflies, and dragonflies) are usually abundant. The drainage is lined with baltic rush, foxtail, and buttercup. Canada thistle and musk thistle are recent additions, having invaded during the past thirty years.

West Pine Canyon Potholes

This is the only source of free water to be found on Moccasin Mesa. Two potholes retain some water. These are not as deep as a series of fifteen potholes at the lower end of the mesa. Woodhouse's toads breed here, and wild horses and deer come to drink. Several Fremont cottonwood trees live in the drainage nearby. The potholes have formed on the turtle-backed white beach deposit of the Cliff House Sandstone.

School Section Spring

This water source runs intermittently from the canyon head down to the confluence of East School Section Canyon. The portion with continuous flow is about a mile down-canyon, where a well was dug by cattle ranchers at the turn of the twentieth century. In fact, four wells were dug in conjunction with plans to establish National Park headquarters in this canyon. The consistent flow does make this water site one of the more important sources of free water for wildlife. The remaining intact hand-dug well is about 2 m deep and was rock lined until the 1996 drought, which represented the driest year on record. That year a desperate bear dug one side of the well out and was able to reach free water. Most years the flow surfaces for a kilometer, and elk, deer, bear, mountain lion, and coyote tracks are common along the waterway. Warblers, Steller's jays, members of the thrush group, red-shafted flickers, mourning doves, and many other bird species are more common here than in areas without free water. In the piñon-juniper woodland the nest sites of great horned owls are almost always near a water site. This may be due to increased prey base, but at least in some cases it is because of the cover provided by the large Gambel oak clones that line these drainage water outcrops. The aqui-barrier at the School Section Spring is a bentonitic layer within the Menefee Shale.

Soda Canyon Spring

Soda Canyon, especially in the drainage below the confluence of East and West Soda Canyons, once supported a natural spring that surfaced for about 500 m. The spring's natural state can no longer be determined, because water pours into West Soda Canyon daily throughout the summer from the Far View sewer lagoon. The sewer effluence began in 1966. Yet the presence of 30-cm-diameter boxelder trees and similar-sized cottonwood trees down-canyon from Battleship Rock indicates a historic natural water source. A spring in upper East Soda Canyon, the site of an old well, runs water at least half of the summer. A record-sized Woodhouse's toad was recorded 16 km down from the canyon head in a rush-lined pool. In 1996 and 1997 the pools and riparian vegetation along the channel sides in this lower portion of the canyon were swept away by intense flooding following a fire.

Fewkes Canyon Plunge Pool

Fewkes Canyon cuts across a southward dip in the Cliff House Sandstone and for that reason has several seeps, most of which tend to be salty and ephemeral. There is also a permanent spring at the base of the upper Cliff House Sandstone at the confluence of Fewkes Canyon and Cliff Canyon. A thin layer of shale serves as the aquifer. This spring is special, for it harbors the only population of maidenhair fern in the region. A second permanent water site, a plunge pool, presently is less than 3 m in diameter and less than 15 cm deep. Surely the Ancestral Puebloans whose cliff dwellings surround the pool dug it deeper to allow for more collection capacity. There are over ten settlements easily within reach of this water, and two hand- and toehold trails attest that inhabitants of mesa-top settlements also visited this area for water. Historically bighorn sheep have been known to drink here. Numerous birds also come to drink in this canyon when free water is scarce on the mesa top.

Little Long House Pothole Series

This series involved eleven potholes at one time, but a park road presently covers or has silted in five of the holes. The lowermost hole holds about 20 liters in a deep crevice, which is shaded and dries up only in years of severe drought. Even though it is precarious to enter the crevice to the water, I have seen evidence that deer, mountain lions, and various birds do so. Like many others, these potholes have formed on the turtle-backed white beach deposit of Cliff House Sandstone.

Red-Spotted Toad Pothole Series

This series is found at the southern end of Chapin Mesa at the confluence of two drainages in the turtle-backed Cliff House Sandstone. A few Fremont cottonwood trees coexist with narrow-leafed cattails. Red-spotted toads breed in these potholes, and the water is also accessible to wild horses and deer. Violet-green swallows are often seen feeding in these potholes. Red midge larvae, diving beetles, water-loving beetles, and dance flies are associated with the site.

Navajo Canyon Echo Cliffs Spring

This canyon, like Soda Canyon, has been altered by wastewater from the Chapin Mesa sewer lagoon. The native spring is found in a narrow portion of Echo Cliffs. Menefee Shale serves as the aqui-barrier. The water surfaces for roughly 500 m in this deep, narrow canyon. The alluvial bottom accumulates sand to 5 m deep, and for the most part the water lies hidden within this layer. In dry years a few plunge pools, directly below tumbled sandstone boulders, are the only free water. A few cottonwood trees, willows, and baltic rush stands are restricted to the channel bottom, which is about 1.5 m wide between nearly vertical sides 5 m high. No oaks line the waterway, and neither birds nor mammals are attracted to the area. Woodhouse's toads and garter snakes have been seen at the upper end of the water site.

Navajo Dike Springs

This is a spring of a different sort, and the hydrology is not fully understood. A mile-long dike runs from the mouth of Spruce Canyon along the face of Echo Cliffs and then along the west slope of Navajo Canyon before it cuts through the end of Long Mesa. Where this dike is found near the floor of Navajo Canyon, we find a spring of continuous flow. The main spring is at the minette cliff base and is sheltered by dense brush. It is not in an expected location, since it is not in the canyon bottom, but wildlife trails will lead the observer to its location. During the 1996 drought a strange thing happened. Wild horses dug a deep bowl into the alluvium (about 2 m in diameter and nearly 0.5 m deep) in the canyon floor about 30 m out from the dike. During July of that year there was not another source of free water for miles. Perhaps because this site is continuously disturbed by thirsty traffic, the only associated riparian vegetation consists of a few non-native tamarisk trees. Tracks of wild horses and cattle, mountain lions, gray foxes, coyotes, skunks, and rock squirrels have been seen around the water's edge.

Long Canyon Drainage Spring

This water site is much like the Echo Cliff spring, except the channel is not so deep; nor is it fed by introduced water. Long Canyon Spring may run 500 m to a kilometer. During the heat of the summer it is reduced to a few isolated pools. A small flow may resume in late summer. The presence of a Menefee Shale bentonite layer, which runs diagonally across the cuesta from northeast to southwest, establishes the aqui-barrier of the Long Canyon Spring.

Rock Canyon Drainage Spring

Rock Canyon Spring has never been known to dry up. Even during the drought years of 1996 and 2000, it ran a trickle, establishing shallow water on rocks and even a few 30-m-deep pools. The water site begins at a 9 m cliff of sandstone, part of the Menefee Shale, where a plunge pool 1 m deep by 9 m in diameter has formed (the largest plunge pool known within our piñon-juniper study area). The aquifer is once again Menefee Shale bentonite. Accuminate cottonwood trees, coyote willows, virgin's bower clematis, alkali buttercups, and herb willows are found along the spring, which has free water for 500 m. This spring is best known for its large breeding population of red-spotted toads.

LIMEY DRAW POTHOLES

This wonderful series of twelve potholes has formed in a shallow drainage within five acres in the turtle-backed white beach deposit of the Cliff House Sandstone. The two potholes on the mesa rim, above a 12 m drop-off, are shaped like a large cream can. Since they are so deep in the sandstone and so shaded, these two holes never dry up. Red-spotted toads breed here. Occasionally summer thunderstorms ravage the potholes and wash a large part of the tadpoles over the cliff. It is possible that a few survive to reach the Rock Canyon Drainage Spring mentioned previously, but the short journey would be treacherous.

HORSE CANYON DRAINAGE SPRING

This water site is similar geologically and biologically to the Rock Canyon Drainage Spring, which is just a half-mile to the east, separated by Wild Horse Mesa. Both water sites are continuously beaten out by cattle. We cannot be sure what the natural state of biota would be without the continuous hoof traffic, but it is likely that there would be more cottonwood and willow growth if the impact were not so high.

HORSE SPRINGS

Horse Springs is the most mysterious, beautiful spring on the Mesa Verde cuesta. It is on the Ute Mountain Ute Reservation in a side canyon near Rock Canyon. The upper Cliff House Sandstone serves as the aquifer, and a thin layer of clay is the aqui-barrier. This spring is unusual in that it is in a large side canyon that drains to the southeast, whereas most Cliff House Sandstone springs are in side canyons draining to the southwest, which more effectively cut across the dip of sandstone. Horse Springs is in a large alcove, so well shaded that it receives sun only during winter months. The sandstone in the back of the alcove is stained with iron, suggesting intrusion along a fault line. Twelve boxelder trees screen the front of the alcove, with herb willows and thick pads of moss lining the back. The alcove is about 9 m deep and 36 m long, and the ceiling is about 6 m high. In the 1960s the Bureau of Indian Affairs installed a concrete trough through which the crystal-clear water is piped. About three gallons per minute of water overflow from the trough, dump into the sandy floor, and run down the drainage through the boxelders before disappearing into the ground. A trail that undoubtedly goes back to prehistoric times runs up the side canyon from Horse Canyon and is now kept open by feral cattle and horses, which probably travel many miles to this water. Another trail has been partly constructed by humans and descends from the mesa top on the west side into the alcove. Tracks and scats of bear and fox are found along this trail. Violet-green swallows feed over the side canyon in the dense boxelder and oak growth.

Ute Canyon Drainage Spring

This entire portion of the cuesta is Menefee Shale, and the included sandstone layer serves as an aquifer 24 km long. The main outcropping of water begins at the confluence of East and West Ute Canyons, and free water flows for about 8 km along a gravelly bottom that is apparently underlain by clay. During summer thunderstorms there is an unusually high sheet wash off the sparsely vegetated clay slopes, and the flow in the drainage can swell to 5 or 10 cfs. The main drainage is sparsely lined for several miles with cottonwoods. Toads breed in the lower end of the outcrop, where the water is only 2 to 5 cm deep. In 1999 sixteen clutches, each with 500 to 1,500 tadpoles, were recorded (Colyer 2000). The water encounters a thick layer of large alluvial gravel 5 km down the canyon, only 200 m above the Mancos River. Here, too, are cottonwood trees, willows, baltic rush, and sedge, supporting breeding toads.

Pernot Canyon Drainage Spring

In a little narrow canyon is another beautiful spring, one of the few associated with Point Lookout Sandstone. Pernot Canyon is cut through the sandstone in a geological syncline, and in this case the sandstone is near the floor of the Mancos Canyon. The Point Lookout Sandstone is not cemented as hard at the south end of the cuesta as it is elsewhere. The spring is up-canyon from a large rockfall that blocks the entry of cattle into the box canyon head; therefore, this is one of the few drainage springs where the Fremont cottonwood trees are reproducing well. Other associated vegetation includes coyote willow, virgin's bower, violet thelypodium, and common reed, all indicating water that is relatively salt-free. The water is free-flowing, at about 2.5 cm depth, for 500 m. There are intermittent pools that hold 15–20 liters. Several small prehistoric cliff dwellings lie in the canyon near the springs.

Wing Springs

This seep, named after a well-known Ute family, apparently flows year-round from the Point Lookout Sandstone, lying over Mancos Shale. It is on the extreme southern end of the cuesta. Historically a 150-m-long pipe was installed in the spring to carry water down the nearly vertical slope of Mancos Shale to a concrete trough. This must have provided water for livestock during the dry summers when the Mancos River, several kilometers away, ceased flowing. A stand of the locally rare helleborine orchid is found in the sandstone layers near Wing Spring.

EPHEMERAL WATER SOURCES AT YUCCA HOUSE AND SURROUNDINGS

Yucca House Monument Springs

Yucca House, a large prehistoric dwelling, was built around a permanent water site that springs from alluvial gravel and today produces about 8 liters per minute. The source of the water is Ute Mountain. Smaller springs are found on both sides of the main dwelling. The water may have been used for limited irrigation on the Mancos shale clay loam during prehistoric times. On this type of soil a little water can go a long way. The water is high in calcium and sodium, with the pH averaging 7.3. The riparian plant community consists of baltic rush, sedges, tall bulrush, broadleaf cattail, and coyote willow. Before South Montezuma Valley received irrigation water from the Dolores River, this spring certainly was sought after by wildlife and people, because all other natural water sources were kilometers away.

Mud Springs

This spring also arises from alluvial gravels and is fed from Ute Mountain waters. Just as at Yucca House, there was a large prehistoric settlement built around this permanent water source. Today surrounding irrigation has altered the area, making it difficult to see the importance that this perennial water once had for the region.

Mitchell Springs

The town of Cortez, once called Mitchell Springs, began its settlement at this excellent spring, named for the first proprietor of a trading post. The water springs out of a Dakota Sandstone fault line on the side of upper McElmo Canyon and drains from a geological basin known as Cortez Basin. The original spring was a few hundred meters long (Maness 2000). Willows, Fremont cottonwoods, sedges, cattails, and baltic rush grow here. House finches, canyon wrens, Scott's orioles, Bullock's orioles, and other lowland birds are associated with this spring. Historically leopard frogs and tiger salamanders bred here. Kit fox, gray fox, red fox, and mule deer have periodically been common along this shallow riparian canyon. A large prehistoric settlement, the Mitchell Springs Site, lies just south of the spring. Before the Dolores Irrigation System brought water to Cortez, the early European settlers came here by horse and wagon to fill water barrels (Maness 2000).

McELMO CANYON TO CROSS CANYON

Graveyard Canyon Potholes

This is a set of eight unusually deep potholes scoured into the Entrada Sandstone. Two of the potholes are nearly bathtub-sized, holding perhaps 90 liters of water when full. Fremont cottonwood, common reed, and narrow-leaf cattail are found along the bedrock channel, indicating water with relatively low salt content. Evidence of a medium-sized prehistoric settlement is found at the cliff base near McElmo Wash.

Goodman Point Spring

This spring is north of Graveyard Canyon, on the Goodman Point Dome. The spring flows at a rate of 3–9 liters per minute from Dakota Sandstone. Currently the water runs eastward down a small canyon, where it is lined with riparian or semiriparian vegetation, including Fremont cottonwood trees, coyote willows, chokecherry shrubs, skunk brush, Gambel oaks, head-high grasses, and recently the noxious weed Canada thistle. One of the largest prehistoric settlements in southwestern Colorado was built around this spring, attesting to its reliability. The site was occupied until about A.D. 1150. Buried masonry can be found throughout the areas. A small ditchlike feature runs from the spring at a grade contouring along the drainage for 200 m, suggesting a prehistoric irrigation ditch. The little valley above the spring now supports vegetation that indicates an elevated water table, and this area may also have been used for agriculture. Early white settlers came to Goodman Point Springs to fill barrels with water, according to oral histories collected in the area. An old road running north and south of the spring was probably built for this purpose. Even in 1996, the driest period documented in sixty-three years of records, this spring continued to flow (Hovenweep Water Site Records).

Yellowjacket Spring

This water site is similar to the Goodman Point Spring. The spring rises at the head of Yellowjacket Canyon where Dakota Sandstone is exposed. Otherwise the area is thoroughly overlain by red loess soil. The flow is similar to if not stronger than Goodman Point Spring. Here, too, there is a very large prehistoric settlement with an unusually high number of kivas, indicating that this site was a focal point for ceremonial activities. The size of this settlement lends additional evidence for the reliability of Yellowjacket Spring.

Sand Canyon Spring

This water site is also similar to Goodman Point Spring. The water rises at the head of Sand Canyon, where Dakota Sandstone serves as an aquifer from the Goodman Point Dome and perhaps also from the general dip of the beds beginning at the upper end along the Dolores Canyon rim. A similar scenario can be envisioned for this area. The Sand Canyon Spring site has a large prehistoric settlement, as might be expected for a good water site in dry arable land.

Cannonball Mesa Springs

This permanent spring in a 21-m-deep side canyon lies at the southern end of Cannonball Mesa. The drainage runs into Yellowjacket Canyon 1 km from McElmo Wash. Dakota Sandstone serves as the aquifer, and the spring originates at the juncture of the sandstone and one of the interbedded clay layers. There is a short stretch of riparian vegetation consisting of Fremont cottonwood trees, coyote willows, and common reeds, but cattle grazing over the past hundred years have obliterated the potential diversity of the natural flora. The flow ranges from 3 to 15 liters depending on annual precipitation, which averages 25 cm. This is the only spring along the 11-km-long stretch of western Cannonball Mesa. A large three-storied prehistoric dwelling was built on the bedrock along the mesa rim, in a U shape around the spring. Currently cattle ranchers depend on the water, which has been piped down-canyon to prevent erosion of a prehistoric Puebloan village.

Cannonball Potholes

At the westernmost tip of Cannonball Mesa, overlooking the confluence of Yellowjacket Canyon and McElmo Wash, are nine shallow potholes up to 2.7 m in diameter and 0.5 m deep. These hold water for several months during the winter and collect rainwater in most summers during August and September. A 0.5-m-high prehistoric masonry fence still in place across the western mesa tip might be related to guardianship of these potholes. Also nearby is a natural slit in the south cliff face that allows passage from McElmo Canyon, perhaps providing access to the water site. These potholes are important breeding grounds for large, pink fairy shrimp, which are numerous here during wet periods.

Calf Cannon Spring

This spring is similar to Cannonball Mesa Spring. Here the water springs from Dakota Sandstone and supports a flora similar to that of other springs in Dakota Sandstone at about 1,750 m. The flow into Calf Creek may be substantial enough for breeding Woodhouse's red-spotted toads. Fremont cottonwood trees and common reeds grow at the site.

Cahone Spring

This is an interesting and important water site at the westernmost end of Cahone Mesa. As with all springs north of McElmo Canyon, the aquifer is Dakota Sandstone. What makes this spring different, however, is a clay layer about 10 m below the mesa rim, where a 30-m-long totally shaded alcove has formed. The entire ceiling weeps and drips water that feeds into the moss-covered sandy bottom. The water then feeds another 6 m downward until it hits the clay-lined canyon bottom. There a small pool is partially kept excavated by thunderstorm water pouring over the lip. This pool is lined by coyote willows, and dragonflies, damselfies, and water striders live in the water. Mockingbirds, black-throated sparrows, gray foxes, and even a yellow-bellied marmot have been known to use the water. Water tests show the water to be nearly neutral in pH and fairly low in minerals (Hovenweep Water Site Records). It is not surprising that a medium-sized prehistoric settlement was built around the spring and that today this is one of the protected units of Hovenweep National Monument.

Hackberry Spring

As the name implies, the rim of this spring in Hovenweep National Monument is lined by about ten beautiful old hackberry trees. Gray foxes and scrub-jays come to feed on the dry, often abundant fruit. The trees and the overhanging alcove make the site wonderfully cool and shaded. The water seeps out of Dakota Sandstone, pools along the back of the alcove, then feeds through the sand into the canyon head drainage, where the riparian floral community is rich. Broad-leaf cattails, herb willows, Fremont cottonwoods, coyote willows, and alkali muhly grass cover the bottom for over 30 m downstream. Violet-green swallows, house finches, canyon wrens, and rufous-sided towhees feed over the lush area. A great horned owl nests on a cliff ledge a few hundred meters away. Hackberry House is the name of the prehistoric settlement on the mesa top just a stone's throw from the spring.

Square Tower Spring

The headquarters of Hovenweep National Monument is just above this spring. The geology is the same as for the two springs above. The flora includes Fremont cottonwoods and coyote willows. Broomrape and coralroot orchids also are able to live in the canyon head due to increased moisture. Because of constant nearby human traffic we do not see much visitation by larger mammals, but surveys of birds have shown that over fifty species (including cliff swallows, long-eared owls, and Cooper's hawks) nest in the canyon. A complex series of prehistoric rooms and towers was built around the spring.

SUMMARY

Although wild and scenic rivers, such as the Dolores River and the Animas River, easily catch our attention as providers of water in a semiarid woodland setting, less visible, smaller, ephemeral sources are equally important. Because they are interspersed in the piñon-juniper woodland, springs, seeps, and bedrock potholes provide water for a multitude of insects, amphibians, birds, and mammals. Intermittent runoff events swell the side canyons, scour the drainages, and replenish the temporary reservoirs. All water is critical in a dry landscape such as the piñon-juniper woodlands of Mesa Verde Country.

REFERENCES CITED

Colyer, M. 2000. Mesa Verde National Park: Water site records. Mesa Verde National Park Resource Management Files.

Maness, B. 2000. Mesa Verde Oral History Program. Report on file at Mesa Verde National Park.

Mesa Verde and Hovenweep Water Site Records. State Soil Test Laboratory, 1980 to 2000.

Yucca House Water Site Records. State Soil Test Laboratory, 1992 to 2000.

THE CLIMATE HISTORY OF SOUTHWESTERN COLORADO

Nolan J. Doesken and Thomas B. McKee

RECENT CLIMATIC CONDITIONS ARE ESPECIALLY FAVORABLE for germination and establishment of piñon and juniper in Mesa Verde Country (Chapter 2). But the oldest trees recorded in the area are many hundreds of years old, and the climates during their establishment may have been different from today's. Archaeologists have recently synthesized available tree-ring, groundwater, pollen, and packrat midden data to reconstruct the climate during the buildup and decline of Ancestral Pueblo populations from A.D. 900 to 1300. This period included several especially persistent droughts (Van West and Dean 2000) that affected the piñon-juniper woodland as well as the potential for agriculture. Past and future climate scenarios and their implications for piñon and juniper success are discussed in Chapter 22. Here we focus strictly on the recent instrumental climate history, the period during which objective measurements of climatic elements (such as temperature and precipitation) have been taken using standard meteorological instruments. Nationally this represents little more than 150 years (Fleming 1990)—a mere blink of the eye in comparison with the time scales over which archaeologists, paleoclimatologists, and geologists have identified and described past climate changes. For the many remote areas of the western United States, such as Mesa Verde Country, instrumental climate histories are even shorter.

Although brief, the instrumental climate record is essential for providing a reference platform from which to describe the present climate. It then becomes possible to compare climates of the past and to imagine climatic conditions of the future. The instrumental climate record also offers the opportunity to evaluate interannual and interdecadal variations that are a challenging but normal part of our climate system. Such patterns provide essential information relevant to the sustainability of piñon and juniper woodlands in Mesa Verde Country.

DATA AVAILABILITY

The Colorado Climate Center at Colorado State University, Fort Collins, routinely monitors climatic conditions throughout Colorado and evaluates current conditions with respect to what has been observed throughout the instrumental record. The earliest instrumental climate records in Colorado were begun in 1852 at Fort Massachusetts in south-central Colorado (Doesken et al. 1991). The United States Army Post at Fort Lewis was the earliest weather station in southwestern Colorado, with observations beginning in 1880 (Bradley and Barry 1973). After 1890 the number of weather stations increased as the newly formed civilian United States Weather Bureau made nationwide climate monitoring a priority to support the growth of agriculture. Volunteer weather observers were recruited, supplied with standard instruments, and trained. Most weather stations were equipped with eight-inch-diameter metal precipitation gauges and self-registering liquid-in-glass maximum and minimum thermometers mounted inside wooden instrument shelters (National Weather Service 1989). Very few changes have been made in this network over the past century, and the data from the NWS cooperative network continue to be the best overall data source for climate monitoring in the United States.

The number of stations increased significantly during the early 1900s but leveled off after 1960. Many stations have come and gone after collecting only a few years to a few decades of data. Ten stations collected data for sixty years and are currently in operation (Table 15.1). All analyses in the following sections are based on data from these long-term sites.

PRESENT CLIMATIC CHARACTERISTICS

Data for the period from 1962 to 1992 were used to identify key features of the current climate of southwestern Colorado. The most consistent feature of the climate is the annual temperature cycle. Monthly averages of the daily maximum and minimum temperatures (Figure 15.1) show the traits common to the interior, continental, and mid-latitude regions, with air temperatures lagging behind the annual solar cycle by a few weeks. On average January is the coldest month of the year, with daytime temperatures typically reaching into the 30s and 40s (° F) across the region and with nighttime temperatures averaging from the single digits to the low 20s. July is normally the warmest month of the year, with highs typically in the 80s and 90s and with lows from the upper 40s to the 60s. Figure 15.2 is based on daily averages and extremes for Cortez and Mesa Verde. The annual temperature wave is close to sinusoidal but has a more gradual temperature rise in the

Table 15.1. Long-term National Weather Service climate stations in southwestern Colorado (currently in operation and with at least sixty years of consistent data)

Station	Elevation (m)	Years of Record for Temperature	Years of Record for Precipitation
Cortez	1,882	64	64
Durango	2,000	98	98
Fort Lewis	2,303	86	92
Fruita	1,357	94	94
Grand Junction	1,467	101	101
Ignacio	1,957	79	79
Mesa Verde National Park	2,156	70	70
Montrose	1,753	91	105
Northdale	2,024	63	63
Norwood	2,127	62	62

Note: Data summaries from 1992.

spring and a more rapid decline in the autumn. Larger day-to-day variations in temperature occur in winter than in summer, related to more frequent changes in air masses and cloud conditions. Day-to-day temperature changes in the Four Corners area, however, are not nearly as great as the variations observed east of the Rocky Mountains.

Mean monthly and annual temperatures generally decrease with altitude in this region, but local topographic effects related to cold air trapping and inversion formation affect this relationship. For example, much of Mesa Verde is unusually mild for its elevation. In general, temperature decreases with elevation are most consistent during the daytime and especially during the summer. A decrease of approximately 5° F per 303 m is typical for summer days. Other factors besides elevation become increasingly important at night and especially during the winter.

Diurnal temperature ranges (day-night temperature differences) are large throughout southwestern Colorado but show considerable local variation (Figure 15.3) based on topography. June has the largest day-night temperature differences, while December typically has the least. The weather station on Mesa Verde has smaller diurnal ranges than any other reporting station in the region, with day-night differences of only about 20° F.

Average annual precipitation over southwestern Colorado ranges from less than 20 cm near Grand Junction to more than 75 cm in the San Juan Mountains. Precipitation generally increases with elevation, but elevation is not the only controlling variable. More than 75% of the annual precipitation falls as rain at lower elevations. Above about 2,576 m the majority falls as snow.

Precipitation patterns are much more complex and much less consistent from year to year than temperature. There are three primary precipitation mechanisms

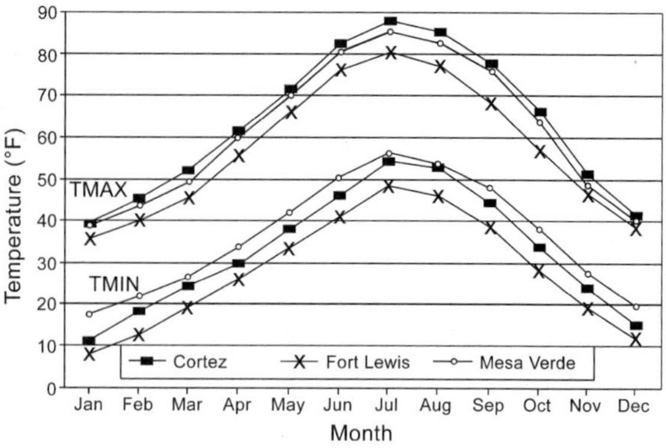

Figure 15.1. *Average monthly maximum and minimum temperatures (°F) for selected southwestern Colorado weather stations—Cortez, Fort Lewis, and Mesa Verde—based on 1962–1992 data.*

influencing southwestern Colorado: large-scale cool-season storm systems that approach from the Pacific (November–April) and are greatly enhanced by orographic lifting, summer convective precipitation that becomes heaviest and most widespread when monsoon moisture spreads into the region from the south (early July–early September), and autumn storm systems that combine tropical air masses and the remains of occasional Pacific hurricanes with early-autumn, mid-latitude storms. Each of these mechanisms operates nearly independently, although some connection to the Pacific El Niño circulation can be traced.

The resulting pattern of mean seasonal precipitation (Figure 15.4) shows a late summer and fall precipitation maximum at most sites. At the lower-elevation sites near the Utah border, October is the wettest month of the year. This seasonal pattern is unique to the Colorado Plateau region. Winter maximums occur nearer the San Juan Mountains. Throughout the Mesa Verde Country extremely dry Junes are a reliable feature of the climate.

Interannual precipitation variability is a key feature of the climate over southwestern Colorado and is greater in this region than in most of the rest of the western United States (Changnon et al. 1991). It is very common to have alternating seasons and years with much-above-average and much-below-average precipitation. A large percentage of any year's precipitation falls from a small number of storms (typical for most arid and semiarid regions). Great variability is produced by a combination of factors related to being south of the primary winter mid-latitude storm track and on the north end of the summer and fall tropical moisture intrusions. Distributions of monthly precipitation are typically skewed in all climates but especially in dry climates with highly variable precipitation. Figure 15.5 shows median versus mean precipitation for selected stations. For many purposes, the median provides a better description of the climate. Medians are less than means for all stations and for nearly all months. Median precipitation for

The Climate History of Southwestern Colorado

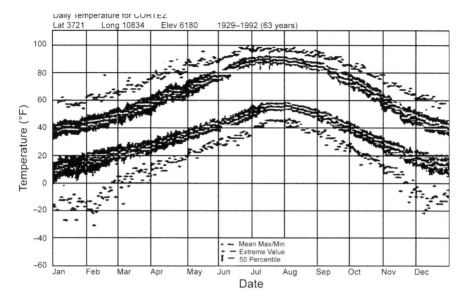

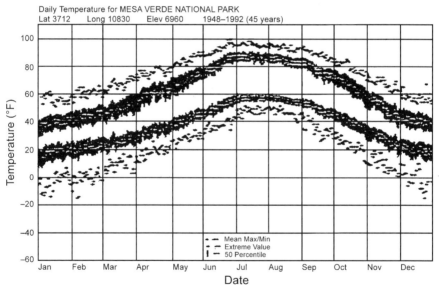

Figure 15.2. Average daily maximum and minimum temperatures, typical variability (vertical lines), and daily extremes for Cortez and Mesa Verde National Park.

a given month is typically about 10% less than the mean. Means and medians are quite similar during July and August, however, and tend to be the most different during September and October.

251

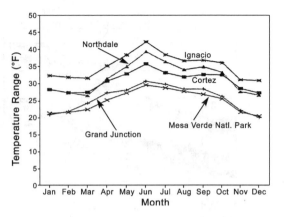

Figure 15.3. Monthly average diurnal temperature range (difference between the average daily maximum temperature and average daily minimum temperature) for selected locations in southwestern Colorado, based on 1962–1992 data.

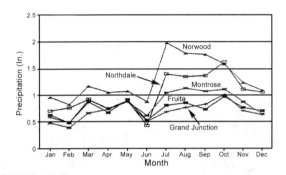

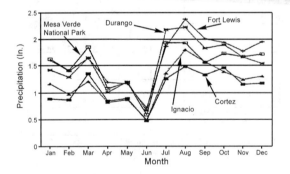

Figure 15.4. Average monthly precipitation in inches for each of five stations with long-term data in Mesa Verde Country, based on 1962–1992 data.

OBSERVED CLIMATIC DIFFERENCES

With barely a century of instrumental observations at our disposal, interannual and interdecadal variations dominate the climate record and challenge any attempts at deducing long-term trends. Furthermore, superimposed on the actual

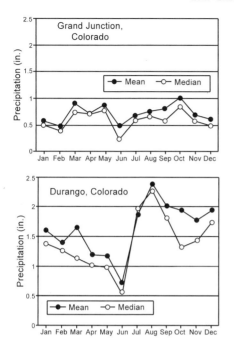
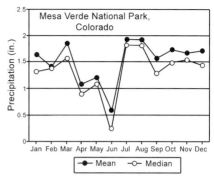

Figure 15.5. Comparison of monthly average precipitation and median precipitation for Grand Junction, Mesa Verde, and Durango, based on 1962–1992 data.

climate signal are variations and discontinuities in the climate record resulting from any of several possible observational issues. As such, we prefer to analyze "climate differences" rather than "climate change." To look at climate differences, the instrumental record was divided into three nonoverlapping 31-year periods: 1895–1925, 1926–1956, and 1962–1992. These periods were chosen arbitrarily but are consistent with available data. We used 31-year periods, rather than the 30-year interval commonly used to establish climate normals, simply for convenience in determining median values.

Only three stations had complete data for all three periods: Grand Junction, Montrose, and Durango. Here we focus only on the Durango station's 31-year average monthly temperatures and precipitation for the three periods (Figure 15.6). Small but observable differences in 31-year monthly temperature averages can be seen. Durango data show 1 to 2.5°F higher average maximum temperatures throughout the year than in the past, but a nonsignificant change in average minimum temperatures is detected. Temperatures at this site (as well as two others in Colorado in which long-term data are available) fail to show the consistent reduction in diurnal temperature range that has been noted for much of North America (Karl et al. 1986).

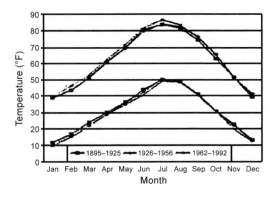

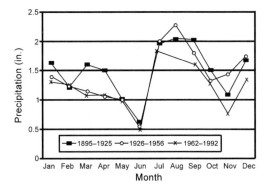

Figure 15.6. Comparison of 31-year average temperatures (top) *and 31-year median precipitation* (bottom) *for Durango for three periods, 1895–1925, 1926–1956, and 1962–1992.*

Not surprisingly, some significant differences in precipitation have been observed. The Durango data show that the wettest period of the instrumental record was 1895–1925. That early period included wet early springs that have not been paralleled in subsequent years. The 1926–1956 period was the driest portion of the record: precipitation was significantly lower from late summer through early winter. The most recent period, 1962–1992, has been characterized by wet summers and wetter early winters (November–December) but drier springs. Changes in the seasonal distribution of precipitation, in combination with subtle temperature changes, could produce significant changes in the overall surface water balance of the region that might be reflected in changes in germination and establishment success of piñon and juniper populations. A comparison of the seasonal distribution of precipitation at the Mesa Verde National Park long-term climate station for the twenty-one individual (not consecutive) wettest and driest years (based on calendar year totals) is shown in Figure 15.7. While the general seasonal distribution of precipitation has been similar in wet, dry, and average years, the wettest years have been characterized by a higher percentage of the annual precipitation falling during the winter months.

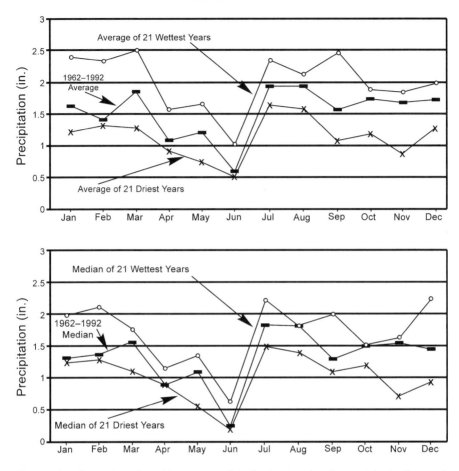

Figure 15.7. Comparison of monthly average precipitation (top) *and median precipitation* (bottom) *for Mesa Verde National Park for the twenty-one wettest and twenty-one driest calendar years, based on 1962–1992 data.*

Frequencies of heavy precipitation events and flooding were not included in this study. Instead a time series of a drought monitoring index developed for Colorado (McKee et al. 1993) called the Standardized Precipitation Index (SPI) is shown in Figure 15.8. The SPI requires only monthly precipitation values for computation but needs consistent and long-term data. Index values of –1 or lower represent periods of drought, while periods above +1 are very wet. The Durango station shares most of the very dry and very wet periods in common with the other Western Slope stations, Grand Junction and Montrose. Overall the worst drought period took place from 1899 to 1903. There was almost no drought

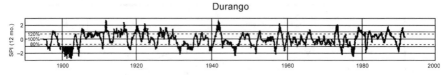

Figure 15.8. Time series of the Standardized Precipitation Indexes (SPIs) computed for consecutive overlapping twelve-month periods for Durango.

from then until the early 1930s. Several droughts have occurred since 1950, but all have been short in duration, including the severe drought of 1976–1977.

The type of information in Figure 15.8 is very helpful in explaining typical characteristics of drought and precipitation variability. Even with 100 years of consistent data, however, it is impossible to assess probabilities for long-duration severe wet or dry episodes with confidence.

SUMMARY

The instrumental record for southwestern Colorado dates back just over a century. Only ten stations have at least sixty consecutive years of temperature and precipitation data. Nonetheless, the data resources are sufficient to define many aspects of the climate of the Four Corners area. Minor temperature differences have been observed from the earliest instrumental records until the present. Most areas show slightly warmer temperatures now than at the beginning of the twentieth century, but with little change in the shape of the annual cycle. In terms of precipitation, southwestern Colorado experiences large interannual variations, which, in turn, have introduced considerable variations in monthly averages over longer periods. The wettest periods in the region occurred from approximately 1904 to about 1930 and during the mid-1980s. Drought was most prevalent from 1889 to 1903 and sporadically but with short duration since the 1930s. Some differences in the seasonal pattern of average precipitation across the region have been observed from the early record until the present. These trends in climatic fluctuation form the backdrop upon which we can begin to interpret fluctuations in piñon and juniper woodland establishment (Chapter 2).

The instrumental record offers a wonderful resource for evaluating the climate. Data users must be aware, however, that instrumental observations do not provide the absolute truth but simply a reasonable representation of the truth. With a relatively small commitment of time and resources, it should be possible to continue and improve consistent instrumental climate monitoring into the future. Every effort should be made to make sure that climate monitoring continues.

REFERENCES CITED

Bradley, R. S., and R. G. Barry. 1973. Secular climatic fluctuations in southwestern Colorado. *Monthly Weather Review* 101: 264–270.

Changnon, D., T. B. McKee, and N. J. Doesken. 1991. Hydroclimatic variability in the Rocky Mountains. *Water Research Bulletin* 27: 733–743.

Doesken, N. J., T. B. McKee, and S. Hersh. 1991. *Colorado Centennial, 1891–1991*. Colorado State University, Fort Collins.

Fleming, J. R. 1990. *Meteorology in America, 1800–1870*. Johns Hopkins University Press, Baltimore.

Karl, T. R., G. Kukla, and J. Galvin. 1986. Relationship between decreased temperature range and precipitation trends in the United States and Canada, 1941–1980. *Journal of Applied Meteorology* 25: 1878–1886.

McKee, T. B., N. J. Doesken, and J. Kleist. 1993. The relationship of drought frequency and duration to time scales. *Preprints, 8th Conference on Applied Climatology,* American Meteorological Society, January 17–22, Anaheim, Calif., pp. 179–184.

National Weather Service. 1989. *Cooperative Station Observations, Observing Handbook No. 2*. United States Department of Commerce, Silver Spring, Md.

Van West, C. R., and J. S. Dean. 2000. Environmental characteristics of the A.D. 900–1300 period in the central Mesa Verde region. *Kiva* 66: 19–44.

PART III

WHAT PROCESSES OF CHANGE ARE OCCURRING IN MESA VERDE COUNTRY?

THE AUTHORS IN THIS SECTION EXAMINE SOME OF THE MAJOR FORCES OF CHANGE in Mesa Verde Country, beginning with physical processes (fire) and ending with the effects of that most ecologically influential species—*Homo sapiens*.

Many people are aware of the large fires that have occurred in Mesa Verde National Park in recent years. Chapter 16 puts these fires into a historical and ecological context by showing that fires have long been a natural part of the Mesa Verde environment. Fires were ignited by lightning, possibly augmented by indigenous humans, and have played a role almost as important as the roles of soil and microclimate in shaping the vegetation patterns. For example, the extensive shrublands at the north end of the Mesa Verde cuesta are largely the result of centuries of fires that favored shrubs but excluded trees. In contrast, the dense piñon-juniper woodlands at the southern end of the cuesta have not burned in the last 700 years. Chapter 17 further examines the ecological role of fire, with a particular emphasis on the insects of Mesa Verde.

The remaining chapters deal with human effects on the natural ecological systems of Mesa Verde Country. Chapters 18 to 20 examine the remarkably diverse ways in which the Ancestral Puebloan people interacted with their environment, including the piñon-juniper woodlands, over a period of many centuries. The woodlands supported the people, who in turn profoundly influenced the woodlands. Finally, in Chapter 21 we see the activities and influences of the most recent human inhabitants of the area: the Euro-American settlers who began arriving in numbers in the late 1800s. Although they have been here a relatively short time, their impact perhaps has been the greatest of any of the people who have ever inhabited Mesa Verde Country.

FIRE HISTORY

M. LISA FLOYD, WILLIAM H. ROMME, AND DAVID D. HANNA

WHEN TRAVELING THROUGHOUT MESA VERDE COUNTRY, the visitor may be struck by a fairly consistent pattern in the landscape in which piñon-juniper woodlands give way (often at slightly higher elevations) to tall, dense shrublands. These shrublands, known as petran chaparral, may or may not be interspersed with scattered conifers. The dynamic interaction between piñon-juniper woodlands and these petran chaparral communities in the Mesa Verde Country has much to do with natural disturbances, yet we know little about the disturbance history and successional dynamics that result in vegetation patterns seen today. We know that fire is a prominent disturbance agent, a phenomenon so graphically displayed during 1996 (Figure 16.1) and 2000 wildfires on the Mesa Verde cuesta, near Hovenweep, and in Disappointment Valley. As in previous chapters, we focus on Mesa Verde because we know more about recent fire events here than elsewhere in the study region; indeed several of the region's large wildfires of the last decade have occurred on the Mesa Verde cuesta.

Since the establishment of Mesa Verde National Park in 1906, the principal goal has been preservation of archaeological and other cultural resources. Therefore, the park has always had a policy of complete fire suppression. Records of fire occurrence have been kept since 1926. From 1926 to 1977 the number of fire starts per year ranged from 0 to 36, with an average of 8 fires/year. Lightning caused 90% of fire starts and humans 10%; 90% of all fires burned less than 0.1 ha. Only 2% of the fires burned more than 4 ha, but these larger fires (Figure

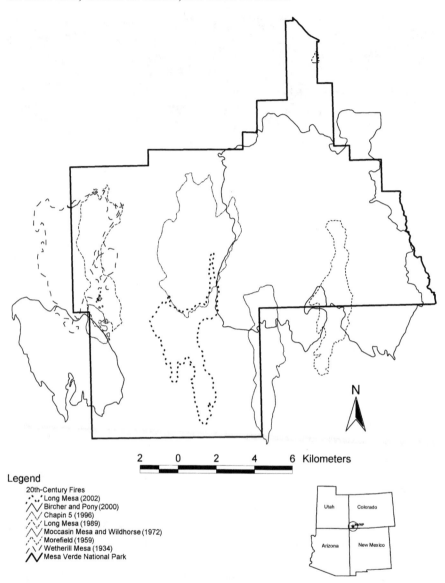

Figure 16.1. Mesa Verde National Park, in the heart of Mesa Verde Country, experienced five landscape-scale fires from 1934 to 1996 (fire outlines are shown).

16.2 and Table 16.1) have accounted for over 95% of the total area burned in the twentieth century, a pattern that is not unique to the Mesa Verde Country (Johnson and Gutsell 1994; Moritz 1997). All of the large fires prior to 1990 occurred during the typical dry period of late June and early July. The recent large fires

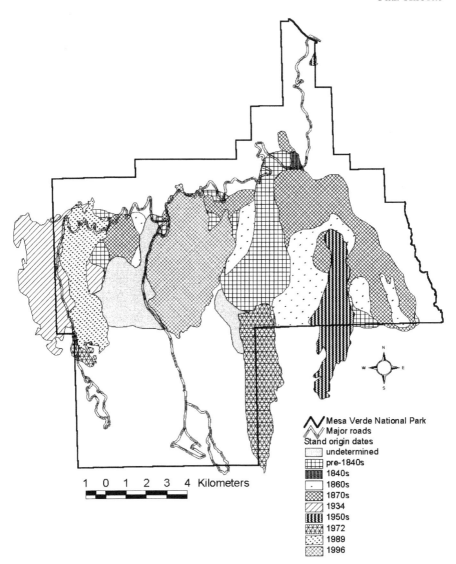

Figure 16.2. The stand origin map of prehistoric fires and historic fires, Mesa Verde National Park (see the text for details of map derivation).

(Chapin 5 fire in 1996, Bircher and Pony fires in 2000, and Long Mesa 2 fire in 2002) all occurred in late July and August, however, a time of year that is typically moist with the summer monsoon but was unusually dry in 1996, 2000, and 2002. Throughout the twentieth century all large fires occurred under conditions of

Table 16.1. Historically documented large fires (> 50 ha) in Mesa Verde National Park since 1931

Fire Year	Fire Name	Extent of Fire (hectares)
1934	Wetherill Mesa	1,934
1959	Morefield	865
1972	Rock Springs	161
1972	Moccasin Mesa	980
1989	Long Mesa	1,800
1996	Chapin 5	1,972
2000	Bircher	10,592
2000	Pony	2,272
2002	Long Mesa 2	1,054

Note: Data are from the park's fire atlas or (since 1996) polygons mapped with GPS.

high winds and prolonged drought. These were severe fires that burned through the crowns of trees and shrubs, killing all or most aboveground vegetation.

Mesa Verde has experienced eight large, intense fires since the 1930s, including four in the last decade (Table 16.1; Figure 16.2). These fires burned through both piñon-juniper and chaparral vegetation and were uncontrollable even with modern firefighting technologies. They destroyed several irreplaceable cultural resources, including an ancient rock art panel, and threatened numerous other cultural features. By monitoring the recent fires and making observations of the older ones, we begin to construct a picture of postfire vegetation recovery.

The natural heterogeneity of the burned Mesa Verde landscape (mesa tops, canyon walls, meadows) supported several different postfire trajectories. Vegetation recovery following the 1989 and 1996 fires depended heavily upon the prefire vegetation types. Piñon pines and the two species of juniper common in our region are fire-sensitive and are easily killed even by low-intensity burns; reestablishment is solely by seed (Fischer and Clayton 1983; Floyd 1982, 1986; Koniak 1985). Where piñon-juniper woodlands burned in Mesa Verde during the twentieth century, piñon and juniper have not become reestablished, for unknown reasons, and these areas have been converted into shrublands (BAER Report 2000; Barney and Frischknecht 1974; Everett and Ward 1984; Koniak 1985). Many of the major shrub species, including Gambel oak *(Quercus gambelii)*, serviceberry *(Amelanchier utahensis)*, snowberry *(Symphoricarpos oreophilus)*, fendlerbush *(Fendlera rupicola)*, and skunkbush *(Rhus trilobata)*, resprout prolifically from roots, rhizomes, or lignotubers after fire kills aboveground portions of the plants (Brown 1958; Clary and Tiedemann 1986; Koniak 1985; Noste and Bushy 1987; Wright et al. 1979). Other shrub species are killed by fire and reestablish solely by seed, including big sagebrush *(Artemisia tridentata)* and black sagebrush *(Artemisia nova)* (Blaisdell et al. 1982; West and Hassan 1985; Wright et al. 1979). Knowing these responses to past fires gives us critical information to use in interpreting vegetation communities for which we have no direct evidence of disturbance, as shown in the following discussion.

Reconstructing long-term fire history in piñon-juniper woodland is challenging because the major species rarely form "fire scars," distinctive, nonlethal injuries that are common in ponderosa pine and other thick-barked trees and that allow precise determination of the year in which a fire occurred. Fire scars expose the wood (xylem) tissue when the living vascular cambium is killed by the fire. As growth of the cambium resumes in the nonburned portion of the stem, the dead portion remains visible; and the year and season of the fire can be precisely determined through dendrochronological sampling. So we searched for other types of evidence for past fires. Written fire records were available only for the last seventy years. Therefore, we needed to devise new methods to determine fire history in landscapes dominated by piñon-juniper and petran chaparral.

In this chapter we describe a method that we developed for reconstructing prehistoric fire history in the petran chaparral communities of Mesa Verde, based on the current age structure of resprouting Gambel's oak. We used this method to detect and map major fires since the 1840s and to estimate turnover times within the shrublands that dominate approximately half of the park area. We also used indirect methods, based on age structure of piñon and juniper, to estimate time since fire and turnover times in piñon-juniper woodlands that cover most of the remainder of the park. The quantitative estimates of fire history in Mesa Verde that we present in this chapter are based on our work through 1998. After this work was completed, wildfires burned the Mesa Verde cuesta in the dry widespread wildfire season of 2000 and 2002. In the final section of this chapter, these recent fires are discussed in light of our recently completed fire history studies.

A METHOD FOR DETERMINING PREHISTORIC FIRE HISTORY IN PETRAN CHAPARRAL

Based on our observations of prolific shrub resprouting following fires, we developed a method that enabled us to reconstruct prehistoric fire history (Floyd et al. 2000). First, we sampled the oldest oak clones throughout the northern portion of the mesa and determined their ages. The dating of postfire oak stem cohorts permits detection of past fires up to about 150 years ago, give or take a decade. We then developed a map showing the decade in which the last severe fire had occurred throughout the shrublands of Mesa Verde National Park. The areas burned each decade were used to estimate the time required to burn an area equal to the entire petran chaparral zone within the park and the expected interval between successive fires at each point in the landscape. Approximately half of the petran chaparral in Mesa Verde had burned during the second half of the twentieth century (Figure 16.3). If this rate of burning continues, it will take about 100 years for an area equal to the entire petran chaparral zone to burn (i.e., turnover time). By constructing a time-since-fire cumulative curve (Floyd et al. 2000), we showed that the slope of the curve (an indication of turnover time) is comparable for the late-twentieth- and late-nineteenth-century periods, so the turnover time for the fire regime of the late nineteenth century also was about 100 years. During the first half of the twentieth century, however, there was only one large fire (in 1934), and the turnover time was longer.

Figure 16.3. Fire in the piñon-juniper zone is particularly intense, as shown here during the Chapin 5 fire in 1996.

Second, we tallied previous fire dates for the sampled stands falling within the perimeters of the 1989 and 1996 fires. By subtracting the date of the previous fire from either 1989 or 1996 (depending on which of these recent fires burned a given stand), we summarized the ages of the stands burned in these recent fires. This gave us a sample of intervals between successive fires at individual points within the landscape. The median age of stands burned by the fires in 1989 and 1996 was about 110 years.

Both of these methods yielded a similar interpretation: any spot within Mesa Verde's shrubland is likely to burn about once every century. In more technical language (a necessary evil because terminology in fire is very confusing), we conclude that the median interval between successive fires at any point in the landscape, and the time required to burn an area equal to the entire extent of the petran chaparral in Mesa Verde (turnover time), is about 100 years under the fire regime that characterized the last half of the twentieth century. Some portions of the chaparral have not burned for over 150 years, but other areas have burned two or even three times during that period. Fire occurrence was significantly reduced during the first half of the twentieth century, but turnover time during the mid- to late nineteenth century apparently was comparable to what we saw in the late twentieth century.

REASONS FOR DISTINCT PERIODS OF FIRE HISTORY IN MESA VERDE SHRUBLANDS

The late-nineteenth-century period may represent a more or less "natural" fire regime in the petran chaparral of Mesa Verde. The main local inhabitants at that time were the Utes; they are known to have done some intentional burning in the general region, but their specific uses of fire are poorly understood (Callaway et al. 1986; Duke 1995; Ellis 1996; Marilyn Colyer, personal communication). Even if the Utes did ignite some fires, the high frequency of lightning ignitions in Mesa Verde and the strong relationship between weather conditions and fire size during the recent period of accurate fire records (Omi and Emrick 1980) suggest that lightning ignitions alone may be sufficient to account for the extensive fires that we documented in the late nineteenth century. European settlers and livestock began arriving in southwestern Colorado in the 1870s (Dishman 1982), but the inaccessible and poorly watered Mesa Verde was not homesteaded until the end of the nineteenth century (Smith 1988).

The early-twentieth-century period of little fire activity probably resulted from two European influences on the ecosystems of Mesa Verde. First, even though people did not attempt to settle this rugged area permanently until the turn of the twentieth century, they were grazing livestock in the Mesa Verde cuesta as early as the 1880s (Smith 1988; Chapter 21). Reports and photos from 1906 to 1935 in the park archives indicate that by the early twentieth century, cattle were numerous in the park, and heavy grazing was occurring in many areas. Heavy livestock grazing removed fine herbaceous fuels that formerly carried fires. The removal of the fine fuels was a major cause for the dramatic drop in fire frequency and fire size that occurred throughout much of the West in the late nineteenth century (Smith 1988). The second important influence during the early twentieth century was an improvement in methods of fire detection and suppression following the large fire of 1934.

Although the park still has a policy of complete fire suppression, it was not successful in preventing extensive fires during the second half of the twentieth century. Despite a lull in fire activity during the first half of the twentieth century, the fire history of the second half-century was not much different from the fire history of the late nineteenth century, when there was no fire suppression at all (Floyd et al. 2000). There may be two reasons for the dramatic increase in fire activity during the last fifty years. First, livestock grazing was phased out in the park from the 1910s through the early 1930s (park records), and plant growth during the long, relatively fire-free period of the early twentieth century may have produced fuel mass and continuity sufficient to carry the large fires of the late twentieth century. Second, there have been several summers during the last half-century when ignitions occurred under severe fire weather conditions, and it has been impossible to control the fires even with modern firefighting technology. This may be related in part to patterns in global-scale weather phenomena, such as the strong variation in the effects of the El Niño–Southern Oscillation system since 1976. This regional climatic pattern has significantly influenced fire frequency and fire effects throughout the Southwest (Swetnam and Betancourt 1998).

FIRE HISTORY IN PIÑON-JUNIPER WOODLANDS OF MESA VERDE

Our fire history information for piñon-juniper woodlands, based on stand ages, is less precise than our spatially explicit reconstruction of shrubland fire history. Nevertheless, it is apparent that intervals between successive fires at any point are generally much longer in the woodlands than in the shrublands. Most of the piñon-juniper woodlands in the southern portion of Mesa Verde contain large, old trees and lack charred snags. We aged one stand, the Glades, that retained evidence of past fire in the form of charred snags and abundant charcoal. This stand was over 270 years old—the fire that created all of the snags and charcoal had occurred back in the 1720s! Thus we assume that stands without such evidence have escaped fire for at least 270 years and may be as old as two stands sampled on Chapin and Wetherill Mesas that were nearly 500 years old. In fact, we suspect that some of the woodland stands in Mesa Verde have not burned since the emigration of the Ancestral Pueblo people 700 years ago, meaning that these are some of the oldest stands in the Four Corners region.

To compare fire regimes in piñon-juniper woodlands and shrublands of Mesa Verde quantitatively, we derived two estimates of turnover time (i.e., the time required to burn an area equal to the total area of woodland in the park) from our measurements of woodland area that has burned in the twentieth century. These estimates were based on two different reference periods and refer to the national park only:

1. In the last century (1898–1998) 1,729 ha of woodland have burned, out of 10,563 ha estimated to have existed in 1898 (i.e., 16% of the original woodland area burned in the twentieth century). If this rate of burning continued, 600 years would be required to burn an area equal to the area of woodland in 1898.
2. In the last half-century (1948–1998) 1,336 ha of woodland have burned, out of 10,170 ha that existed in 1948 (i.e., 13% in 50 years). At this rate, 380 years would be required to burn an area equal to the area of woodland in 1948.

The second estimate (ca. 400 years, based on the period 1948–1998) probably is our best estimate of the "natural" fire turnover time in woodlands of Mesa Verde, because our more precise data from the shrublands indicate that the fire regime of the mid- to late nineteenth century (which we regard as the "natural" fire regime for this area) was similar to that of the late twentieth century.

FIRE HISTORY AND VEGETATION PATTERNS IN MESA VERDE

Piñon-juniper woodlands and petran chaparral in the northern portion of the Colorado Plateau and southern Rocky Mountains (including Mesa Verde) appear to be characterized by a fire regime of infrequent but severe (stand-replacing) fires. Nearly all of the historically documented fires of the twentieth century that occurred in Mesa Verde either went out naturally before burning more than several square meters or spread uncontrollably through the crowns of trees and shrubs, covering hundreds or thousands of hectares. Extensive low-intensity surface fires spreading through surface fuels without killing the trees generally have not been observed in this area; and the dominant tree species, Colorado piñon pine (*Pinus edulis*) and Utah juniper (*Juniperus osteosperma*), are rarely found with fire scars. Mean fire intervals and turnover times appear substantially shorter in the chaparral vegetation than in the piñon-juniper woodlands. In Mesa Verde we estimated a "natural" fire turnover time of approximately 100 years in shrublands, compared with about 400 years in woodlands.

The differences that we detected between shrublands and woodlands cannot be explained by differences in frequency of fire starts. In Mesa Verde more lightning ignitions occur each year in piñon-juniper woodlands than in shrublands (Omi and Emrick 1980). Regardless of where they start, however, most fires extinguish naturally before burning more than an individual tree or several square meters of ground. Most of the large fires observed in the twentieth century started in piñon-juniper woodlands, then spread into shrublands, where they burned a large area.

We suggest two possible reasons for greater occurrence of extensive lethal fire and hence shorter turnover times in the shrublands of Mesa Verde. First, there are differences in the fuel complexes of the two major vegetation types (Bruner and Klebenow 1979; Floyd-Hanna et al. 1996). Even though piñon-juniper woodlands have very heavy fuel loads capable of tremendous heat release under certain weather conditions, the horizontal continuity or connectedness of fuels in this vegetation type is low. Fine fuels (twigs and needles) and herbaceous cover are generally sparse, and there are many patches of rock or bare soil up to several meters in width containing little or no fuel interspersed between the comparably sized patches of thick trees, brush, and heavy downed woody material (Barney and Frischknecht 1974). Thus fires in piñon-juniper woodlands tend to be limited to individual patches of fuel and fail to spread over a large area except under conditions of extreme drought and sustained high winds—conditions that allow the fire to spread through the tree canopy, such as occurred in 1989, 1996, and 2000 (Bruner and Klebenow 1979; Omi and Emrick 1980). In contrast, the shrublands contain dense patches of shrubs underlain by leaf litter, and the spaces between the shrub patches are filled with forbs and grasses that create a nearly continuous light fuel layer. Heat release in the shrublands generally is lower than in the piñon-juniper woodlands, but extensive fire spread can occur under dry conditions without the need for high wind.

The second reason for more extensive fires in shrublands than in woodlands, at least in Mesa Verde, may be related to the landscape position of the two vegetation types. The woodlands are largely restricted to the southern portion of the top of the cuesta, with cliffs and sparsely vegetated slopes to the south and west. Fires that started to the west and south would be unlikely to spread into the woodlands because the cliffs create barriers to fire spread, and fires ignited to the north or east would not tend to move into the woodlands because the prevailing winds are from the west and south. Thus fires that started within the woodland itself tend to be the ones that burned the woodland, and then only under high-wind conditions. In contrast, the shrublands are located in the higher northern and eastern portions of the cuesta, with natural fire barriers only on the north. Hence fires starting almost anywhere on the cuesta (especially in the southern and western portions) could spread over a large portion of the shrublands. The importance of natural barriers to fire spread is further suggested by our observation that many of the small patches of piñon-juniper woodland within the extensive shrubland portion of Mesa Verde are located just north or east of a small cliff or steep south-facing slope with little vegetative cover.

ASSESSING THE FIRES OF THE 2000s

In the hot, dry summers of 2000 and 2002, Mesa Verde went up in flames. In 2000 two separate large fires together burned more area within the park than all of the fires of the previous century combined. The Bircher fire was ignited by lightning to the north and east of Mesa Verde on July 20, escaped control, and burned into the park's eastern flank. Over nine days it burned most of the eastern third of Mesa Verde National Park—a total of 10,592 ha inside and outside of the park. The Bircher fire was barely extinguished when lightning ignited another fire in the Ute Mountain Tribal Park on the west side of the park on August 2. This second large fire, the Pony fire, burned into the southwestern portion of Mesa Verde National Park and reached a final size of 2,272 ha before it was finally extinguished ten days later. Only two years later 1,054 ha burned on Chapin Mesa, damaging several homes and numerous park facilities as the fire swept through from July 29 through August 1. This fire was misnamed the Long Mesa fire although it never reached Long Mesa. To distinguish it from the 1989 Long Mesa fire, we call it Long Mesa 2 here.

The occurrence of two enormous fires in a single summer caused us to re-evaluate our interpretations of the natural—or unnatural—fire regime in Mesa Verde based on the fire history of the previous 150 years. At first glance it appeared that the magnitude of the fires of 2000 was unprecedented in the park's ecological history. It also seemed that our earlier conclusion of a more or less natural fire regime still prevailing in the Mesa Verde ecosystem was unfounded.

These initial impressions led us to reevaluate our fire history data and interpretations. In the process we discovered that our previous analyses had focused on too small an area to understand the patterns and long-term processes of large, severe fires in piñon-juniper and petran chaparral ecosystems of southwestern Colorado.

Of necessity, we had restricted our previous fire history analyses to the boundaries of Mesa Verde National Park itself, simply because no data existed for areas outside the park. We had recognized all along, however, that the park covers less than one-half of the larger natural landform that we call Mesa Verde. The portions of Mesa Verde that lie outside the national park are covered by similar vegetation and landforms. We obtained satellite imagery of the entire Mesa Verde cuesta, including both Mesa Verde National Park and the Ute Mountain Ute Tribal Park, and mapped the extent of piñon-juniper woodlands and petran chaparral using methods developed earlier for a vegetation map of the national park (Floyd-Hanna et al. 1996). Based on our detailed studies within the national park (described earlier in this chapter), we assumed that all areas of chaparral on the top of the Mesa Verde cuesta represented the locations of severe fires at some time during the last 150 years. We further assumed that areas now covered by piñon-juniper woodland had not burned within the last 150 years. From this information, we estimated the fire history of the entire cuesta and evaluated the 2000 and 2002 fires within that larger context.

Our primary concern in this assessment was with the ancient piñon-juniper woodlands that support distinctive plant and animal communities and that have recovered extremely slowly, if at all, after fires of the last century (Table 16.2). We determined that the total area of piñon-juniper woodland on the cuesta at the beginning of the twentieth century was 43,578 ha. From 1901 to 2002 a total of 6,624 ha of woodland burned (most of it in 2000, but also substantial amounts in 1934, 1959, 1972, 1989, 1996, and 2002). This represents about 15% of the woodland that existed in 1901. Following the reasoning presented earlier in this chapter, at this rate of burning it would take about 660 years to burn an area equal to the entire woodland portion of the Mesa Verde cuesta. If we restrict our analysis to the last half-century (1951–2002), we compute a turnover time of about 375 years. As we have already explained, we think that the period 1951–2002 is a better reflection of the "natural" fire regime in the Mesa Verde ecosystem.

The interesting result of this new analysis is that we obtained about the same estimate for the entire Mesa Verde cuesta as we had obtained earlier for just that portion of the cuesta lying within the national park. Both analyses converged on an estimate of about four centuries for the natural fire turnover time in piñon-juniper woodlands. Thus two final conclusions emerge from this new analysis. First, we must be very sensitive to scale effects in all ecological studies (Lewin 1992). This is especially true when we deal with ecological processes that occur over very extensive land areas or very long periods, such as fire in the Mesa Verde ecosystem. Second, we think that our original conclusion is still sound: the current

Table 16.2. Piñon-juniper woodland on the Mesa Verde Plateau burned in twentieth-century fires

Year	Area of Burned Woodland (ha)	Area of Woodland Remaining (ha)
pre-1934	—	43,578
1934	750	42,828
1951	15	42,813
1959	409	42,404
1972	966	41,438
1989	90	41,348
1996	741	40,607
2000-Bircher	1,834	38,773
2000-Pony	1,096	37,677
2002	723	36,954
Total	6,624	

Note: Information was derived from a vegetation map made around 1950, a fire atlas and field surveys, and fire records of more recent fires kept in archive files at Mesa Verde National Park.Table 17.1. Species diversity, abundance, and number of insect species in burned and unburned piñon-juniper habitat in Mesa Verde National Park.

fire regime on the Mesa Verde cuesta is *not* outside the range of natural variation (Cole 2000; Landres et al. 1999). Admittedly, the fires of the 2000s were frighteningly large and severe. From the perspective of an individual human lifetime they were unprecedented. From the perspective of ecological time (centuries to millennia), however, they were not all that unusual.

So we conclude that the fires in 2000 were not "unnatural" in terms of fire size and severity. Nonetheless, there remain reasons for concern about these recent fires, related mainly to invasion of alien species within the burned areas, global-scale changes in the earth's climate that are forecast for the next century, and land-use changes around the protected areas within Mesa Verde and the Ute Mountain Ute Tribal Parks. These concerns are discussed in Chapter 23.

FIRE IN MESA VERDE AND IN THE REGION

Our discussion of fire ecology in this chapter has focused on Mesa Verde—both the national park and the prominent landform that dominates so many views in southwestern Colorado. We believe that most of our conclusions (based on detailed studies in Mesa Verde) also can be extrapolated to the other landscapes surrounding Mesa Verde that are within the scope of interest of this book. We must caution, however, that fire ecology research in the surrounding areas is very limited and that some of the ecological features of Mesa Verde may be unique and not representative of other places. We urgently need additional research to test the applicability of ecological concepts developed in Mesa Verde to other parts of southwestern Colorado and the Colorado Plateau.

Two critical features of Mesa Verde may produce important ecological differences from the surrounding areas. First, the top of the cuesta is relatively warm and wet compared with other areas of piñon-juniper vegetation. The cuesta receives high precipitation because of its high elevation and its projection above the surrounding land surface. It also has warm temperatures because of its southward-dipping slopes. Consequently, Mesa Verde has a relatively long growing season for a piñon-juniper woodland, and this may influence rates of plant growth, fuel accumulation, and plant responses after fires. Lightning strikes also may be more frequent here than in surrounding lowlands. These factors have not been systematically evaluated for their possible influence on fire history and fire effects, so we simply do not know how much difference they make in Mesa Verde's fire regime. They may be very important, however, and we encourage further scientific investigation into their effects.

The second unique feature of Mesa Verde is its distinctive landform. The cliffs and sparsely vegetated slopes along the southern and western edges of the cuesta probably have always acted as partial barriers to fire spread. Therefore, as we have explained, past fire intervals in piñon-juniper woodlands lying just above the cliffs may be substantially longer than intervals in other woodlands exposed to fires spreading from adjacent areas. Petran chaparral in Mesa Verde also may experience longer fire intervals than chaparral in other, more exposed locations. In particular we hypothesize that chaparral located adjacent to ponderosa-pine forests in the nearby foothills of the San Juan Mountains burned more frequently prior to modern fire control policies than did the somewhat protected chaparral of Mesa Verde. Ponderosa-pine forests were subject to low-intensity fires every ten to twenty years until about 1880 (Swetnam and Baisan 1996), and many of these fires must have spread into adjacent chaparral. We need additional research into fire history to determine the specific ways in which Mesa Verde's fire regime is either unique or representative of the larger surrounding area.

Although this book focuses on southwestern Colorado, we urgently need more studies on fire history and fire effects in piñon-juniper woodlands throughout the Colorado Plateau. Some public land managers in this region are considering aggressive prescribed fire programs in piñon-juniper vegetation in the belief that these ecosystems were characterized historically by frequent fire and that the relative paucity of fire in the last century is anomalous. This belief is supported by studies in ponderosa-pine forests and in piñon-juniper woodlands elsewhere in the West (e.g., in the Great Basin region). Our findings in Mesa Verde, however, suggest that piñon-juniper woodlands of the Colorado Plateau may have been shaped over the centuries by a fundamentally different kind of fire regime than woodlands elsewhere in the West (Romme et al. in press). If Mesa Verde is in fact representative of the rest of the Colorado Plateau, then the low incidence of fire in many piñon-juniper woodlands of the Four Corners region during the twentieth century may be entirely natural. Aggressive burning of a vegetation type that rarely burned in the past would not "restore" natural conditions but could in fact damage such systems severely (Cole 2000). Our research has demonstrated that the piñon-juniper woodlands of Mesa Verde are ancient forests that have developed in the absence of fire for several centuries. A key question for southwestern fire ecologists is whether Mesa Verde is unique in this respect or whether similarly ancient piñon-juniper woodlands exist elsewhere across the Colorado Plateau. At the moment we cannot answer this question.

Finally, when discussing fires in old-growth woodlands of Mesa Verde, we cannot ignore the possible impact on archaeological sites. To address concerns about the effects of recent fires on the cultural resources of Mesa Verde National Park, in 1993 a group of archaeologists, fire scientists, and managers convened a workshop to identify cultural resources at risk from wildfires as well as strategies for reducing losses to future fires (Romme et al. 1993). Highly vulnerable resources identified by this group included the cliff dwellings for which Mesa Verde is world famous, culturally scarred trees, wooden structures from the historic period, and park administration buildings. One of the recommended strategies for reducing future losses to wildfire was to obtain better information on the frequency, spatial extent, and ecological effects of past fires. As summarized in this chapter, we have begun development of these data.

Since that meeting in 1993, four large wildfires have threatened numerous cultural resources. Of particular regret was the spalling of sandstone from Battleship Rock, one of Mesa Verde's finest petroglyph faces, during the Chapin 5 fire (BAER Report 2000). The cultural assessment that followed the 2000 Bircher fire identified the presence of 1,086 previously documented archaeological sites, and we know from past fires that additional sites will be revealed now that the vegetation is removed. The Pony fire probably had an impact on many of the 387 known sites as well. The effects of fire in the piñon-juniper woodlands of the Mesa Verde cuesta are thus compounded by archaeological losses. While fire is a natural part of the woodland development, it is important to consider the breadth of postfire effects, which in this case may threaten prehistoric and historic treasures.

SUMMARY

Recent uncontrollable wildfires in Mesa Verde and surrounding areas have threatened significant cultural and natural resources. Understanding the long-term role of fire in piñon-juniper woodlands and petran chaparral is challenging, because written records are limited and the major species rarely form fire scars. Nevertheless, our research (and that of others) has demonstrated some of fire's important natural influences on vegetation dynamics and overall ecological function throughout Mesa Verde Country. In response to our increasing knowledge about fire, resource managers throughout the Southwest are revising fire management plans to accommodate the natural role of fire in these ecosystems.

In this chapter we have described a method for dating and mapping the spatial extent of past fires in shrublands dominated by Gambel oak (*Quercus gambelii*), based on detecting a prominent cohort of stems that resprout from lignotubers after fire kills aboveground portions of the plants. We have determined that the turnover time in petran chaparral of Mesa Verde National Park is about a century, whereas the median fire interval and turnover time in piñon-juniper woodlands, based on current tree age structures and twentieth-century fire history, are about 400 years. Many woodland stands contain trees more than 500 years old and probably have not burned for many centuries. Most fires are very small, but the majority of the cumulative area burned during a decade or a century is burned by only one or a few very large fires. These very large fires tend to be severe: they kill all or most of the aboveground vegetation. Petran chaparral recovers very quickly after fire, because most of the major shrub species resprout prolifically; but recovery of burned piñon-juniper woodlands requires centuries.

The fire frequency and fire effects that we have observed in the last 100 years do not appear to differ significantly from pre-1900 fire frequency and fire effects, especially when viewed at the scale of the entire Mesa Verde cuesta. Large fires in the last decade, however, have occurred more frequently and later in the summer season than did large fires earlier in the twentieth century. We cannot yet say whether this last decade of severe fire effects is simply a part of the normal long-term variation in this system or reflects the beginning of a more ominous shift toward a new fire regime as a consequence of global climatic changes or other causes.

REFERENCES CITED

BAER (Burned Areas Emergency Rehabilitation). 2000. *Accomplishment Report*. National Park Service, Mesa Verde.

Barney, M. A., and N. C. Frischknecht. 1974. Vegetation changes following fire in the pinyon-juniper type of west-central Utah. *Journal of Range Management* 27: 91–96.

Blaisdell, J. P., R. B. Murray, and E. D. McArthur. 1982. *Managing intermountain rangelands: Sagebrush-grass ranges*. USDA Forest Service General Technical Report INT-134.

Brown, H. F. 1958. Gambel oak in west-central Colorado. *Ecology* 39: 317–327.

Bruner, A. D., and D. A. Klebenow. 1979. *Predicting success of prescribed fires in pinyon-juniper woodland in Nevada*. USDA Forest Service Research Paper INT-219.

Callaway, D., J. Janetski, and O. C. Stewart. 1986. Ute. In W. L. D'Azevedo (ed.), *Handbook of North American Indians*, vol. 11: *Great Basin*, pp. 336–367. Smithsonian Institution, Washington, D.C.

Clary, W. P., and A. R. Tiedemann. 1986. Distribution of biomass within small tree and shrub form *Quercus gambelii* stands. *Forest Science* 32: 234–242.

Cole, D. N. 2000. Paradox of the primeval: Ecological restoration in wilderness. *Ecological Restoration* 18: 77–86.

Dishman, L. 1982. Ranching and farming in the lower Dolores River Valley. In G. D. Kendrick (ed.), *The river of sorrows: The history of the lower Dolores River Valley*, pp. 23–41. USDI National Park Service and Bureau of Reclamation. U.S. Government Printing Office, Denver.

Duke, P. 1995. Working through theoretical tensions in contemporary archaeology: A practical attempt from southwestern Colorado. *Journal of Archaeological Method and Theory* 2: 201–229.

Ellis, R. N. 1996. The Utes. In R. Blair (managing ed.), *The western San Juan Mountains: Their geology, ecology, and human history*, pp. 225–233. University Press of Colorado, Niwot.

Everett, R. L., and K. Ward. 1984. Early plant succession on pinyon-juniper controlled burns. *Northwest Science* 58: 57–68.

Fischer, W. C., and B. D. Clayton. 1983. *Fire ecology of Montana forest habitat types east of the Continental Divide.* USDA Forest Service General Technical Report INT-141.

Floyd, M. E. 1982. Interaction of piñon pine and *Quercus gambelii* in succession near Dolores, Colorado. *Southwestern Naturalist* 27: 143–147.

———. 1986. Inter- and intra-specific variations in piñon pine reproduction. *Botanical Gazette* 147: 180–188.

Floyd, M. L., W. H. Romme, and D. Hanna. 2000. Fire history and vegetation pattern in Mesa Verde National Park, Colorado, USA. *Ecological Applications* 10(6): 1666–1680.

Floyd-Hanna, M. L., A. W. Spencer, and W. H. Romme. 1996. Biotic communities of the semiarid foothills and valleys. In R. Blair (managing ed.), *The western San Juan Mountains: Their geology, ecology, and human history*, pp. 143–158. University Press of Colorado, Niwot.

Johnson, E. A., and S. L. Gutsell. 1994. Fire frequency models, methods and interpretations. *Advances in Ecological Research* 25: 239–287.

Keeley, J. E. 1992. Recruitment of seedlings and vegetative sprouts in unburned chaparral. *Ecology* 73: 1194–1208.

Koniak, S. 1985. Succession in pinyon-juniper woodlands following wildfire in the Great Basin. *Great Basin Naturalist* 45: 556–566.

Landres, P. B., P. Morgan, and F. J. Swanson. 1999. Overview of the use of natural variability concepts in managing ecological systems. *Ecological Applications* 9: 1179–1188.

Leopold, A. 1924. Grass, brush, timber, and fire in southern Arizona. *Journal of Forestry* 22(6): 1–10.

Lewin, S. A. 1992. The problem of pattern and scale in ecology. *Ecology* 73: 1943–1967.

Moritz, M. A. 1997. Analyzing extreme disturbance events: Fire in Los Padres National Forest. *Ecological Applications* 7: 1252–1262.

Noste, N. V., and C. L. Bushy. 1987. *Fire response of shrubs of dry forest habitat types in Montana and Idaho.* USDA Forest Service General Technical Report INT-239.

Omi, P., and L. Emrick. 1980. Fire and resource management in Mesa Verde National Park. Contract CS-1200-9-B015. Unpublished report, on file at Mesa Verde National Park.

Romme, W. H., M. L. Floyd-Hanna, and M. Conner. 1993. Effects of fire on cultural resources at Mesa Verde National Park. *Park Science* 13(3): 28–30.

Romme, W. H., M. L. Floyd-Hanna, and D. D. Hanna. In press. Ancient pinyon-juniper forests of Mesa Verde: A cautionary note about thinning and burning for restoration. In *Proceedings of the conference on Fire, Fuel Treatments, and Ecological Restoration: Proper Place, Appropriate Time,* Colorado State University, April 2002. USDA Forest Service General Technical Report RMRS-GTR.

Smith, D. A. 1988. *Mesa Verde National Park: Shadows of the centuries.* University of Kansas Press, Lawrence.

Swetnam, T., and J. L. Betancourt. 1998. Mesoscale disturbance and ecological response to decadal climatic variability in the American Southwest. *Journal of Climate* 11: 3128–3147.

Swetnam, T. W., and C. H. Baisan. 1996. Historical fire regime patterns in the southwestern United States since A.D. 1700. In C. D. Allen (technical ed.), *Fire effects in southwestern forests: Proceedings of the Second La Mesa Fire Symposium, Los Alamos, 1994,* pp. 11–32. USDA Forest Service General Technical Report RM-GTR-286.

Tiedemann, A. R., W. P. Clary, and R. J. Barbour. 1987. Underground systems of Gambel oak (*Quercus gambelii*) in central Utah. *American Journal of Botany* 74: 1065–1071.

West, N. E., and M. A. Hassan. 1985. Recovery of sagebrush-grass vegetation following wildfire. *Journal of Range Management* 38: 131–134.

Wright, H. A., L. F. Neuenschwander, and C. M. Carlton. 1979. *The role and use of fire in the sagebrush and piñon-juniper plant communities.* USDA Forest Service General Technical Report INT-58.

EFFECTS OF FIRE ON INSECT COMMUNITIES IN PIÑON-JUNIPER WOODLANDS IN MESA VERDE COUNTRY

Deborah M. Kendall

Large wildfires have torn through the piñon-juniper woodlands on Mesa Verde four times in the last decade (Chapter 16). In the wake of each intense, stand-replacing fire, few residual plants or animals survive. Insects, key players in the health of piñon-juniper ecosystems (Frischknecht 1975), are no doubt dramatically affected. Insect communities in grasslands (Bulan and Barrett 1971), boreal forests (McCullough et al. 1989), chaparral (Moore et al. 1979), and sandhills (McCoy 1987) have been altered by fires. Studies of fire effects on insects in piñon-juniper woodlands, however, have not been published. At Mesa Verde National Park the large fires of the last decade have provided the opportunity to assess insect succession; here the focus is on old-growth piñon-juniper woodlands that burned in the 1989 Long Mesa fire.

Patterns of insect recolonization of burned habitats involve mobility of the insect species, the successional profile of returning vegetation, and the vegetational composition of returning plants. Plant species diversity and ground cover, nutritional quality of the foliage, and chemical deterrents are all relevant to reestablishing insects. These plant characteristics influence the population dynamics, species diversity, and trophic structure of the returning insect community.

Fires at Mesa Verde in the last decade have been stand replacing, where most of the aboveground biomass and leaf litter are destroyed. This might have both positive and negative effects on insects, by eliminating niches but also opening the habitat to greater insect mobility and potential for recolonization.

Figure 17.1. The tachinid fly, typical of herbaceous weedy habitats, occurs in postfire vegetation communities at Mesa Verde National Park. Illustration by Agnes Suazo.

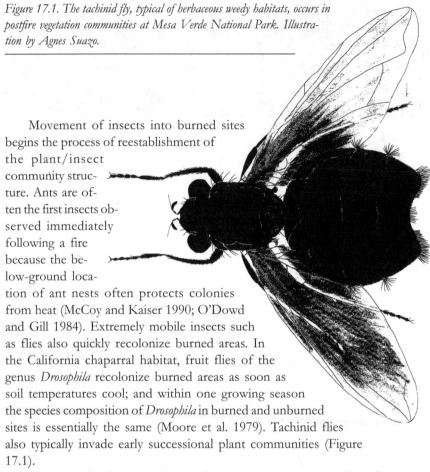

Movement of insects into burned sites begins the process of reestablishment of the plant/insect community structure. Ants are often the first insects observed immediately following a fire because the below-ground location of ant nests often protects colonies from heat (McCoy and Kaiser 1990; O'Dowd and Gill 1984). Extremely mobile insects such as flies also quickly recolonize burned areas. In the California chaparral habitat, fruit flies of the genus *Drosophila* recolonize burned areas as soon as soil temperatures cool; and within one growing season the species composition of *Drosophila* in burned and unburned sites is essentially the same (Moore et al. 1979). Tachinid flies also typically invade early successional plant communities (Figure 17.1).

Patterns of early succession in the plant community greatly influence the recolonization patterns of insects. Other studies suggest that early successional plant communities support large populations of herbivorous (plant-feeding) insects that feed on a wide variety of plant species. Generalist insects may expand on early postfire plant communities, perhaps because they consist of high plant species diversity. Insect species diversity commonly increases in the first postfire growing seasons as well, probably due to changes in plant community structure (McCoy 1987; Mushinsky 1985). The early successional communities offer numerous feeding and enemy-free spaces that allow rapid increases in insect herbivore populations.

Nutritional quality of early successional plant species also influences the population dynamics of herbivores, which in turn affects predaceous and detritivorous insects. Early successional plants are usually high in nitrogen and support rapid increases in insect herbivore populations (Bach 1990).

Other studies have shown that if perennials or woody shrubs become dominant as vegetation matures over time after fire, they offer greater numbers of enemy-free spaces and tend to support predaceous (feeding on insects and other animals) and detritivorous (feeding on decayed organic material) insects. Older successional plants are often trees and shrubs that contain large quantities of herbivore-deterrent compounds such as tannins and phenolic glycosides (Hemming and Lindroth 1999). Plant deterrent compounds are also important in determining herbivore population dynamics and species diversity.

Given this background from other areas, we began a study of postfire recolonization patterns in unburned and burned piñon-juniper woodlands following the 1989 Long Mesa fire in Mesa Verde National Park.

INSECTS FOLLOWING FIRE IN MESA VERDE

The Long Mesa fire in July 1989 burned roughly 1,800 ha of piñon-juniper woodland, mountain shrublands, and sagebrush habitats. Two study sites were chosen to examine the effects of fire on insect and plant communities. Both sites consisted of old-growth piñon-juniper woodlands; one site burned in the 1989 fire, and the other remained undisturbed.

In burned and unburned piñon-juniper woodlands, insects were collected by pit trap and sweep net methods every two weeks from May to August in 1991 and 1992. Ants were eliminated from the collection because their numbers depend on adjacency to ant nests (that might survive fire under the soil surface) and do not reflect recolonization patterns. Insect species diversity and abundance were calculated. Alternative measures of diversity that account for relative abundance, such as the Shannon-Weaver Index (Shannon and Weaver 1963), are available; but here the diversity (number of species) and abundance (number of sampled insects of all species) are considered for simplicity.

In order to characterize the preburn plant community, we measured the percent coverage (Daubenmire 1959) of all vegetation. In addition, the diameter and height were recorded for all living shrubs and trees.

Insects exhibited specific patterns of recolonization three years after the Long Mesa fire. Relative to unburned woodland, the burned areas supported a greater number of insect species, a higher abundance of all insects combined, and a similar proportion of herbivores and predators (Tables 17.1, 17.2, and 17.3).

Early postfire insect assemblages in Mesa Verde also included a larger proportion of generalists (insects that exploit a wide range of plant species) than in the unburned site (Tables 17.2 and 17.3). The insect community in the burned piñon-juniper habitat also exhibited a more diverse trophic structure, including herbivores, predators, detritivores, and flower visitors, such as bees and wasps that feed on pollen and nectar (Table 17.2).

How can we interpret such a dramatic increase in insect populations after fire? As we have already described for other habitats, a greater number of feeding

TABLE 17.1. Species diversity, abundance, and number of insect species in burned and unburned piñon-juniper habitat in Mesa Verde National Park

	Burned	Unburned
Number of insects	453	56
Number of species	33	19
Trophic niches:		
Herbivores	95.8%	98.2%
Predators	2.4%	1.8%
Detritivores	1.8%	0

Note: Data were collected in 1991 and 1992.

TABLE 17.2. Insect species in burned piñon-juniper habitat in Mesa Verde National Park

COLEOPTERA (BEETLES)
Carabidae (Ground Beetles)
 Harpalus desertus
 Harpalus fraternus
 Piosoma setosum

CERAMBYCIDAE (LONG-HORNED BEETLES)
 Gnathachaeops pratensis
Cleridae (Checkered Beetles)
 Trichodes ornatus
Coccinellidae (Ladybird Beetles)
 Hippodamia convergens
Silphidae (Carrion Beetles)
 Nicrophorus guttula
Tenebrionidae (Darkling Beetles)
 Eleodes obscurus

DIPTERA (FLIES)
Conopidae (Conopid Flies)
 Physocephala texana
Tachinidae (Tachinid Flies)
 Peleteria spp.
Therevidae (Therevid Flies)
 Ozodiceromya spp.

HEMIPTERA (TRUE BUGS)
Lygaeidae (Seed Bugs)
 Lygaeus reclivatus
 Lygus spp.
Nabidae (Damsel Bugs)
 Nabis spp.
Pentatomidae (Stink Bugs)
 Codophila remota
 Holcostathus spp.
 Thyanta spp.

HOMOPTERA (LEAFHOPPERS)
Cicadellidae (Leafhoppers)
 Achorotila spp.

HYMENOPTERA (BEES, WASPS)
Andrenidae (Andrenid Bees)
 Andrena spp.
Anthophoridae (Digger Bees)
 Melissodes spp.
Bombidae (Bumblebees)
 Bombus nevadensis
Halictidae (Halictid Bees)
 Agapostemon spp.
 Ashmeadiella spp.
 Diclictus spp.
 Halictus spp.
Megachilidae (Leaf Cutter Bees)
 Anthidium spp.
Mutillidae (Velvet Ants)
 Dasymutilla vestita
Sphecidae (Sphecid Wasps)
 Amnophila azteca
 Podolonia communis

Note: Data were collected in 1991 and 1992.

TABLE 17.3. Insect species in unburned piñon-juniper habitat in Mesa Verde National Park

COLEOPTERA (BEETLES)	HEMIPTERA (TRUE BUGS)
Cleridae (Checkered Beetles)	Lygaeidae (Seed Bugs)
Cymatodera spp.	*Lygaeus reclivatus*
Coccinellidae (Ladybird Beetles)	*Lygus* spp.
Hippodamia convergens	Pentatomidae (Stink Bugs)
Mysia spp.	*Codophila remota*
Curculionidae (Weevils)	*Holcostathus* spp.
Magdalis lecontei	*Thyanta* spp.
Elateridae (Click Beetles)	HOMOPTERA (LEAFHOPPERS)
Ctenicera spp.	Cicadellidae (Leafhoppers)
Meloidae (Blister Beetles)	*Achorotila* spp.
Meloe spp.	
	HYMENOPTERA (BEES, WASPS)
DIPTERA (FLIES)	Sphecidae (Sphecid Wasps)
Otitidae (Picture-Winged Flies)	*Podolonia communis*
Tritoxa spp.	
Tachinidae (Tachinid Flies)	
Peleteria spp.	

Note: Data were collected in 1991 and 1992.

and enemy-free niches were very likely provided by early postfire plants. In general, when old-growth piñon-juniper in Mesa Verde Country burns, the pattern of plant reestablishment depends on the understory (Chapter 16). If shrubs and perennial grasses make up the understory, they will resprout during the first growing season after fire. In other habitats a plethora of forbs including *Polygonum sawatchense*, *Penstemon linarioides*, *Lupinus ammophilus*, and *Chenopodium fremontii* germinate during the first postfire year. The first plants to colonize this particular site were native perennial grasses and herbaceous forbs (Table 17.4). The burned piñon-juniper woodland was characterized by a greater number of forbs than the unburned site (Tables 17.4 and 17.5). Trees and woody shrubs were absent, because fire killed all the aboveground vegetation in 1989 (Chapter 16). New foliage, stems, petals, pollen, and nectar supported a greater diversity and number of insects that are predominantly herbivores (95.8%) in the burned woodland. Also, bare ground exposure (47.6%) may have contributed to ease of mobility by cursorial insects such as ground beetles (Table 17.2).

Plants in the burned piñon-juniper habitat may also have been rich sources of nitrogen. It is well-known that some plant species in burned sites assimilate greater quantities of nitrogen from the soil and exhibit greater aboveground biomass production than they do in unburned habitats (Blank et al. 1994). The high nitrogen content of foliage in turn often leads to increases in herbivore populations (Forkner and Hunter 2000). There was a clear increase in insect populations while

Table 17.4. Plant species percent coverage in burned piñon-juniper woodland

	Percent Coverage
Bare Ground	47.6
Grasses	2.4
Herbaceous Forbs	
Alyssum alyssoides	2.1
Arenaria congesta	3.6
Carduus nutans	2.8
Erigeron etoni	10.9
Heterotheca villosa	8.5
Hymenoxys acaulis	2.8
Lomatium dissectum	6.3
Opuntia spp.	0.4
Penstemon linarioides	2.6
Senecio spp.	2.3

Note: Data were collected in 1991 and 1992.

Table 17.5. Plant species percent coverage in unburned piñon-juniper woodland

	Percent Coverage
Bare Ground	64.8
Grasses	14.4
Herbaceous Forbs	
Allium spp.	0.1
Ipomopsis aggregata	0.9
Senecio spp.	0.1
Thelopodium spp.	0.4
Yucca baccata	0.1

Note: Data were collected in 1991 and 1992.

feeding on postfire successional communities (Table 17.1), and increased nitrogen content of the foliage may be one explanation (although it was not measured in this study).

The unburned piñon-juniper site supported perennial grasses, forbs, shrubs, and trees, with a composition that is fairly typical of the middle-elevation portions of the mesa tops (Chapters 2 and 3). Dominant woody species included *Pinus edulis* (piñon pine), *Juniperus osteosperma* (Utah juniper), *Prunus virginiana* (chokecherry), *Quercus gambelii* (Gambel oak), and *Amelanchier utahensis* (Utah serviceberry). The understory was predominantly bare ground (64.8%); herbaceous forbs were uncommon, but perennial native grasses were more abundant than on the burned site (Tables 17.4 and 17.5).

Chemical differences in the unburned (older) foliage might be in part responsible for the fact that abundance and diversity of insects were lower in this unburned piñon-juniper woodland than in the burned site (Table 17.1). Most insect herbivores are generalists that feed on a wide variety of herbaceous forbs, a lifeform conspicuously absent from the unburned site. It appears that the older vegetation of the unburned piñon-juniper woodland does not support a large number of insect herbivores. Rather, insects feeding on late successional plants of unburned woodlands are often specialists that are less likely to be affected by the secondary plant compounds, such as tannin, terpenes, and phenols produced by the host plants.

SUMMARY

Insects are critical to maintaining the health of piñon-juniper ecosystems, yet fire effects on these important populations are not well-known. Three years after the 1989 Long Mesa fire in Mesa Verde National Park, burned piñon-juniper woodlands supported primarily herbaceous forbs and early successional insect species. The plant community was dominated by annuals that are very likely rich in nitrogen content but low in herbivore feeding deterrents. The early successional insect community exhibited greater breadth of insect species and overall higher abundance. Increased postfire insect diversity may be attributed to an increase in microhabitats for feeding and hiding, and one explanation for higher insect abundance is increased nitrogen content of plants growing in fire-burned soils.

REFERENCES CITED

Bach, C. E. 1990. Plant successional stage and insect herbivory: Flea beetles on sand-dune willow. *Ecology* 71: 598–609.

Blank, R. R., F. Allen, and J. A. Young. 1994. Growth and elemental content of several sagebrush-steppe species in unburned and post-wildfire soil and plant effects on soil attributes. *Plant and Soil* 164: 35–41.

Bulan, C. A., and G. W. Barrett. 1971. The effects of two acute stresses on the arthropod component of an experimental grassland ecosystem. *Ecology* 52: 597–605.

Daubenmire, R. 1959. A canopy-coverage method of vegetational analysis. *Northwest Science* 33: 43–64.

Forkner, R. E., and M. D. Hunter. 2000. What goes up must come down? Nutrient addition and predation pressure on oak herbivores. *Ecology* 81: 1588–1600.

Frischknecht, N. C. 1975. Native faunal relationships within the piñon-juniper ecosystem. In *The piñon-juniper ecosystem: A symposium,* pp. 55–65. Utah State University, Logan.

Hemming, J. D., and R. L. Lindroth. 1999. Effects of light and nutrient availability on aspen: Growth, phytochemistry, and insect performance. *Journal of Chemical Ecology* 25: 1687–1714.

McCoy, E. D. 1987. The ground-dwelling beetles of periodically-burned plots of sandhill. *Florida Entomologist* 70: 31–39.

McCoy, E. D., and B. W. Kaiser. 1990. Changes in foraging activity of the southern harvester and *Pogonomyrmex badius* (Latreille) in response to fire. *American Midland Naturalist* 123: 112–123.

McCullough, D. G., R. A. Werner, and D. Neumann. 1989. Fire and insects in northern and boreal forest ecosystems of North America. *Annual Review of Entomology* 43: 107–127.

Moore, J. A., C. E. Taylor, and B. C. Moore. 1979. The *Drosophila* of southern California: I. Colonization after a fire. *Evolution* 33: 156–171.

Mushinsky, H. R. 1985. Fire and the Florida sandhill herpetofaunal community, with special attention to responses of *Cnemidophorus sexlineatus*. *Herpetologica* 41: 333–342.

O'Dowd, D. J., and A. M. Gill. 1984. Predator satiation and site alteration following fire: Mass reproduction of alpine ash, *Eucalyptus delegatensis*, in southeastern Australia. *Ecology* 65: 1052–1066.

Shannon, C. E., and W. Weaver. 1963. *The mathematical theory of communication.* University of Illinois Press, Champaign.

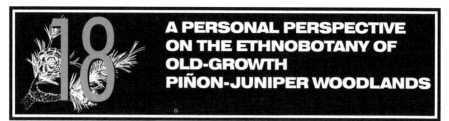

18. A PERSONAL PERSPECTIVE ON THE ETHNOBOTANY OF OLD-GROWTH PIÑON-JUNIPER WOODLANDS

WILLIAM J. LITZINGER

WHEN I WALK IN THE STILLNESS OF A DENSE, MATURE STAND of piñon and juniper trees, I am drawn to details of each tree's individual life history as recorded in its size, shape, pattern of growth, sex, and maturity. I look for evidence of significant events in each tree's history, such as fire scars, wind damage, disease, and attack by predators. As a botanist and ecologist, I observe other plants in the forest and assess them in similar ways. I have come to understand these as important events and processes in the lives of the plants themselves.

Sometimes I wonder what I would be looking for if I were trying to subsist in the piñon forest environment. Humans have walked in piñon forests for thousands of years, gathering plants for food, fuel, and utility (Chapters 19 and 20). I wonder how ancient foragers came to know the plants around them. Through the lens of their unique cultural experience and subsistence needs, did ancient foragers see these plants in different ways than I do?

Having been fortunate to watch piñon woodlands as they change with seasons and years, I have come to know the trees and how they respond to their environment as individuals. Recognition of each one engenders within me great respect and caring. My relationship with the trees also gives me a sense of stewardship and responsibility for the forest as a whole. Perceiving the finer details of the natural world, as did the ancient indigenous foragers, involves detection of color, form, and texture. Indeed it is possible that I have seen some of the same individual plants indigenous foragers saw centuries ago. Although our cultural viewpoints and relationships with piñon trees are vastly different, I believe we share a deep sense of kinship with these plants and these landscapes. I have an overwhelming sense that indigenous foragers probably saw the piñons much as I do today; therein I find my connection with indigenous foragers, for I know they must have felt the same sense of caring, respect, and stewardship toward these trees.

CULTURAL IMPORTANCE

Every cultural group that has lived within the natural range of the piñons now or in the past has used them and in some cases relied on them as major sources of food and fuel. For thousands of years people have been using the piñon pine in the Southwest. Remains of seed coats and wood charcoal are found in archaeological sites from the earliest times. Piñons provide for a number of human needs: the seeds are used for food; the wood for fuel and construction material; the resin as a medicine and as an adhesive for mending, construction of implements, and sealing water bottles; and the pollen as an important element in religious ceremonies.

All the piñon species have significant nutritional value. *Pinus cembroides* (Mexican piñon) is the richest in protein and the lowest in starch, *Pinus monophylla* (single-leaf piñon) the lowest in protein and starch, and *Pinus edulis* (Colorado piñon) the richest in fats. *Pinus edulis* has the thinnest seed coat and appears to be the most preferred as an edible. The amino acid composition of the piñon kernel protein contains seven of the nine essential amino acids in adequate amounts. Especially rich are tryptophane and cystine (Lanner 1981), two amino acids often in low concentration in other indigenous plant foods, particularly maize (*Zea mays* L.) and wild grass seed.

R. B. Woodbury and E.B.W. Zubrow (1979) argue that the large production of piñon seeds for food provided early peoples in the Southwest with a buffer that allowed them to experiment with agriculture. An abundance of piñon seed and wood fragments and remains of many other native plants found in the archaeological sites of the Ancestral Puebloan "Basketmakers" (ca. A.D. 200 to 700) show that they intensified their reliance on agriculture while still gathering a wide range of wild plants. Because of the marginality of agriculture in the arid Southwest, this pattern persisted into historic and recent times among Pueblo and Navajo farmers.

A dramatic and widely recorded decline of piñon remains in archaeological sites dates to the late Ancestral Puebloan occupation of Mesa Verde Country, just prior to the catastrophic abandonment of the area. While some piñon remains are found in late Ancestral Pueblo sites (Figure 18.1), forest clearing was obviously intensive and widespread. It is likely that only small remnant stands remained at the time of abandonment of the area around A.D. 1300. Indeed, D. G. Wycoff (1977) documents that it took at least 350 years for piñon forests to recover after the Pueblo abandonment.

Figure 18.1. A cliff dwelling, home to Ancestral Puebloans during the 1200s, in Mesa Verde National Park. Illustration by Susie Harvin.

R. M. Lanner (1981) has summarized the extensive ethnographic record of piñon use in the greater Southwest and Great Basin. There is evidence that *Pinus edulis* was so important that it was known well beyond its natural range. It was an item of trade among various indigenous peoples of the Southwest (such as the Zunis, Hopis, and Navajos) and as far as Mexico, where it was sold in markets in preference to local native species (Benavides in 1630, as cited in Robbins et al. 1916:41).

PREDICTING THE PIÑON NUT HARVEST

In the Great Basin, where agriculture did not take a strong hold, the Shoshones and Paiutes relied extensively on the piñon cone crops (Figure 18.2). Julian Steward (1938) suggested that the social organization of the Shoshones and Paiutes is based in part on the availability and procurement of piñons and other wild foods. D. H. Thomas (1972, 1973) thought that certain powwows and fandangos were held for the purpose of predicting the timing of the piñon harvest. Family groups would gather into seasonal villages near the harvest sites.

Figure 18.2. Piñon pine seeds (produced in female cones) have provided an excellent but ephemeral food source for human populations throughout the Southwest. Illustration by Noni Floyd.

Because piñons do not produce seed crops every year, relying on them as a food has some risks. Seed production in piñons fluctuates between large crops, or masts, alternating with unproductive years. For every ten years, one to three years may have large seed crops, while during the intervening years there is little or no seed production. Piñons have two-and-a-half-year cycles of seed maturation; therefore, potential crops are visible during the growing season prior to harvest (Floyd and Kohler 1990). Predicting the piñon harvest has importance culturally and ecologically. When there is a mast, the people do well to organize and concentrate their effort so they can harvest a crop large enough to store for the unproductive years.

The predictability of piñon seed production is a complicated biological problem. Because of this and because ethnobotanical studies tend to be superficial and concentrate on the uses of plants and not the perception of the plants as natural organisms, we lack precise information about piñons that would help us better evaluate Steward's and Thomas's hypotheses about the role of piñon in indigenous culture.

In reviewing the botanical and ecological literature on piñon forests, two modern viewpoints emerge. First is the view that piñons are a sporadic resource with commercial potential in mast years. There is no recognition of the potential of yield over the long term. The second viewpoint reflects the perception of the arid southwestern woodlands as not being as productive as they could be. The piñon forests are seen as wastelands that need to be removed and replaced with grasslands that can be used for cattle grazing. Removal of trees has created tension and some violence among some Native Americans (who consider the woodlands to be communal resources important for long-term survival) and some Anglo and Hispanic Americans (who value the piñon woodlands for short-term profit manipulation). Much of present-day management of the piñon woodlands, by means of chaining, frequent burning, herbicide applications, and the seeding of non-native grasses, has caused deterioration of the environment occupied by the piñon (Gifford and Busby 1975). Recent fire-related fuel reduction efforts threaten the integrity of the woodland as well (M. L. Floyd and W. H. Romme, unpublished data). Clearly, short-term manipulation is not compatible with this fragile, long-term-yielding ecosystem.

There is a critical need for more biological and ecological data about piñons to prevent future deterioration of the piñon woodlands and allow sensible use of this resource. As G. P. Nabhan (2000) points out, because traditional ecological knowledge is often highly sensitive to critical relationships and interactions, it is important for ecologists to begin to integrate this knowledge to help in understanding natural phenomena. C. S. Fowler (2000) argues further that humans not only recognize and passively take advantage of the diversity of plants but also manipulate the pattern of diversity in an active manner that is advantageous to both plants and humans. This manipulation has long been recognized in our relationship with domesticated plants and plants that occupy anthropogenic habitats. At present there is scant information on the manipulation of native plants in the piñon forests by indigenous peoples. No doubt a few living Shoshone, Ute, and Paiute people have this information. The knowledge held by indigenous people is important, but their relationship with these trees also represents a stewardship of nature that we can share as a value.

IDENTIFYING CULTURALLY MODIFIED TREES

Even though the use of the piñon forests by indigenous people may have diminished or even ceased in some areas, it still may be possible to ascertain evolutionary aspects of long-term human interaction with the piñons. Because of the life expectancy of piñons (greater than 350 years) it is possible to reconstruct the age diversity of the forest in the recent past. This information may be used to test specific hypotheses. For example, based on M. E. Floyd's (1983) observations, it is reasonable to ask whether the Paiutes, Utes, and Shoshones manipulated piñon forests to maximize yield by selectively encouraging highly productive middle-aged trees for seed production and removing older trees for construction and firewood. This effect would be expected to be even more pronounced in piñon populations located near sedentary agricultural peoples such as the Hopis or Zunis, who because of their greater populations have increased construction and fuel needs and at the same time increased demand for piñon seeds. The intensification of these contrasting needs—the benefits of mast crops for food reserves vs. the demand for fuel wood—would be expected to result in an exploitation pattern that created piñon forests of only young and middle-aged trees. This is only one approach to addressing the effects of Native American utilization of piñons, but unique stand structures and other features such as branch removal and shaping should be evident and easily observed in living trees.

SUMMARY

It is important that we shift away from the prevalent viewpoint in modern Anglo culture that the piñon forests are a wasteland in need of improvement. This point of view has caused overexploitation and destruction of the piñon forests for short-term economic gain. Our attempts as biologists and ecologists to gain a more complete understanding of the piñon woodland will benefit from a greater appreciation and understanding of the traditional values of indigenous cultures regarding the piñon. People who have intimate contact with the piñons and manipulate the populations to maximize long-term resource use create a stewardship relationship with these plants and the natural world. An understanding of how humans have successfully and sustainably interacted with the piñon can help provide a model for managing the piñon-juniper woodlands and other mature ecosystems for long-term yield. This model of land use would aim to incorporate efficient fuel wood and seed use with maintenance of the health and vitality of the piñon-juniper ecosystem.

REFERENCES CITED

Floyd, M. E. 1983. Dioecy in five *Pinus edulis* populations in the southwestern United States. *American Midland Naturalist* 110(2): 405–411.

Floyd, M. L., and T. A. Kohler. 1990. Current productivity and prehistoric use of the piñon (*Pinus edulis,* Pinaceae) in the Dolores Archaeological Project Area, southwestern Colorado. *Economic Botany* 44(2): 141–156.

Fowler, C. S. 2000. Ethnoecology: An introduction. In P. E. Minnis (ed.), *Ethnobotany: A reader,* pp. 13–16. University of Oklahoma Press, Norman.

Gifford, G. F., and F. E. Busby. 1975. *The piñon-juniper ecosystem: A symposium.* Utah Agricultural Experiment Station, Logan.

Lanner, R. M. 1981. *The piñon pine: A natural and cultural history.* University of Nevada Press, Reno.

Nabhan, G. P. 2000. Interspecific relationships affecting endangered species recognized by the O'odham and Comcáac cultures. *Ecological Applications* 10(5): 1288–1295.

Robbins, W., J. P. Harrington, and B. Freire-Marreco. 1916. *Ethnobotany of the Tewa Indians.* Bureau of American Ethnology, Bulletin 55. Smithsonian Institution, Washington, D.C.

Steward, J. H. 1938. *Basin-Plateau aboriginal sociopolitical groups.* Bureau of American Indian Affairs, Bulletin 120. Smithsonian Institution, Washington, D.C.

Thomas, D. H. 1972. Western Shoshone ecology: Settlement patterns and beyond. In D. D. Fowler (ed.), *Great Basin cultural ecology: A symposium,* pp. 135–154. Publications in the Social Sciences. Desert Research Institute, Reno, Nev.

———. 1973. An empirical test of Steward's model of Great Basin settlement patterns. *American Antiquity* 38: 155–176.

Woodbury, R. B., and E.B.W. Zubrow. 1979. Agricultural beginnings, 2000 B.C.–A.D. 500. In A. Ortiz (ed.), *Handbook of North American Indians,* vol. 9: *Southwest.* Smithsonian Institution, Washington, D.C.

Wycoff, D. G. 1977. Secondary forest succession following abandonment of Mesa Verde. *Kiva* 42(3–4): 215–231.

19 SOME ETHNOBOTANICAL USES OF PLANTS FROM PIÑON-JUNIPER WOODLANDS

Marilyn Colyer

The Ancestral Puebloan people who once roamed Mesa Verde Country and Native American people who currently live here have used the plants of the piñon-juniper woodlands in myriad ways. We have evidence of prehistoric uses from seeds and pollen found in human coprolites. We have also found stores of seeds and dried fruits. Fragments of wood or fiber artifacts have shown us that plants were extensively used for clothing, tools, housing, food, and medicine. Ethnobotanical studies of living people document intensive use of native plants in the recent past and suggest that a few wild plants are still sought after today. Much of the information in this chapter comes from diffuse sources (my neighbors, acquaintances, and friends) who have shared their knowledge of specific plant uses with me. It is difficult to document these sources; some are confidential, but most of the information presented here was garnered over time, tried, and found to be useful. Information specific to uses of medicinal plants was provided by numerous people, most of whom are no longer living, as part of research conducted by Dr. H. D. Harrington and myself (Harrington 1967).

Perception of the natural history of plant species must be especially sharp in those who collect ethnobotanical specimens (as discussed in Chapter 18). They must know the location of each special plant and when it is best to harvest the part desired. I have gone out with Utes, Navajos, and those of Spanish descent, generally elderly women, to learn about uses of plants in the piñon-juniper woodland. Their knowledge of the intricate ecology of each species is astounding, and they travel paths that probably have been passed down for many generations. They have related methods of harvest, storage, and preparation.

Several factors concerning the time of harvest must be considered. First, crops are inconsistent. Some depend on deep winter moisture, some depend on late summer moisture, and some are wiped out by frosts in late spring. Deep, plentiful winter moisture will produce good fruit crops in the mountain shrub species and the prickly pear cactus and also promote excellent root storage in perennial forbs. In contrast, late summer moisture is necessary for a crop of annual succulents, such as purslane. The expectations are never the same from year to year, and sometimes the fruit production is most unexpected. For example, 1999 was quite wet and 2000 was the driest precipitation year in the 72-year record at Mesa Verde National Park. Forbs and grasses showed little growth and almost no seed production in 2000, but the mountain shrub fruit crop was quite good. This is probably because the roots obtained adequate winter moisture in 1999 and because there was no late spring frost in 2000. Spring frosts can completely obliterate the crop of acorns, fleshy fruits of shrubs, and even yucca fruits.

Piñon-juniper woodlands are perhaps among the most provident zones in the Southwest. Where they are intact, with a well-developed understory, and where moist side canyons and riparian lands exist, we find a vast array of plants that have been used for food, fiber, medicine, dyes, and even spiritual purposes. I discuss only a few of these plants in this chapter. Chapters 18 and 20 focus on the many documented uses of piñon and juniper; therefore I concentrate on understory species. I use common names in the discussion; scientific names are presented in Table 19.1.

PLANTS OF THE UPPER ELEVATIONS OF MESA VERDE

In the higher elevations of the Mesa Verde cuesta as well as in side canyons at lower elevations the mountain shrublands intergrade with the woodland. Gambel oak, chokecherry, squawapple, skunkbush, and buffaloberry are often confined to drainages or rocky outcrops where there is relatively high moisture accumulation. Gambel oak has many ethnobotanical uses and is the most widespread of the mountain shrubs. It is a white oak; the acorns of this group do not require leaching to remove tannic acid. The Utes, and certainly the prehistoric peoples, could easily gather large quantities of acorns. The Utes told me that they ground hulled acorns into flour and made a breadlike food from the flour. Gambel oak wood also has many uses. Oak stems have a naturally grown curve at the base when growing on steep slopes. These stems were flattened laterally, and maybe fire-hardened, into hockey stick–shaped digging sticks. Oak was also preferred for hafted axe handles because it was tough and flexible, so that it could be wrapped around the stone axe or hammer head. I believe that the stem was steam-treated to increase flexibility, and when the wood cooled and dried, it retained the newly

TABLE 19.1. Common and scientific names of ethnobotanical plants from Mesa Verde Country

Group	Common Name	Scientific Name
UPPER ZONE	Gambel oak	*Quercus gambelii*
	Chokecherry	*Prunus virginiana*
	Squawapple	*Peraphyllum ramosissimum*
	Skunkbush	*Rhus aromatica* subsp. *trilobata*
	Buffaloberry	*Shepherdia argentea*
	Biscuit-roots	*Cymopterus purpureus*
		Cymopterus constancei
		Cymopterus fendleri
		Cymopterus bulbosus
		Lomatium grayi
		Lomatium dissectum
	Bead potato	*Lomatium triternatum*
	Datil yucca	*Yucca baccata*
	Narrow-leaf yucca	*Yucca harrimaniae*
LOWER ZONE	Flex-stem sego lily	*Calochortus flexuosus*
	Mariposa lily	*Calochortus flexuosus*
	Fleshy-fruit prickly pear	*Opuntia phaeacantha*
	Muttongrass	*Poa fendleriana* subsp. *longiligula*
	Ricegrass	*Achnatherum hymenoides*
	Purslane	*Portulaca oleracea*
	Yellow beeweed	*Cleome lutea*
	Pink beeweed	*Cleome serrulata*
AFTER FIRES	Coyote tobacco	*Nicotiana attenuata*
	Pigweed	*Amaranthus* sp.
	Knotweed	*Polygonum sawatchense*
	Wolfberry	*Lycium pallidum*
FOOD AND MEDICINES	Goosefoot	*Chenopodium Fremonti*
	Yarrow	*Achillea lanulosa*
	Bitterbrush	*Purshia tridentata*
	Yerba immortal	*Asclepias asperula*
	Grommel	*Lithospermum ruderale*
	Wormwood	*Seriphidium tridentata*
		Seriphidium novum
		Artemisia bigelovii
		Artemisia figida
		Artemisia ludoviciana
		Oligosporus dracunculus subsp. *glaucus*
	Mormon tea	*Ephedra viridis*
	Snakeweed	*Gutierrezia sarothrae*
	Wild clematis	*Clematis ligusticifolia*
	Oregon grape	*Mahonia repens*
	Broomrape	*Orobanche fasciculatum*
	Evening primrose	*Oenathera pallida*
	Wild onion	*Allium macropetalum*
	Wild potato	*Solanum jamesii*
	Moonflower	*Datura wrightii*
	Ground cherry	*Physalis fendleri*
	Cottonwood	*Popalus angustifolia*
	Willow	*Salix exigua*
	Dogbane	*Apocynum* sp.

formed shape. Another use of oak is as a source of tannic acid for tanning leather. The enlarged stem galls, found in certain infected clones, are a rich source of tannic acid.

Utah serviceberry rivals oak in its density and widespread nature. The small serviceberry fruit is often infected with a rust, and we have no evidence that people have eaten this fruit extensively. The wood provided roofing material and tools such as hunting bows. The common occurrence of serviceberry in archaeological sites is probably due to the usefulness of the hard, common wood.

Chokecherries are still used for food, and some people make a flavorful wine from the juices. Until modern times the Utes ground the entire berry, including seeds, and formed little cakes that were sold to the Navajos as blood cleansers and food. In the 1960s three people died from eating the freshly ground fruits because they had not allowed the mash to dry (which dissipates the high prussic acid content).

Squawapple has also been used prehistorically and is still gathered today by some people. This miniature apple is very sour and high in pectin and can be delicious when mixed with other fruit. Little is known of early uses of this fruit.

Skunkbush berries, also known as pink lemonade berries, are found in tight clusters of small berries with large seeds. The wonderful thing about this fruit is that it dries on the bush and can be harvested throughout the winter. The Navajos relish the ground fruit, mixed with sugar and eaten in a pudding that they call "cheel-chin." I know a number of people who steep the berries for tea, sweeten it a little, and drink the pretty pink brew. Even though the fruit is also used by wildlife, in a good year there are usually a few clusters remaining at the tops of the bushes.

Buffaloberry is a shrub usually confined to riparian habitats. Fruit production is rare because of late frosts, but when crops are present the berries can be abundant, lending a crimson cast to the silver-leafed bushes. The berries, although small, were harvested by early European settlers after the first fall frost when they could be knocked onto a cloth and gathered in great quantities. My grandmother made a pink vinegar from buffaloberries that was very important for the preservation of other foods. The berries are somewhat sour but are edible fresh or preserved whole. The Utes ate them on the spot or mixed them with other fruits and meat, forming pemmican.

Another common member of the mountain shrub complex, fendlerbush, appears to have been especially selected for items that require extremely hard wood. Hard wooden arrow points and awls were made of fire-hardened fendlerbush stems. This species, the densest of the mountain shrubs, was used more often than any other shrub for these purposes. In addition to the extreme hardness, the stems are smooth and straight-grained and therefore less vulnerable to fracture.

Of the forbs in the upper elevations of Mesa Verde, perhaps the most important are the biscuit-roots, four species of *Cymopterus* and three species of

Figure 19.1. Lomatium triternatum, *a biscuit-root common in Mesa Verde, is prized for its edible root. Illustration by Pat Oppelt.*

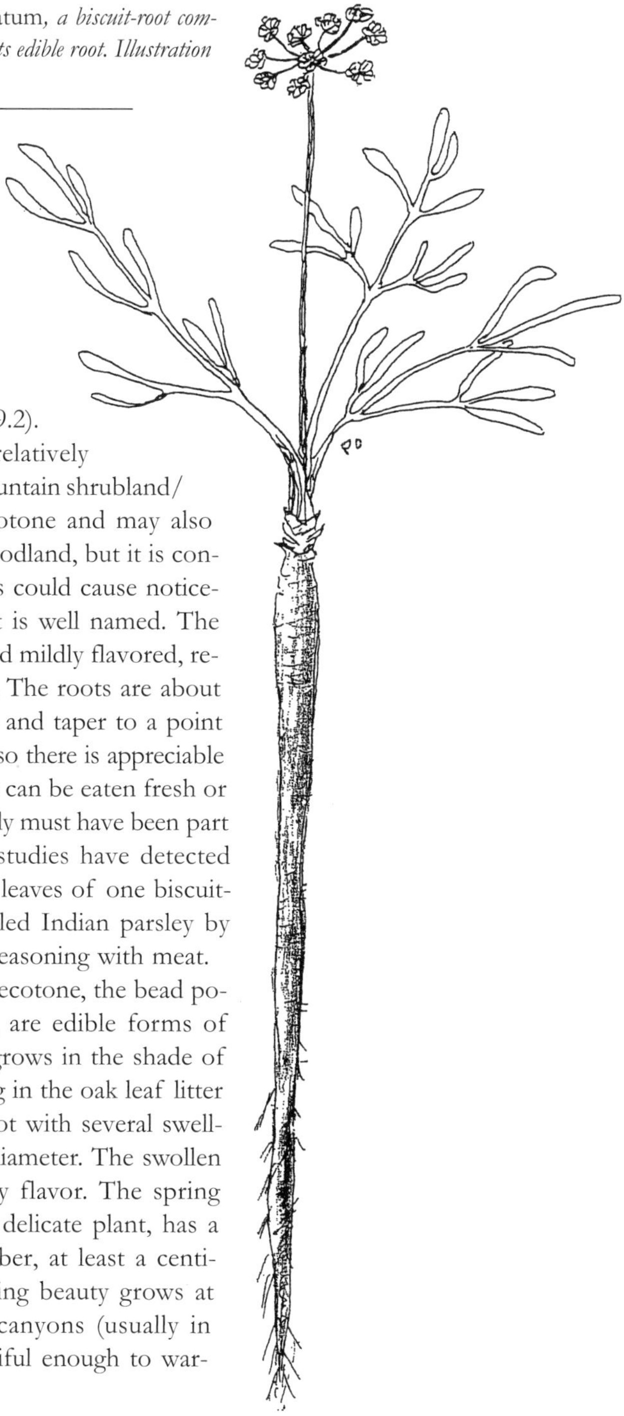

Lomatium (Figures 19.1 and 19.2). Presently these plants are relatively common throughout the mountain shrubland/piñon-juniper woodland ecotone and may also be common in the dense woodland, but it is conceivable that annual harvests could cause noticeable reductions. Biscuit-root is well named. The long roots are nonfibrous and mildly flavored, reminding one of pure starch. The roots are about 1 cm in diameter at the top and taper to a point nearly 22 cm within the soil, so there is appreciable food in each root. The roots can be eaten fresh or cooked; although they certainly must have been part of the prehistoric diet, no studies have detected root cells in coprolites. The leaves of one biscuit-root, *Lomatium grayi*, are called Indian parsley by Apaches and are used as a seasoning with meat.

Two other roots of the ecotone, the bead potato and the spring beauty, are edible forms of stored starch. Bead potato grows in the shade of oakbrush, and a little digging in the oak leaf litter will turn up a horizontal root with several swellings, up to a centimeter in diameter. The swollen nodes have a pleasant nutty flavor. The spring beauty, for such a tiny and delicate plant, has a surprisingly large starchy tuber, at least a centimeter in diameter. The spring beauty grows at canyon heads and in side canyons (usually in meadows) and can be plentiful enough to warrant the digging.

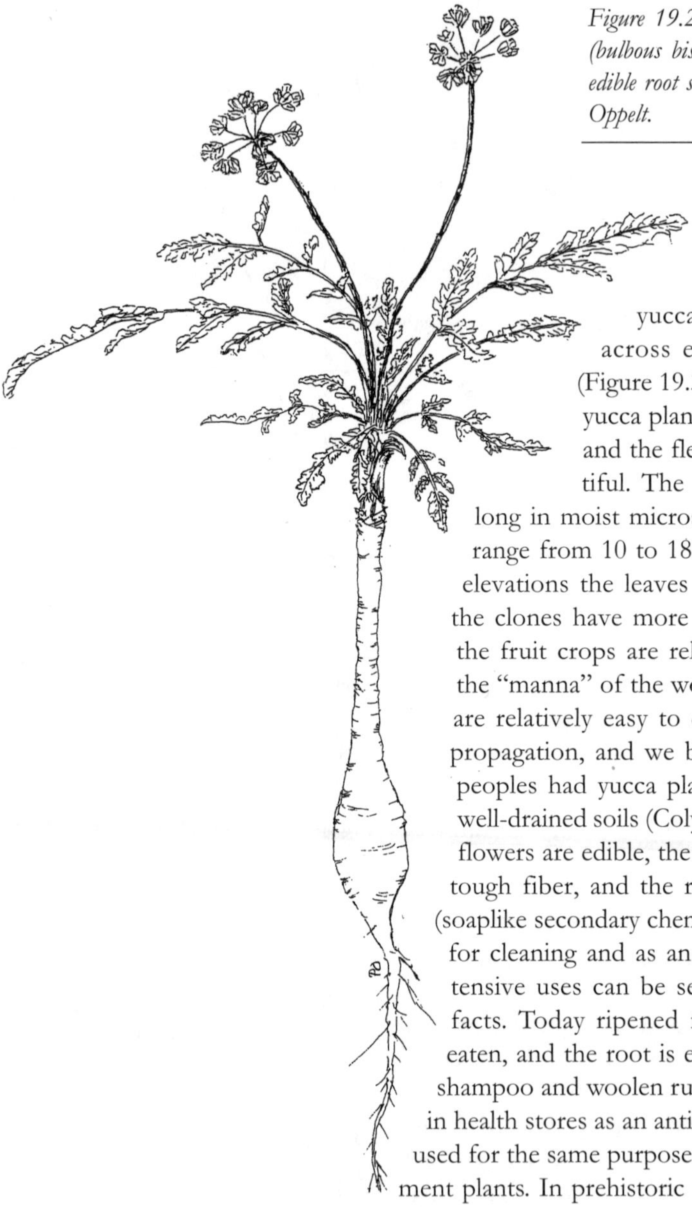

Figure 19.2. Cymopterus bulbosus *(bulbous biscuit-root), showing detail of edible root structure. Illustration by Pat Oppelt.*

Datil or broad-leafed yucca is widely distributed across elevations in our area (Figure 19.3). At higher elevations yucca plants tend to grow bigger, and the fleshy fruit is more plentiful. The leaves can be a meter long in moist microsites, and the fat fruits range from 10 to 18 cm long. In the lower elevations the leaves are only half as long, the clones have more plants, and in general the fruit crops are relatively small. Yucca is the "manna" of the woodland. The rhizomes are relatively easy to dig and transplant for propagation, and we believe that prehistoric peoples had yucca plantations on sites with well-drained soils (Colyer 1976). The fruit and flowers are edible, the leaves have a long and tough fiber, and the root contains saponins (soaplike secondary chemicals) that can be used for cleaning and as an antibiotic. Yucca's extensive uses can be seen in prehistoric artifacts. Today ripened fruits are occasionally eaten, and the root is extracted for a superior shampoo and woolen rug cleaner. Yucca is sold in health stores as an antibiotic and is reportedly used for the same purposes in some sewage treatment plants. In prehistoric times the fruit was extensively harvested; the seeds were removed and stored as dried slices or patties or strung on yucca stems. Cordage made from the long fibers was used widely. We find fine and patterned cordage sandals, baskets, and even watertight canteens made of yucca covered with piñon pitch. The cordage was used to repair pottery and make heavy ropes. Matted fibers were formed into pouches. We find "quids" of the fibers that may represent a medicinal use for

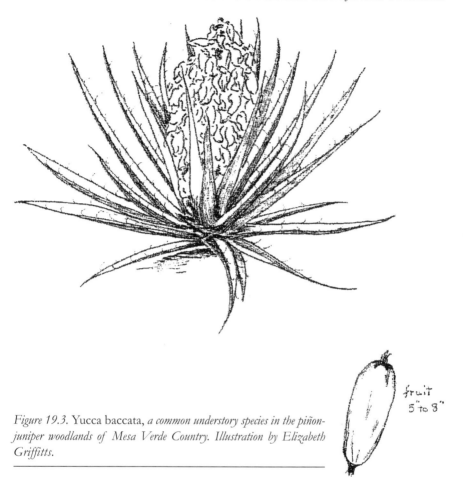

Figure 19.3. Yucca baccata, *a common understory species in the piñon-juniper woodlands of Mesa Verde Country. Illustration by Elizabeth Griffitts.*

digestive ailments. The frayed tips of the leaves were used as brushes for painting pottery.

PLANTS OF THE LOWER ELEVATIONS OF MESA VERDE

All wild onions are edible, but the most abundant species is *Allium macropetalum*, which grows thick as grass in the Blue Hills horizon of the Mancos Shale. Sometimes in April the onions are so prevalent that their odor can be detected for a considerable distance. The little bulbs, over a centimeter in diameter, grow just below the surface of the ground and can easily be dug up. The Utes call this species "sweet onion" and say that they eat it with bread. The flavor is mild when eaten fresh or cooked. Other onion species with larger leaves, found in the shade of oak, were used as a potherb: the entire plant was cooked and eaten.

Sego lilies also have an edible bulb. In this case, the bulb is deep and hard to dig, but the flavor can be appreciated. The flex-stem sego lily is abundant after wet winters and springs and is found in the Mancos Shale floodplains. The other two species are present on the red loess soils but are generally more scattered and would be more difficult to harvest in quantity. Sego lilies reportedly saved the lives of early European settlers during years of domestic crop failure.

The bulbous biscuit-root occurs in lower elevations in Mesa Verde Country. It has a swelling on the root 10 to 15 cm below the ground surface. This plant is most common in clay soils derived from the Mancos Formation.

The fleshy-fruited, large "padded" prickly pear (*Opuntia phaeacantha*) is found on rimrock or in sandy soil throughout the woodland, forming large clones (Figure 19.4). In some years it is plentiful, with fruits as large as a domestic lime. The small glochid spines must be removed by rubbing or searing. About 80% of the prehistoric coprolites studied during the Wetherill Mesa Archeological Project contained prickly pear seeds and glochids. The glochids may have also come from the cactus pads, which were used by some Native Americans until quite recently. This plant is also easy to propagate by transplanting pads during moist times, and it is likely that the prehistoric people maintained prickly pear plantations.

Most of the native grasses do not lend themselves to human use, but two exceptions are muttongrass and Indian ricegrass. Muttongrass, the most common grass in a healthy piñon-juniper woodland in Mesa Verde Country, begins its growth in April. The young grasses, including the new heads, are tender, taste sweet, and do not have prickly awns. We find grass quids in the archaeological remains. Similarly, Indian ricegrass has fairly large seeds and could serve as a grain. We have found ricegrass seed hulls in coprolites as well as grass stems wrapped in bundles with burned ends. Ethnobotanical reports tell of Native Americans harvesting this grass by heating it over a slanted rock above a fire and causing the seeds to pop free. It is likely that both of these grasses were plentiful enough for harvest.

One of the most important wild plants used by people is purslane. This annual appears after summer rains and can grow rapidly to form mats up to 25 cm in diameter. The rapid growth is succulent and has been designated as having a high nutritional value, having a high protein content, being rich in Vitamins A and C, and being very digestible. Purslane grows in sandy loams in warmer bottomland and can become the dominant ground cover in late summer, provided there is ample rainfall. The vivid green growth with red understems taking over bottomlands can be quite beautiful. We know that the prehistoric peoples used the plant as a potherb, and we have even found large storages of the tiny black seeds. Early European settlers regularly harvested purslane and dried it for winter food.

Beeweed seeds were found in over half of the prehistoric coprolites studied from Mesa Verde archaeological sites. Two beeweed species are found in the lower elevations. Pink beeweed reaches 1.5 m tall, and yellow beeweed is generally about 45 cm tall. It is interesting that purple beeweed was prolific in the postfire

Figure 19.4. Prickly pear cactus (Opuntia polycantha) *produces fleshy edible fruit. Illustration by Elizabeth Griffitts.*

vegetation community on Wetherill Mesa following the Pony fire in 2000. Both species can be plentiful in the right conditions, such as in disturbed sandy soils free of interspecific competition with good spring or summer rains. The Utes were still harvesting the young plants for potherbs in the 1960s. Another reported use of this plant is for pottery paint. Ancestral Puebloans utilized iron-rich minerals for their early pottery designs. The later pottery was painted with organic or carbon paint. Beeweed and tansy mustard were boiled to form carbon paint.

PLANTS THAT FLOURISH AFTER FIRES

Several plants stand out after fires in the piñon-juniper woodland. These are annuals whose seeds lie in the ground for decades or maybe even centuries waiting for the heat, the smoke, or the nutrient-enriched soil that comes with fire. Soil nitrogen, phosphorus, and potassium increase significantly after fires, and this fertilization effect persists for about five years (Colyer 1990–1999).

Fire enhances populations of two species of *Chenopodium* (goosefoot). While they are spindly, single-stemmed specimens before fires, they can become the dominant species after fire, characteristically developing into bushes 1.5 m tall! In the second year after the fire, goosefoot reaches about 0.5 m tall and is sparsely distributed. By the third year it has disappeared, except for millions of hard black seeds waiting in the ground. During the first year following a fire, goosefoot leaves, stems, and seeds are edible and were harvested by prehistoric peoples. Ample evidence (such as seed hulls in coprolites and caches of stored seeds) suggests that this plant was an important food. These small black seeds are often identified as "cheno-ams," for the seeds of *Chenopodium* and *Amaranthus* (pigweed) are nearly indistinguishable. We also find *Amaranthus* following fires and in sandy alluvium in bottomlands.

Another plant that changes from a long-waiting straggler to a robust proliferator of large-grained seeds after fire is piñon knotweed. This little annual persists by coming up from year to year in gopher mounds and along drainages and game trails. But with the exposed landscape and enriched soil that a fire provides, it also becomes a multistemmed, bushlike plant, reaching 20 cm tall. The stems are loaded with relatively large, fusiform, hard-coated seeds. Some of the knotweed plants will drop nearly a quart of seeds on the ground in a pile at the end of the first summer after a fire. This is another species that we consistently found to be abundant in prehistoric human coprolites. Knotweed was a semicultigen, growing in association with corn, beans, and squash. Perhaps people practiced farming associated with burning.

A third fire-facilitated plant that remains only a year or two is *Nicotiana attenuata* (wild tobacco). This species is rarely found and if present is usually in disturbed sites. In these instances it is quite small compared with the 2-m-tall "bushes" seen after fire. Wild tobacco was documented in burned mature piñon-juniper woodlands following several fires in Mesa Verde National Park (Floyd, personal communication, 1996). Wild tobacco has a high nicotine content.

ORIGINS OF ETHNOBOTANICAL PLANTS

According to Dr. Joseph Winters (New Mexico State University), many of the ethnobotanical species of the potato-tomato family, Solanaceae, may have been brought to the Four Corners from Mexico. Today five species remain. Many more species of the Solanaceae were being used for food in lands farther to the south; and potatoes, tomatoes, chilies, and eggplants are commonly used by modern peoples. One reason why Dr. Winters suspects the recent arrival of these plants in the Four Corners area is their common association with archaeological sites.

Ground cherries are found growing in bottomlands and near archaeological sites. During the first half of the twentieth century ground cherries were a common weed in modern bean fields that replaced prehistoric gardens and houses. Ground cherries have a very small tomato-like fruit, encased in a husk. The fruits have been used as preserves, mashed into a pudding, and eaten fresh. Ground cherry seeds were found in over half of the human coprolites from Mesa Verde archaeological cliff dwellings. Wolfberry (known locally as Moki bush, referring to the association with prehistoric sites) is sometimes found in dense, large stands covering several acres. It does best at the lower elevations and particularly likes the rock piles of fallen structures. I have often seen it covered with pale cream-colored tubular flowers but rarely with much fruit. The fruit left an acrid flavor in my mouth that lasted for hours. It is said that this fruit is eaten with a white clay that may contain calcium carbonate or perhaps a gypsum clay or a kaolinite. The white clay probably neutralizes solanic acid in the wolfberry. Could it be that the name "wolfberry" is derived from this tantalizingly orange fruit that bites?

Another plant of the Solanaceae family is the wild potato. I know of only three populations of wild potato in the Four Corners. These are in sandy alluvium and are maintaining good numbers even though pocket gophers commonly dig and eat the tubers, which reach diameters of several centimeters. There may be ten to fifteen tubers per plant, easily retrieved from sandy soil. They taste and look like potatoes.

Moonflower is another plant of the same family. This long-lived narcotic plant is found only at the lower elevations and often shows cultural affiliation. Reportedly moonflower has been used during ceremonies to produce an altered state of mind. Possibly coyote tobacco was used in a similar way. Both plants belong to the Solanaceae family.

MEDICINAL PLANTS

Much of what is known about prehistoric use of medicinal plants comes from studying living cultures. The list of plants found in the woodland with documented medicinal use could be quite lengthy. In this section I discuss only a few that are well documented.

Yarrow is found in many of the piñon-juniper understory communities. This plant has legendary medicinal uses that bring relief to the wounded and ailing. The scientific name *Achillea* comes from the fact that in Greek mythology Achilles carried this plant to battle to apply to wounds. Yarrow preparations make wounds close and heal.

Pitch from the piñon tree has been reported to be an excellent treatment for open sores and even toothaches. The turpentine acts as an antibiotic, and the resin seals out air and pathogens.

The tannins and quercin in oak act as an effective astringent, and a decoction from the inner bark will bring relief for insect bites, burns, toothaches, and sore throats. All of these problems probably afflicted prehistoric and historic peoples.

I learned from Ute tribal members that they used bitterbrush as a tissue relaxant for a woman after childbirth. A tea was made of the inner bark, to be consumed while the medicine man performed a dance four times around the new mother. The same tea is one of the few solutions with a native source that is recommended and safe to use as a soothing eyewash.

External and internal parasites have been problems for people for centuries. Three plants are available to treat these afflictions. Extracts of larkspur roots kill external parasites such as head lice, fleas, and ticks. Piñon lousewort and fleabane, powdered and applied, can do the same. *Artemisia* spp. (sagebrush) teas, taken over a two-week period, will cause the expulsion of round worms—hence the common name "wormwood."

Several plants have ethnobotanical uses associated with reproduction. Yerba immortal, a milkweed, is well documented as an abortive. Grommel was used historically by the Navajos in the 1800s, when no more than three children were allowed in a family. Caches of grommel seeds found in archaeological sites may have been collected for use as an abortive.

Mormon tea is one of the best known blood cleansers found in the Southwest. It contains ephedrine, which acts as a decongestant. Many early settlers regularly used a tea made of this plant as a spring tonic.

Cottonwoods and willows are well-known for their anti-inflammatory properties. Chewing the young stems can reduce inflammation, bring down fevers, and reduce muscle aches. The active compounds are salicin and populin. One man of Mexican descent always kept bundles of snakeweed hanging on his porch to take as a cure for headache. Wild clematis also constricts blood vessels around the brain and treats cluster headaches.

Several plants have been used for heart medication. Dogbane, in the hands of an experienced person, can be used to increase the heart rate; an overdose can be fatal. Oregon grape root is still collected, processed, and used to regulate the heart safely.

Juniper cones (or "berries") and twigs are effective for urogenital problems such as kidney infection, cystitis, and vaginal infections. The tea is consumed for a week or two and then discontinued.

Broomrape is taken in its entirety as a sedative. It is highly prized in China, where it has been used for centuries. This plant is parasitic on shrubs such as sagebrush and bitterbrush and is only common during wet springs. The root can be up to 0.8 m in length.

Evening-primrose tea relaxes the vagus nerve and controls coughing. The roots and leaves are steeped for a tea. It is thought that the high level of potassium nitrate is the effective material. There are eight species of evening primrose in the woodland, and it can be locally common. Sometimes it is associated with prehistoric sites, which may indicate that it was cultivated for use in an earlier time.

SUMMARY

In closing, I would like to emphasize that this discussion is only an introduction to the many uses of plants known in Mesa Verde Country. This type of knowledge is often couched in spiritual or legendary frameworks and not readily passed from native peoples to botanists. Rarely is it well documented or published. I close with a few general sources that outline more of the uses for plants from the piñon-juniper zone.

REFERENCES CITED

Colyer, Marilyn. 1976. Yucca production in thirty-six habitat-types. Report on file, Mesa Verde National Park Natural Resources.

———. 1990–1999. Long Mesa burn and Chapin #5 burn, soil nutrient analysis, 1990 through 1999. Report on file, Mesa Verde Natural Resources.

Coon, Nelson, and Louise Hyde. 1976. *The Rodale herb book*. Rodale Press, Emmaus, Pa.

Cummings, Linda Scott. 1992. Anasazi diet: Variety in Hoy House and Lion House—Coprolite record and nutritional analysis. Unpublished manuscript.

Foster, Steven, and James Duke. 1990. *Medicinal plants of the central and eastern United States*. Peterson Field Guide. Houghton Mifflin Company, Boston.

Harrington, H. D. 1967. *Edible native plants of the Rocky Mountains* (with contributions by Marilyn Colyer). University of New Mexico Press, Albuquerque.

Minnis, P. E. 1985. *Social Adaptation to Food Stress: A Prehistoric Southwestern Example*. University of Chicago Press, Chicago.

———. 1991. Prehistoric diet in the northern Southwest: Macroplant remains from Four-Corners feces. *American Antiquity* 54: 543–563.

Moore, Michael. 1979. *Medicinal plants of the Mountain West*. Museum of New Mexico Press, Santa Fe.

———. 1992. *Los Remedios: Traditional Herbal Remedies of the Southwest*. Red Crane Books, Santa Fe.

Steiger, M. A. 1977. Anasazi diet: The coprolite evidence. M.A. thesis, Department of Anthropology, University of Colorado, Boulder.

Willy, Helen. 1976. *Wild Foods Cookbook and Field Guide*. Workman Publishing Company, New York.

Winter, J. C. 1978. Anasazi agriculture at Hovenweep, I: Field systems. *Contributions to Archaeological Studies, Albuquerque Center for Anthropological Studies* 1: 83–97.

ORAL HISTORY SOURCES

I have interviewed over 120 local people concerning varied natural resource historic conditions, including uses of native plants for food, medicine, and other purposes. This was done partially in conjunction with preparation of materials for Harrington 1967. The following people generously shared their knowledge of plant uses:

Elders Group with Fred Niborn, Ute Mountain Tribe, 1967
Camillo Torez, Dolores, Colorado, 1968
Martina Sandoval, Navajo heritage, Sanders, Arizona, 1970
Ethyl Garlinghouse, Apache heritage, Mancos, Colorado, 1975
Rowlene Vecinti, Apache heritage, Chama, New Mexico, 1978
Mary Verlaria, Italian heritage, Dolores, Colorado, 1969

THE ANCESTRAL PUEBLOANS AND THEIR PIÑON-JUNIPER WOODLANDS

Patricia Robins Flint-Lacey

For over a thousand years, from about A.D. 1 to 1300, the piñon-juniper woodland of the Mesa Verde region was home to a group of farming people whom archaeologists now call the Ancestral Puebloans (formerly Anasazis). With their stone tools, they transformed parts of the Mesa Verde National Park area from forests and woodlands to farming plots. Archaeologists know that they raised corn (*Zea mays*), beans (*Phaseolus vulgaris*), and squash (*Curcurbita pepo*) and utilized the wild seeds, roots, and greens of the woodland environments. Rabbits, deer, rodents, and other forest animals were hunted for food. The piñon and juniper wood itself was used for tools, house-building material, and fuel.

Archaeologists find evidence of the Ancestral Pueblo people in the form of artifacts, abandoned homes, and activity areas. The farms were more like what we would think of as gardens, located in areas where the soil was fertile and water was available. Evidence from soil samples surrounding sites and from food remains in hearths indicates that for the duration of the occupation the people continued to rely in part on the wild resources of the woodland. Until about A.D. 900 they found a balance between clearing some of the trees and understory shrubs for their gardens and keeping the forest woodland for the animal habitat and wild plants it nourished.

Finally, though, by A.D. 1300 the Ancestral Pueblo people left the Mesa Verde region for a combination of reasons. A series of droughts, increasing population, more intensive use of the land with consequent soil deterioration, mesa-top deforestation, and reduced harvests all contributed to the abandonment of their homes in the piñon-juniper woodland (Osborne 1976; Schoenwetter and Dittert 1968).

In this chapter I focus specifically on the artifacts, homes, and activity areas found in Mesa Verde National Park that show a clear connection to the piñon-juniper woodlands. These are the things of the past that we can see and touch. It is from this material culture that archaeologists infer ways in which the Ancestral Pueblo people made a living on the Mesa Verde cuesta from A.D. 550 to 1300.

WOODEN ARTIFACTS

Artifacts are the tangible evidence for the Ancestral Pueblo use of the piñon-juniper woodland. The wood of the trees and the shrubs entered into every aspect of life, from digging sticks to dishes, from cradleboards to wooden planks, from wooden flutes to snowshoes, from juniper jar rests to tinder, from oak basket foundations to juniper bark (bast) bundles, from arrows to axe handles, and from weaving tools to prayer sticks.

Digging Sticks

Digging sticks were farming tools used for planting and weeding corn, beans, and squash. Sixteen digging sticks made of serviceberry (*Amelanchier utahensis*) and mountain mahogany (*Cercocarpus montanus*) were found on the Wetherill Mesa Survey (Hayes 1964). Digging sticks found in Mug House were made of serviceberry (Ives et al. 1997), Gambel oak (*Quercus gambelii*) (Hayes 1998), mountain mahogany (Floyd and Kohler 1990), juniper (*Juniperus*) (Cattanach 1980), and four-winged saltbush (*Atriplex canescens*) (Rohn 1971).

Containers

Three shallow wooden dishes were recovered from Wetherill Mesa (Hayes 1964). One of the dishes, at Site 1370, was made of Douglas fir (*Pseudotsuga menziesii*). One, at Site 1892, was constructed by splitting pine and hollowing the inside. The dish at Site 1365 was made of piñon (*Pinus edulis*); it "almost exactly fits a bin depression at the foot of the metate and served the purpose of collecting the ground meal" (Hayes 1964:125).

An aspen (*Populus tremuloides*) bark cylinder was discovered at Mug House (Rohn 1971), and a cottonwood (*Populus* sp.) bark bucket was found at Site 643 on Chapin Mesa (Rohn 1977). The bucket was made from a long strip of cottonwood bark folded in the middle and spirally whipped along its sides with split strips of yucca leaves. It was 36 cm high with a maximum opening of 24 cm by 18 cm. A wooden scoop was also discovered on Chapin Mesa (Rohn 1977). Two hoops of oak were found that were probably prepared for the foundation of a ring basket (Osborne 1980). Willow (*Salix* sp.) was also used as a ring foundation for twilled yucca (*Yucca* sp.) baskets (Figure 20.1) (Osborne 1980).

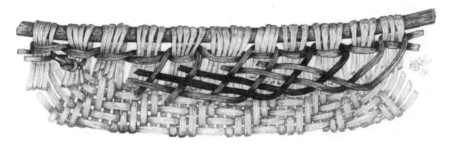

Figure 20.1. Drawing of exterior of fragmentary twilled yucca strip basket from Long House, Mesa Verde National Park, Colorado. Drawing by Marilyn Colyer, Mesa Verde National Park Research Center, Wetherill Mesa Long House Report Plates Folder 9.

CRADLEBOARDS AND PLANKS

Other wooden items found on Wetherill Mesa were the head of a cradleboard and a lapboard (Hayes 1964). Another small pine board, at Mug House, was probably the side of a cradleboard (Rohn 1971). A large ponderosa pine (*Pinus ponderosa*) plank discovered at Mesa Verde was probably from the east wall lintel in the kiva at Fire Temple. It was 3.4 m long, 0.49 m wide, and 0.08 m thick (Nichols 1965).

FLUTE, SNOWSHOE, AND JUNIPER BARK (BAST)

A wooden flute with four holes was found at Spring House on Mesa Verde (Nordenskiold 1979). A snowshoe was found at Ruin 9 (Nordenskiold 1979: Plate 48). Juniper bark (bast) was used for jar rests (Osborne 1980:322, Figure 390b). Juniper bark (bast) found in loosely tied bundles may also have been used as tinder. A bundle was discovered at Long House wrapped around newly made awls, presumably for the protection of the points (Figure 20.2).

ARROWS

All of the arrows excavated in Long House were made in two parts: a cane shaft and a wooden arrow point (Osborne 1980). The shafts of the arrows were made of locally available cane (*Phragmites communis*), approximately 6 mm in diameter. Similar shafts were found at Mug House, where some foreshafts were made of mock orange (*Philadelphus* sp.) and four-winged saltbush (*Atriplex canescens*) (Rohn 1971). The wooden arrow points were made of the shrubby hardwoods of the piñon-juniper woodland: serviceberry, four-winged saltbush, and fendlerbush (*Fendlera rupicola*) (Osborne 1980).

Figure 20.2. Juniper bark (bast) bundles from Long House, Mesa Verde National Park, Colorado. Photo by Fred E. Mang Jr. Mesa Verde National Park Research Center Negative 15629. Figure from Long House Report photo did not appear in published document Figure 465, 570A 4th file.

HAMMER AND AXE HANDLES

Sticks of oak were used as handles for hammers and axes in Long House and Mug House (Osborne 1980; Rohn 1971) (Figure 20.3). A. H. Rohn (1971:210) found a "partially peeled branch of oak about 92 cm long and 1 cm thick, wrapped twice around the groove so that the two ends extend parallel for about 30 cm from each side of the head. Just below the head, the ends are wrapped four times with a yucca-fiber cord tied in a square knot. Wear marks indicate that the ends were once wrapped also near the butt of the handle." A hafted stone axe was also found on Chapin Mesa at Site 519 (Rohn 1977).

WEAVING AND SPINNING TOOLS AND PRAYER STICKS

A rod made of poplar (*Populus*) may have been part of a loom (Osborne 1980). The spinning and weaving arts were further indicated by a spindle whorl, a rectangular disk made of juniper, 8.1 cm long by 6.9 cm wide by 9 mm thick

Figure 20.3. Hammerhead with oak handle from Mug House, Mesa Verde National Park. Handle is about 30 cm long. Photo by Fred E. Mang Jr. Figure 246 Mug House Report.

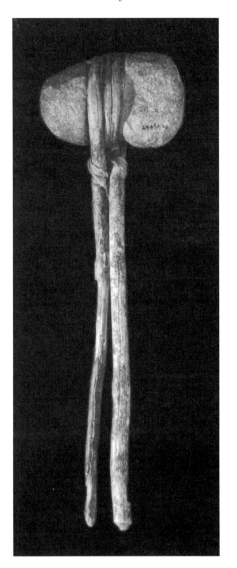

(Osborne 1980). Another spindle whorl, made of ponderosa pine, was found at Mug House (Rohn 1971). Battens for loom weaving, made of juniper and mountain mahogany, were discovered at Mug House along with wooden awls made of fendlerbush, mountain mahogany, and chokecherry (*Prunus virginiana*) (Rohn 1971). Carved sticks of serviceberry (*Amelanchier utahensis*), skunkbush (*Rhus trilobata*), willow, and juniper were thought to be prayer sticks called pahos (Osborne 1980).

INFERENCES FROM THE WOODEN ARTIFACTS

The wooden artifacts indicate that the Ancestral Pueblo people had a full life on Mesa Verde. The artifacts served as tools, containers, musical instruments, and footwear. The people of Mesa Verde enjoyed flute music, prayed, hunted with arrows, planted their seeds and cultivated the ground with digging sticks, made thread and wove fabric for clothing, and used snowshoes.

HOUSE BUILDING MATERIAL

The Ancestral Pueblo people incorporated the resources of the piñon-juniper woodland into their homes. Wooden wall pegs, doorjambs, window lintels, ladders, steps, and roof beams are found in most cliff dwellings (Cattanach 1980; Fiero 1988; Rohn 1971). In Long House in Mesa Verde National Park major portions of the cribbed juniper log roof of Kiva Q were found in place (Cattanach 1980). In the cribbing technique, logs from 2.5 to 3 m (8 to 10 feet) long were first laid from one upright stone support (pilaster) to another. Then a second layer of logs

Figure 20.4. Kiva Q original roofing at Long House, Mesa Verde National Park, Colorado. Roof beams are about 2 to 3 m long. Mesa Verde National Park Research Center, Wetherill Mesa Long House Report Plates Folder 1.

was laid from the midpoint of each log in the first layer (Figure 20.4). In another part of Long House jacal walls "consist of a series of vertical poles—one piñon and five juniper—extending from the cave floor to ceiling. A great number of brush stems, placed vertically, were packed between the main uprights and secured with horizontal oak members and yucca ties. The whole structure was packed with adobe" (Cattanach 1980:35).

Wood was also used for a stockade around Two Raven House (occupied between A.D. 900 and 1100) at Mesa Verde. One piece of the stockade was identified as juniper with an outer ring dated to A.D. 978 (Hayes 1998). Similar stockades have been found in the Mesa Verde region (Kuckelman 1988; Rohn 1975). At Dobbins Stockade 5MT8827 the stockade enclosed an area of 317 m^2 (Kuckelman 1988).

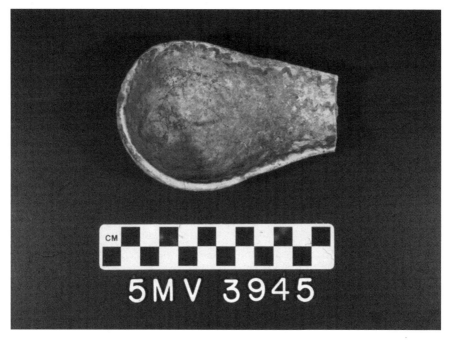

Figure 20.5. Black-on-white ceramic ladle from pottery firing kiln, Site 5MV 3945, Mesa Verde National Park, Colorado. Photo by Julie Bell, Catalog #60138.

FUEL

Piñon and juniper wood was used as fuel to keep warm, to provide light, to cook food, and to "fire" the beautiful black-on-white pottery of the Colorado Plateau (Ford 1984). Charred wood fragments of piñon and juniper were found in seven pottery firing kilns excavated in Mesa Verde National Park (Ives et al. 1997; Matthews 1994) (Figure 20.5).

ANCESTRAL PUEBLO FARMING WITHIN THE PIÑON-JUNIPER WOODLANDS

Deforestation may have been one of the results of the Ancestral Pueblo farming in the Mesa Verde region, as increasing amounts of piñon and juniper were harvested for fuel and home construction materials (Floyd and Kohler 1990). Pollen analysis on Wetherill Mesa showed a relative decrease in juniper and piñon pollen during the Ancestral Pueblo occupation and an increase in weedy edible plants such as goosefoot (*Chenopodium* sp.) and amaranth (*Amaranthus* sp.), grasses, and plants like sagebrush (*Artemisia* sp.) (Martin and Byers 1965). Because of the locations from which the pollen samples were taken, however, there still could have been parts of the mesa top that supported areas of piñon-juniper woodland (Martin and Byers 1965). After the Ancestral Pueblo people abandoned the region, the increase in tree pollen has been interpreted as "secondary forest succession onto previously disturbed, cleared, and cultivated lands" (Wycoff 1977:215). The oldest trees alive today on the cuesta are up to 600 years old, probably germinated in these old fields after abandonment (Chapter 2).

North of Mesa Verde in the Dolores area, corn pollen from about A.D. 600 was found with high levels of fine charcoal fragments, implying that "settlers may have been using large fires to clear the land for farming" (Petersen et al. 1987:18). Areas of sagebrush (*Artemisia tridentata*) may have been cleared before the woodland, because the soil is generally more fertile there (Petersen and Scott 1987). Sagebrush is found in "well-drained, coarse alluvium in the canyon bottoms . . . and the deep and relatively well-drained loess-derived soils between the canyons" (Petersen 1987:18). Clearing the land may have changed the distribution of animals in the region. Gardens would have attracted rabbits, rodents, deer, and other mammals just as they do today (Flint and Neusius 1987; Neusius 1984; Seme 1984).

After A.D. 900 on Mesa Verde, water-control devices such as check dams, hillside terraces, and reservoirs altered the water flow and helped to stop some soil erosion (Rohn 1963). The check dams would have increased the available moisture for crops and native weedy plants growing behind the dams in the rich alluvium. Ethnographic Hopi and Zuni fields are usually located in areas where they will receive runoff water (Winter 1976). In pollen studies at Hovenweep to the east of Mesa Verde National Park, "the consistent occurrence of cultigen and weed pollen with water control structures . . . increases the probability that the devices were actual farms" (Winter 1976:22).

Despite their efforts at soil- and water-control devices, the Ancestral Pueblo people had abandoned their homes in the Mesa Verde region by A.D. 1300. The depletion of the surrounding piñon-juniper woodland may have been a factor. Analysis of human coprolites from Step House, Hoy House, and Lion House indicates a decreasing reliance on forest food such as piñon nuts and wild animals (Stiger 1979). Abandonment of habitation sites was also probably related to the decreased agricultural potential of the land and the final effects of recurring drought (Cordell 1975). The Ancestral Pueblo people left their homes in the Mesa Verde region and migrated south (Osborne 1976).

SUMMARY

Wooden artifacts, tools, house building material, and fuel are the direct physical evidence of the Ancestral Pueblo interaction with the piñon-juniper woodland. Through these material things we glimpse people living in all seasons of the year, adapting to the seasons, using snowshoes, weaving cloth for clothing, using digging sticks for cultivating gardens, and making handles for their stone tools.

Pollen analysis of the areas surrounding archaeological sites suggests that farming in small gardens included the encouragement of native "weedy" plants that would have been good to eat. The food of the gardens also encouraged rabbits, rodents, deer, and other mammals to come in close for the hunters. Some woodland areas remained intact, because the pine and juniper pollen is never missing from the record. Water was certainly a concern for the farmers. There is ample evidence of water-control devices to conserve water behind stone check dams in drainages, to slow water runoff with hillside terraces, and to construct reservoirs. More of these features appear in the archaeological record after A.D. 900.

The Ancestral Pueblo people left a remarkably long record of life in the piñon-juniper woodland of the Mesa Verde region. Scientific study of these remains continues. New evidence from undisturbed archaeological sites (including pollen samples, artifacts, and architecture) will further clarify our picture of these peoples of the past. The evidence we now have indicates effective and creative interaction to sustain their farming life in the piñon-juniper woodland for over a thousand years. After A.D. 900, however, the increasing population, the concomitant intensified use of land and water, and recurring drought seem to have gradually depleted the resources of their environment to the point that by A.D. 1300 these farming people had simply left the land and their homes in the Mesa Verde region.

REFERENCES CITED

Cattanach, G. S. 1980. *Long House, Mesa Verde National Park, Colorado.* Publications in Archeology 7H. Wetherill Mesa Studies. National Park Service, Washington, D.C.

Cordell, L. S. 1975. Predicting site abandonment at Wetherill Mesa. *Kiva* 40(3): 189–202.

Fiero, K. 1988. Balcony House: A history of a cliff dwelling, Mesa Verde National Park, Colorado. Draft. Mesa Verde Museum Association, Inc., Mesa Verde National Park, Cortez, Colo.

Flint, P. R., and S. W. Neusius. 1987. Cottontail procurement among Dolores Anasazi. In K. L. Petersen and J. D. Orcutt (comps.), *Dolores Archaeological Program: Supporting studies—Settlement and Environment,* pp. 257–273. USDI Bureau of Reclamation, Denver.

Floyd, M. L., and T. A. Kohler. 1990. Current productivity and prehistoric use of the piñon (*Pinus edulis,* Pinaceae) in the Dolores Archaeological Project Area, southwestern Colorado. *Economic Botany* 44(2): 141–156.

Ford, R. I. 1984. Ecological consequences of early agriculture in the Southwest. In S. Plog and S. Powell (eds.), *Papers on the Archaeology of Black Mesa, Arizona,* Vol. II, pp. 127–138. Southern Illinois University Press, Carbondale and Edwardsville.

Hayes, A. C. 1964. *The Archeological Survey of Wetherill Mesa, Mesa Verde National Park, Colorado.* Archeological Research Series 7A. National Park Service, Washington, D.C.

———. 1998. *Two Raven House: Mesa Verde National Park, Colorado.* Archeological Research Series 7. Mesa Verde Museum Association. Mesa Verde National Park, Colorado.

Hayes, A. C., and J. A. Lancaster. 1975. *Badger House Community, Mesa Verde National Park. Publications in Archeology 7E: Wetherill Mesa Studies.* National Park Service, Washington, D.C.

Ives, G. A., et al. 1997. Mesa Verde Waterline Replacement Project Phase III: Archeological and historical studies. Draft. Division of Research and Resource Management. Mesa Verde National Park, Colorado.

Kuckelman, K. A. 1988. Excavations at Dobbins Stockage (Site 5MT8827): A Pueblo II habitation. In K. A. Kuckelman and J. N. Morris (comps.), *South Canal: Volume I,* pp. 46–107. Four Corners Archaeological Project Report Number 11. Complete Archaeological Service Associates, Cortez, Colo. Prepared for Cultural Resource Program: Bureau of Reclamation Upper Colorado Region, Salt Lake City. Contract No. 4-CS-40-01650. Delivery Order No. 16.

Martin, P. S., and W. Byers. 1965. Pollen and archaeology at Wetherill Mesa. In D. Osborne (comp.), *Contributions of the Wetherill Mesa Archeology Project. Memoirs of the Society for American Archaeology* 19: 122–133.

Matthews, M. 1994. *Macrobotanical remains from seven kilns excavated at Mesa Verde National Park, Colorado.* San Juan College Cultural Resources Management Program Report 93-SJC-014. San Juan College, Farmington, N.Mex.

Neusius, S. W. 1984. Garden hunting and Anasazi game procurement: Perspectives from Dolores. Paper presented at the 49th Annual Meeting of the Society for American Archaeology, Portland, Ore.

Nichols, R. F. 1965. A large hewn plank from Mesa Verde, Colorado. In D. Osborne (comp.), *Contributions of the Wetherill Mesa Archeological Project. Memoirs of the Society for American Archaeology* 19: 51–56.

Nordenskiold, G. 1979. *The Cliff Dwellers of the Mesa Verde: Their Pottery and Implements.* Translated by D. L. Morgan. Reprinted: Rio Grande Press, Glorieta, N.Mex. Originally published: P. A. Norstedt and Soner, Stockholm, 1893.

Osborne, C. M. 1980. Objects of perishable materials. In G. S. Cattanach (ed.), *Long House, Mesa Verde National Park, Colorado,* pp. 317–367. Publications in Archeology 7H. Wetherill Mesa Studies. National Park Service, Washington, D.C.

Osborne, D. 1976. Slow exodus from Mesa Verde. *Natural History* 85(1): 38–45.

Petersen, K. L. 1987. Vegetation classification for the Dolores Project Area, southwestern Colorado and southeastern Utah. In K. L. Petersen and J. D. Orcutt (comps.), *Dolores Archaeological Program: Supporting Studies—Settlement and Environment,* pp. 7–25. USDI, Bureau of Reclamation, Denver, Colo.

Petersen, K. L., V. L. Clay, M. H. Matthews, and S. W. Neusius. 1987. Implications of Anasazi impact on the landscape. In K. L. Petersen and J. D. Orcutt (comps.), *Dolores Archaeological Program: Supporting studies—Settlement and environment,* pp. 145–184. USDI, Bureau of Reclamation, Denver, Colo.

Petersen, K. L., and L. J. Scott. 1987. Modern surface transect of pollen samples. In K. L. Petersen and J. D. Orcutt (comps.), *Dolores Archaeological Program: Supporting Studies—Settlement and Environment,* pp. 29–38. USDI, Bureau of Reclamation, Denver, Colo.

Rohn, A. H. 1963. Prehistoric soil and water conservation on Chapin Mesa, southwestern Colorado. *American Antiquity* 28(4): 441–455.

———. 1971. *Mug House, Mesa Verde National Park, Colorado.* Archaeological Research Series 7D. National Park Service, Washington, D.C.

———. 1975. A stockaded Basketmaker III village at Yellow Jacket, Colorado. *Kiva* 40(3): 113–119.

———. 1977. *Cultural Continuity and Change on Chapin Mesa.* Regents Press of Kansas, Lawrence.

Schoenwetter, J., and A. D. Dittert Jr. 1968. An ecological interpretation of Anasazi settlement patterns. In B. J. Meggers (ed.), *Anthropological Archeology in the Americas,* pp. 41–66. Anthropological Society of Washington, Washington, D.C.

Seme, M. 1984. The effects of agricultural fields on faunal assemblage variation. In S. Plog and S. Powell (eds.), *Papers on the Archaeology of Black Mesa, Arizona,* Vol. II, pp. 139–157. Southern Illinois University Press, Carbondale and Edwardsville.

Stiger, M. A. 1979. Mesa Verde subsistence patterns from Basketmaker to Pueblo III. *Kiva* 44(2–3): 133–144.

Winter, J. C. 1976. *Hovenweep 1975.* Archeological Report No. 2. San Jose State University, San Jose, Calif.

Winter, J. C., and W. J. Litzinger. 1976. Floral indicators of farm fields. In J. C. Winter, *Hovenweep 1975,* pp. 123–168. Archeological Report No. 2. San Jose State University, San Jose, Calif.

Wycoff, D. G. 1977. Secondary forest succession following abandonment of Mesa Verde. *Kiva* 42(3–4): 215–231.

"ONLY MAN IS VILE"

Duane A. Smith

Isolated in the southwestern corner of Colorado, near the spot where four states touch, lies the beautiful land that became Mesa Verde National Park. For well over five millennia men and women have crisscrossed, lived, hunted, and worked in this region. Before their coming, this quiet land and its animal inhabitants saw only the changing of the seasons.

Although humans have been there for only a smidgen of geological time, their impact has been far greater than their numbers or their time of residency. That impact gradually transformed the land—slowly most of the time, rapidly at fewer times. And more often than not, natural processes were modified for short-term gains, with little attention to long-term consequences. In his "Missionary Hymn," Anglican bishop and hymn writer Reginald Heber understood this human influence. Writing in 1819, he scolded:

> Though every prospect pleases,
> And only man is vile.[1]

Such pessimism perhaps hits too hard, but wherever humans walked in the Four Corners region, they left behind the wake of having been there.

The only constant is change, as the old saying goes. The Four Corners area has been evolving ever since creation; all organisms have contributed their share. Then came the Anasazis, now known as Ancestral Puebloans, and they too changed the land where they lived. One of the most intriguing theories about why they abandoned the region speaks to this clearly. By the end of the thirteenth century they had gathered, hunted, and farmed the land for more than a thousand years. Their impact on the natural resources—land, animals, plants, and timber—had started to take its toll. Even under their primitive farming methods the land suffered. Apparently they had removed most of the forest on the mesa by the late thirteenth century, and their impact on wild game could not have been much less. With a larger population than ever before and changing climatic conditions (becoming warmer and drier), the Mesa Verde people must have faced an environmental crisis.

With no solution in sight, they packed up and moved away, leaving behind mysteries that have yet to be unraveled. Granted, an environmental crisis may be only part of the equation, but it remains a very viable part.

The Utes, who followed them into the region a century or so later, were more hunters and gatherers than farmers. With their low population density, in small bands, they probably had less impact on the land than those who came before and after them. The Utes moved about seasonally and resided in Mesa Verde Country when the Spanish arrived in the eighteenth century. By the 1760s, Spanish explorers, miners, and missionaries had penetrated the area from their base along the Rio Grande.

The La Platas and the San Juan Mountains held out the promise of mineral treasures and the promise of Utes as potential converts to Christianity and providers of profits for traders. Sometime during these decades Mesa Verde received its name. From a distance it did look like a green table. Who might have named it has been lost to history; yet by the second half of the eighteenth century that name (along with La Plata, San Juan, Río Animas, and other geographical designations) had achieved general use.

Mesa Verde, however, did not hold a guarantee of mineral wealth. Nor did it harbor profits for traders or converts for missionaries. The Utes, for some unknown reason that the Spanish did not understand, avoided the area. Spanish explorers, such as Father Francisco Alansio Domínguez and Father Francisco Silvester Vélez de Escalante in 1776, saw it from a distance and even commented on the "ruins" in the area. On August 13, for instance, the party camped near an "ancient times small settlement," the ruins of which reminded Escalante of "the same type of those of the Indians of New Mexico." Interestingly, he also noted that "there is here [Dolores River valley] everything a medium-sized town might need for its subsistence, products from well-watered land, pasture, timber, and firewood."[2]

Neither the Spanish nor the New Mexicans made much of an impression on Mesa Verde or the immediate surrounding region, except by leaving place-names. These New Mexico visitors journeyed through, looking for things that did not seem to be found there. In the nineteenth century what became known as the "Old Spanish Trail" passed nearby (somewhat following the route of Domínguez and Escalante); but, except for occasional "pot hunting," little caught the attention of the travelers on their way to the West.

Change came in the new century, accelerating as the years rolled by. Generally unappreciated at the time, the steady westward march of the United States provided the solution to permanent settlement. While Domínguez's party traveled through the region hoping to blaze a trail to the California missions, colonists back on the East Coast were equally busy declaring independence. Until the purchase of the Louisiana Territory from France in 1803, they were hemmed in east of the Mississippi River. How quickly these frontiersmen moved west can be seen by the fact that William Becknell established winter quarters (1824–1825) somewhere in or near the present Mesa Verde National Park.

Given credit for being "the father of the Santa Fe Trail," Becknell had been the first to travel from Missouri to the romantic and potentially profitable community of Santa Fe in 1821. Mexico, having declared its independence from Spain, made the fatal mistake of opening its northern territories to its expansionist neighbor. Spain had originally closed the door, which had never been very far open anyway. After Zebulon Pike's 1806–1807 expedition showed that such an adventure was easily possible for the Yankees now living on the banks of the Mississippi River, Spanish officials feared the worst. Americans could come in larger and larger numbers, overwhelming the territory.

Fur trappers also probed the area from their bases at Taos and Abiquiu. This was the southern fur trade, less important economically than the central and northern Rockies trade. While the beavers were not completely decimated, as happened in some regions, the San Juan Mountains' populations decreased. The few beavers that existed in the Mancos and Dolores Rivers apparently disappeared, trapped out in the 1820s and 1830s.

Meanwhile, leaving the Santa Fe trade, Becknell turned for several seasons to trapping, trading, and exploring. That explains why he camped for a winter (1824–1825) in what became southwestern Colorado and what attracted him to the ruins. Becknell found "an abundance of broken pottery . . . well baked and neatly painted" and many small stone houses, "some one story beneath the surface of the earth." He speculated that "this was probably a town where the ancient Mexican Indians resided." He could find no further information from the "Spaniards, who seldom visit this part of the country." All this he described in a letter to the Franklin, Missouri, *Missouri Intelligencer* (June 25, 1825), giving its readers their first description of the wonders and enigmas of this region. Escalante praised the land, and Becknell described the mysterious ruins, thereby touching upon the two lures that eventually brought visitors and settlers.

Still, this land remained "terra incognita" well into the 1870s. Frank Fossett in his *Colorado . . . Tourist's Guide* (1878) pictured the extreme southwestern corner of Colorado as the site of the "ancient ruins of towns and cities, built by an extinct race of people."[3] There seemed to have been agricultural settlements in the "fertile river valleys," cave dwellings nearby, and cliff houses or fortresses.

Even then, as Fossett wrote, change had been coming for a quarter of a century. It did not happen overnight. At the conclusion of the Mexican-American War in 1848, this area became part of the United States. A few explorers toured through, but an event in far-off California in January 1848 set off a series of circumstances that finally opened the region to settlement: the discovery of gold.

Following in the footsteps of the Forty-Niners searching for California's mother lode came the Fifty-Niners to Clear Creek in the Pike's Peak country. Within a year prospectors scrambled into the heart of the San Juan Mountains, relentlessly pursuing their golden dreams. Weather, mountains, scarcity of gold, and the Utes drove them out—but not for long. Back they came in 1869 and 1870; within a few years little camps and permanent settlements took root.

At the mouth of La Plata Canyon (bisecting the mountains of the same name) John Moss established Parrott City. In the year of statehood, 1876, it became the county seat of La Plata County, which stretched over the entire southwestern corner of the centennial state. Moss also established a "ranch" on the Mancos River to winter his stock and along the way called attention to the ruins in the area. Visitors started coming. Photographer William H. Jackson was one of the first, and his photographs called attention to the region as his earlier ones had done for Yellowstone National Park. He actually rode all the way around Mesa Verde without exploring it, probably because of the problems in going up the canyons cluttered with trees, brush, and the downed material of the ages. The ruins in Mancos and McElmo Canyons caught his attention.

Others slowly journeyed into the region in and around Mesa Verde, attracted by photographs, drawings, and articles that whetted their curiosity. Some found "cliff houses" in the canyons. They also found that the Utes were sometimes hostile to their ventures, that travel was long and tedious, and that accommodations were often primitive.

At this threshold of the coming of a new age for southwestern Colorado, a debate raged about whether this represented a desert or another western "garden of Eden," awaiting only the coming of the ambitious, hardworking pioneers. George Crofutt straddled the fence in his 1885 *Grip-Sack Guide*. This country "is no means a total desert," as some writers represented it, although the region between the Mancos and Dolores Rivers "generally is dry and sterile." Along the streams, however, "are found grass covered meadows" and "broad belts" of rich bottomlands.[4] Only time would tell.

Parrott City would never have sole dominance over this vast region. Just a few miles east of the tiny camp, the contemporary agricultural village of Animas City slowly grew. Then the Denver and Rio Grande Railroad gave birth to Durango on the Animas River in September 1880; within months it became the largest town on Colorado's Western Slope, topping 2,000 inhabitants. The next summer the tracks and trains finally arrived. Now visitors, investors, and settlers could come with an ease little imagined a decade before. Even closer to Mesa Verde, near where Moss's ranch had been, the permanent settlement of Mancos appeared. By the 1880s it had become a ranching and outfitting point for prospectors, tourists, and settlers. A short day's ride away on the Dolores River, Big Bend served ranchers. With their grasses and milder winter climate, these valleys offered better rangeland than that in the high mountains. A host of mining communities in the La Platas and San Juans provided a potential market.

Among those who came to settle and ranch in the Mancos Valley was the Benjamin Wetherill family in 1880. They settled slightly south and west of the "thriving little village," as the *Durango Morning Herald* (November 12, 1887) described Mancos. Stretching in all directions were the community's "farms," as the *Herald* enthusiastically depicted them. They were more like ranches, however, with subsistence farming. Each one faced long odds of isolation, altitude, and at that time lack of accessible nearby markets. The railroad, when it arrived, would ease some problems. As these pioneers cleared the land, cut timber, diverted water to their pastures and fields, and built roads, they—like the "extinct people" before them—started to transform the land.

The Wetherills illustrated how this happened; for them, as for their contemporaries, this all represented progress, the "march of empire," as many Americans lovingly called it. Benjamin and his five sons, variously described as working on a farm or ranch, were hailed for their "industry and hard labor." "Their large farm is enclosed by a substantial board fence" and divided into fields "ranging from four to ten acres." Everything about the ranch, with its large and well-filled barns and a strong and compact log house, gave "evidence of thrift and comfort." At Durango's agricultural and mineral display in 1887, they received praise for their "fine wheat on exhibition."[5] Their ranch house, fields, corrals, hay and straw stacks, and barns looked much like those of their neighbors. Agricultural settlement had come to stay, along with the miners.

For the boys, ranching held more interest than humdrum farming, and it was natural for them to winter their cattle along the more sheltered and milder Mancos Canyon. Unlike some of their neighbors, this Quaker family befriended their Ute neighbors and treated them fairly. In return, the Utes let them graze their cattle on the reservation south of Mancos. There the Wetherills wintered their cattle and had a winter camp.

At the same time, they learned from the Utes about "the old people" deep in the canyons along the river. Unfettered by the "spirits of the dead," the Wetherills searched for them during their leisure time and while trailing cattle that drifted into those canyons along the Mancos. The result was the "discovery" of what later were named Cliff Palace and Spruce Tree House in December 1888.

A new era dawned for southwestern Colorado, Mesa Verde, archaeology, and every nearby community. From this point, tourism and growth would change the region more swiftly and more permanently than anything had since the first humans walked along the river bottom and climbed the mesas.

For instance, to the west of Mesa Verde, Cortez gained a toehold in 1886, placing settlements on three sides of Mesa Verde. No southwestern Colorado settlement started out with more handicaps, including the usual isolation and no railroad. Its semiarid location lacked an adequate nearby water supply. Without water, the "lifeblood" of the west, Cortez had little hope of growing into a viable community.

Securing that viability took precedence over everything else. A large-scale irrigation system tapping the Dolores River seemed a natural and simple solution. Only a narrow ridge separated the river from the Montezuma Valley, and a tunnel could quickly resolve that. Work on it had started in February 1886, even before Cortez's beginnings.

It proved one thing to scheme and plan, quite another to tunnel 5,400 feet through the ridge from the Dolores River to the Montezuma Valley. Facing money problems, the original Montezuma Valley Water Supply reorganized. Then a potential rival appeared: the Dolores Land and Canal Company, which blasted a 4,000-foot cut through the divide. Not enough potential buyers existed for them both to succeed; so, fortunately, they merged before bankruptcy claimed either one of them. By 1890, with water gushing into the valley, over 100 miles of canals had been built, along with diverting dams and a partly constructed storage reservoir. Cortez's future looked brighter.

The new *Montezuma Journal* (April 23, 1888) expressed the feelings of the home folks as the life-saving water approached: "The event should receive that public and enthusiastic recognition by the people of Montezuma Valley, that its close association with their future interest and prosperity warrants. Let's celebrate, and let it be a novel one by all taking WATER." The water arrived—three months later than planned, but few cared. Their town was saved. Its hopes had triumphed.

Already these pioneers had unintentionally modified their environment by planting non-native trees, which Cortez home owners kept alive, as one resident remembered, by putting "all wash water, even such slop as not too greasy on the trees to keep them alive" until the irrigation water arrived.[6] These non-native species included elms, ashes, Russian olives, and various fruit trees. They looked nice when they survived, provided shade, and produced fruits in favorable years, including peaches, pears, and apples.

With less tribulation, Durango and Mancos had done the same thing. They wanted to re-create the life they had left behind: how better to give that appearance than with familiar grasses, bushes, and trees? That these old familiar varieties crowded out native plants and took more valuable water than native species (and might not be suited for the elevation or climate) troubled few people. Unbeknownst to them, they also brought along some reasons for the allergies that they assumed they had left behind.

With irrigation water, the Montezuma Valley offered more farmland than the La Plata or Mancos Valley. Still, even with the water, much of the homesteaders' effort would be aimed toward dryland farming, which came into vogue in the 1890s and early twentieth century. There would never be enough water to irrigate the valley; nonetheless, where it went, agriculture blossomed. It would take longer for the great plain to be converted into farms, with huge pinto bean fields, among other crops, replacing the sagebrush, scrub oak, and native grasses. Not until the 1920s, amid a general agriculture depression, would some of the outlying acreage come into production.

Ranchers continued grazing their cattle, particularly along the Dolores, La Plata, and Mancos Rivers, just as the Wetherills had done. Sadly, like pioneers from time immemorial, they tended to overstock and overgraze the range. That had been an American tradition: use the natural resources and move on. Somewhere to the west, they could make a new start on virgin land. But time was running out; few fresh possibilities existed.

Neighboring white settlers pressured to open this "fertile and salubrious" land not being utilized "correctly" by the Utes. As a result the resident Ute bands were split. Those favoring individual allotments became the Southern Utes. They occupied the eastern part of the old reservation. Those favoring communal ownership became the Ute Mountain Utes, and their land bordered Mancos, Mesa Verde, and Cortez to the south. One of the last land rushes occurred in 1899, when Southern Ute lands along the Animas, Florida, and Los Pinos Rivers and bordering mesas were opened to homesteading.

Meanwhile, white settlement had not only surrounded Mesa Verde on three sides but had finally crept into the canyons and onto the mesa top itself. Like the Wetherills, others saw potential there and also in the neighboring valleys.

Starting in 1889, the Mesa Verde area was increasingly used as free, uncontrolled open range for stock. The practice did not stop with the eventual creation of the park. Until August 1911, sheep and cattle grazed without permit or restriction.

It is understandable why ranchers moved into Mesa Verde. The canyons provided shelter in the winter. Those mesa fingers tend southward toward the broad canyon of the Mancos River, where water is available in all seasons. When the snowmelt came in the late winter and early spring, the grasses grew and the flowers blossomed—an ideal situation for cattle coming off a winter range. In some ways these months display this captivating country at its best, before dry May and June arrive.

While ranchers and homesteaders gained a Mesa Verde toehold, another Wetherill-spurred development started affecting the mesa's environment: tourism. As news spread of their findings, the Wetherills quickly found out that tourists could be a more lucrative proposition than ranching/farming. Their Alamo Ranch in the "wilderness of sage-brush and pinon-pine," as one visitor described it, quickly emerged as a jumping-off point for trips to the ruins. With the completion of the Rio Grande Southern Railroad to Mancos in 1891, tourism took on a new meaning and mode. Mancos gained the initial advantage as the "gateway to the ancients."[7]

That railroad itself had an impact on the environment. Road rights-of-way had to be surveyed; land had to be cleared and graded; railroad buildings and yards had to be constructed; and, with the coming of trains, fires caused by sparks from engines had to be squelched. These were the harbingers of a variety of environmental changes to come.

Realizing the bonanza at their doorstep, other entrepreneurs from Mancos, Durango, and Cortez tried to copy the Wetherills, with varying degrees of success. A triangular rivalry developed among the three communities. Cortez soon dropped by the wayside, but Durango and Mancos quickly became involved in an urban fray over their respective aspirations.

It was a three-day trip from Mancos: one in, one touring, one out on horseback. Tourists traveled down the Mancos River, then up the canyons to the ruins. The numbers slowly increased, particularly after the Mesa Verde exhibit at the World's Fair in Chicago in 1893. Visitors originally camped out near water sources, then set about to pot-hunt for souvenirs amid the ruins. The next logical step occurred when a Mancos rival of the Wetherills, Charles Kelly, built a cabin near Spruce Tree House to improve the stay. With a spring nearby, it provided tourists with at least a rough, frontier type of comfort. Each step damaged the local ecology.

The initial environmental impact seemed harmless enough. Trees were cut for firewood; rattlesnakes, which had long called the ruins home, were killed; rough trails were blazed from one site to another; random shots were taken at birds and other critters; and litter now marked the tourists' path. Some of these things threatened the ruins and their heritage, perils that inspired a dedicated group of women to stage a decade-long fight to preserve the sites. It resulted in victory when conservationist president Theodore Roosevelt signed the national park bill in June 1906. That created a whole new perspective on those trying to make a living within the suddenly drawn park boundaries.

The early superintendents faced a daunting multitude of issues and problems. A fair share of these concerned the land and environment. Time was of the essence.

Something else that the government had little expected came to the forefront. Coal was known to exist within Mesa Verde National Park, mostly on the west and northwest sides. Cortez relied on that coal. The railroad never reached Cortez to open its market to the much larger and more significant coal fields at Durango and Mancos's neighbor, Hesperus. Coal transportation by wagon was a costly, time-consuming effort. As early as the 1890s, visitors reported coal outcroppings, including the famous Gustaf Nordenskiold, who wrote in his *Cliff Dwellers of the Mesa Verde* about coal being "found in fairly thick seams."[8] Small mines opened just about the time Roosevelt established the park.

Figure 21.1. Early-day visitors stayed at the Kelly Cabin and picked up "souvenirs" throughout the ruins. Mesa Verde National Park. Courtesy, Jean Bader.

By 1900 the long-silent Mesa Verde already whispered with the voices and sounds of tourists and ranchers who pastured their cattle among the ruins and canyons. And the repercussions had only started. For nearby Coloradans, the park's attraction represented one huge potential dollar sign (Figure 21.1). The *Mancos Times* (July 6, 1900), for example, suggested that little beyond "a good wagon road" needed to be built into the heart of the region to increase "Cliff Dwelling Travel at least 10 fold."

The *Denver Times* (August 11, 1907) concurred: the "torturous" trails ought to be replaced by good roads (Figure 21.2). Eventually hotels and "creature comforts" should be provided for the weary wayfarer. "Then indeed will the Mesa Verde national park be second to none in the country; not even the far-famed Yellowstone park."

What impact did those early visitors have on Mesa Verde? The first permanent superintendent, Hans Randolph, wrote the secretary of the interior in September 1907: "[T]here having been no restraint on campers in the past it does not require much imagination to realize how very dirty it is." He wanted to employ two workers for six weeks to clean up the "very disorderly condition—tin cans, papers and all kinds of rubbish." That cleanup was followed by orders stating that

Figure 21.2. By the 1920s, Mesa Verde was popular, and the human impact multiplied: a tour conducted in 1929. Courtesy, Mesa Verde National Park Archive.

in the future, "no rubbish would be allowed to accumulate about camps or in the vicinity of the principal ruins."[9]

While that was going on, the superintendent and rangers also had problems with the ranchers. In March 1908 they advised cattle ranchers that their animals would "not be permitted to graze" within the park boundaries. A plan was already under way to prepare regulations for "the impounding of the animals."[10] Plans did little good; only action would suffice.

This action, of course, ran full force into what westerners considered their vested right to use all the natural resources within their reach whether they owned them or not. The government's conservation practices had already angered westerners for over a decade as Uncle Sam set aside forest preserves, soon to become national forests. Coloradans stood in the forefront in fighting this intrusion on their "natural" rights, even to the point of threatening and shooting at rangers. Now it seemed that Washington planned to go even further—not a pleasant prospect for residents of Montezuma and La Plata Counties.

Meanwhile, homesteaders had already filed on land before the park had even been established. From 1900 to 1903, four homesteaders had applied for homesteads of 160 acres each. In 1900 William and Albert Prater and Mabyn Morfield settled in the canyons named after them. Three years later Ellen Waters filed for a homestead in what was soon to be named Waters Canyon. They all had more to do with ranching and water sources than with farming. What their status in the park would be remained undetermined.

One of the first things the superintendents did was to try to improve the dangerously scarce sources of water, particularly as tourism increased. A survey found that fourteen wells had been dug before park establishment (some for tourist use, a few for cattle). The principal source, a spring near Spruce Tree that Kelly utilized, had been used for over a decade. In 1908 a dam was built at the head of Spruce Tree Canyon to store water for "pack and saddle stock" and to "increase the domestic water supply." With only eighty visitors that year, that amount sufficed; but in less than a decade problems appeared.

Some of these issues were addressed in the park's 1908 *Laws and Regulations*. For example, the owners of patented land within the park "are entitled to full use" of their property. The superintendent was authorized to designate camping places; and all garbage and refuse "must be deposited in places where it is not offensive to the eye or would contaminate any water supply." Visitors "are forbidden to injure or disturb mineral deposits, ruins, wonders, relics and the like."[11] Vandalism had already become a problem for park officials and spurred them into action, as it had done with the women earlier.

The problem of what to do with the homesteaders in the park could not be resolved so neatly by the 1908 regulation. The overriding issue was whether private ownership should be allowed in the park. But it also involved whether cattle were overgrazing the canyon bottoms, the use of water, and the impact on timber and other resources.

The first move to limit the problem came in 1911. The park banned sheep (those legendary banes of western cattle ranchers) "unless [the superintendent knew] of some good reason to the contrary." The Waters family had 600 sheep grazing in the canyon, and this had brought the issue to a head. Waters claimed that he moved his animals "every three days thus preventing the sod being destroyed or damaging the grazing tracts." Park officials concurred, but the rule came down from Washington, and sheep roamed free only another year. Because cattle were allowed to graze in national forests, the superintendent could allow ranchers and Utes to graze horses and cattle at a fixed rate per annum. Proceeds went into the Mesa Verde National Park fund.

Another minor environmental problem occurred when visitors brought dogs into Mesa Verde. This dilemma was finally resolved with a 1913 "doggie warning." Dogs taken into the park must "be prevented from chasing animals and birds or annoying passersby." They either had to ride in wagons or be "led behind them," then be "kept within the limits of the camps when halted."[12] The punishment for disobeying was that dogs found at large would be killed. That seemed

331

unfair when the owners were to blame. Their feline antagonists were not permitted in the park at all.

Cats and dogs aside, Superintendent Thomas Rickner reported that in 1914 three people (all landowners in the park) had permits to graze cattle, up to a total of 1,335 head. They were required to aid rangers in maintaining order and to "guard against fires."[13] Considering the amount of available grazing land, that appeared to be too many cattle. It was not long before the effects on the grassy meadows at the canyon's floor became noticeable. World War I (with its all-out effort to defeat the barbarian Hun) put further pressure on the park to allow more cattle and sheep to graze. They needed the animals in the war effort to supply troops and "starving Europe."

While all this happened, the interest in coal continued unabated. One of the initial things Randolph did after his appointment was to go with a "mineral expert" to "thoroughly inspect" the park to find out if any "valuable mineral deposits" occurred there. Fortunately for Mesa Verde's future, only coal seemed to offer any profitable potential. The U.S. Geological Survey came to the same conclusion in a report stating that "the appearance of coal is good and without doubt it can be profitably mined, where conditions are favorable."[14]

Several small mines operated on the park's western boundary. Now the question arose as to whether individuals or the federal government owned them, or if they should be operated at all. The inhabitants of Cortez fretted. These mines provided jobs, supplied some of the community's fuel needs, and helped reduce local coal imports. They asked their congressman, Edward Taylor, to look into the matter and open the park to coal mining. Eventually he sponsored a bill that passed Congress after some debate, only to be promptly vetoed by President William Howard Taft. He argued that the bill did not give the leasing money to the park, which would be "a radical departure from other park legislation."

Coal mining went on without authorization, as Hans Randolph advised the secretary of the interior (the National Park Service would not be created until 1916). Other congressional attempts to permit mining failed, until the secretary finally granted permission in June 1910 to authorize leases, provided they "shall not include any of the prehistoric ruins" or hinder public access.[15] Fortunately, the coal outcroppings were a long way from the major ruins or any place where the public typically wandered.

Coal mining could now legally be carried on in the park. Within six months two leases were approved and official mining started. Neither mine ever moved beyond what could generously be classified as small. Slightly more than 600 tons were mined the first year; the figure jumped to 1,300 tons in 1913. Matters were not helped when the grade turned poor, with streaks "unfit to use." Transportation from mine to market continued to be difficult and time-consuming. Operations limped into 1919 with a 228-ton production, hardly a decent day's work for a successful coal mine. This tonnage was definitely too small to reduce costs or turn a profit; so mining stopped, while ranching continued in the canyons and mesas.

That was the situation in 1919, after the world had been saved for democracy and other good things by the "boys over there." Amid the excitement of the time, a major development occurred that shaped the park's destiny. Since Mesa Verde's creation, the superintendency had been a political appointment, like far too many other jobs. Such appointments had caused their share of controversy, and the appointees basically saw little wrong with leasing coal mines or running cattle within the park. After all, the lessees were neighbors and friends who appreciated the opportunity. Now the park service appointed a professional archaeologist, Jesse Nusbaum, as superintendent. Nusbaum, who assumed the position in May 1921, saw very little right with the idea of leasing.

Nusbaum (thirty-four years old) had initially been to Mesa Verde to help with a survey of the ruins in 1907–1908. One of his first projects now became the effort to improve the quality of visits by tourists. This included widening the roads for cars, improving the "distressingly bad" sanitary arrangements, replacing inept staff, improving the museum, identifying ruins, and moving the park headquarters out of Mancos into Mesa Verde. His first winter in the park brought back memories of the 1890s. The road was closed by snow, and it required three days by pack and saddle horses to make the round-trip to Mancos for mail and supplies.

Settled in and with some immediately pressing problems resolved, Superintendent Nusbaum turned to the matter of grazing permits. Ranchers overgrazed the park, he wrote; "prodigious overgrazing of the range" had taken place. Looking back through the history of grazing permits, he found that since 1907, no more than 850 head had ever been authorized. With the absence of any ranger to ride the range to check on the total number, he claimed that in some years nearly 5,000 head had been in the park. Nusbaum may have exaggerated a little here; but, in his defense, there was also confusion in Mesa Verde's records. Records showed 2,008 cattle in the peak year for grazing (1918). The owner of three tracts of patented land, including five wells and windmills, "completely monopolized water sources and grazing." It even turned out that one of the previous superintendents and his son-in-law had run large numbers of cattle within Mesa Verde.

Writing to the park service in December 1922, Nusbaum remarked that farmers living in the mesa area considered it their right (as long as they did not disturb the ruins) "to run cattle and sheep and cut timber." The boundaries were not marked, and funds had never been given to do so; the Department of Justice had refused to prosecute two cases because signs had not been posted.

In the background, the political realities presented themselves. From talking to previous superintendents, Nusbaum had learned that any action affecting "grazing permits was extremely hazardous, very much like professional suicide." That did not deter him. The problem appeared plain enough, but the answer not so simple.

In 1922, new grazing leases and regulations included no grazing on Chapin Mesa (the site of the visitor center and Spruce Tree House, among other ruins) and required repair of any damage to roads caused by cattle. The park service charged twenty cents a head on the park's west side (less water) and forty cents on the east side. The superintendent even offered to pay half the cost for five miles of "drift fence" (along the main road) if the permittees would "stand the same amount." The fence went up, with the ranchers doing the building in lieu of cash. "Then, thank God, no more cattle cutting up roads." Nusbaum further adopted a plan gradually to phase out grazing by 20% each year.[16]

Another problem presented itself. Nusbaum was fed up with the cattle grazing near the "so-called mummy lake." He believed stock ranchers rather than the original Mesa Verde inhabitants had conducted water into it. In an April 1922 report he complained that the cattle "invariably watered there every season." They littered the site, which he genteelly described as "unpleasant to eye and nose."[17]

Step by step, an effort was made to purchase privately owned land within Mesa Verde National Park. The 1920s agricultural depression, with sharply declining cattle prices, both helped and hindered the success of this effort. Several homesteaders were very willing to sell the land but not so willing to reduce their herds in the park because of the low prices. The superintendent continued with his plan and had nearly succeeded in removing the cattle by 1927.

Nusbaum intended to end grazing in 1926 but ran up against a widow who managed to convince the government that she should have one more year. That was the last obstacle. As the National Park Service director bluntly wrote Nusbaum on April 20, 1927, "we are not in the cattle business."[18] Congressman Edward Taylor helped clear up the final landownership question by securing $5,000 to purchase the last 320 acres of patented land. Grazing by permit ceased within the park boundaries.

That ended one phase of the grazing problems but did not solve the issue completely. In the years that followed, cattle and horses wandered in from the Ute Mountain Ute Reservation despite the protests of the park service. This problem decreased with the creation of a Ute park bordering Mesa Verde to the south.

This all lay in the future when Nusbaum turned his attention to the coal mines. He solved this issue with much less emotion. The meagerness of the deposits smoothed the way; only one lessee tried mining in the 1920s, without success. The improved roads and trucking to Cortez that made transporting coal to the community easier and cheaper also eased the way to a solution. A 1920 congressional act further strengthened Nusbaum's hand. It excluded prospecting and mining within national parks.

By carefully investigating boundary lines, Nusbaum found that the entrance to the small mine intermittently operating had been on park land, but the mine itself was located on Ute land. He promptly shifted the "small" royalties that had accrued to the tribe. Thus the coal mine issue died away, but it had raised a significant question about mineral resources in national parks.

Eventually, in 1931, a bill passed Congress to simplify park administration, "to provide for the uniform administration of the national parks." It dealt specifically with clauses in the original organic acts that were inconsistent with conservation and park usage. Section one prohibited mining in Mesa Verde and Grand Canyon National Parks. A long-needed precedent had been established. The coal-mining issue proved particularly important. Mesa Verde had been a significant test case. While there would be pressure in the years after 1931 to open park lands to mining, Mesa Verde has never been one of the target areas—its very limited mineral potential has generated no interest.

The short-sighted but understandable mistake of allowing grazing and mining within Mesa Verde luckily produced no permanent damage or unsightly scars. Mesa Verde National Park had been blessed. The outcome could have been much different.

To have allowed grazing and mining would have compromised the integrity of this park and have been incompatible with the park's mission and preservation. The lesson learned has not been lost locally. Mesa Verde has not been violated since then. This curbing and then denial of what many westerners thought was their basic right to utilize nearby natural resources fell before the twentieth century's interest in conservation and preserving national parks. Along with it came increased involvement of the federal government in land and park issues. Times change, and old attitudes had to give way.

In the decades since then, nature has reclaimed the locations of mining activity. Short of physically touring the coal sites, their long-ago presence is not noticeable. By purchasing the homesteads and closing the open range, a tremendous step was taken to restore Mesa Verde's natural environment. That allowed nature, with the help of the park service, to heal the wounds of overgrazing and homesteading. The grass grew back, although many non-native plants took root and may never disappear.

Today's visitor can little imagine how this once overgrazed and often trashed landscape looked to previous generations. Most of the trash of the past century has been cleaned up by the park service over the years, and the sites have been allowed to return to nature. Maybe Bishop Heber was wrong. Man is not as "vile" as he pictured. Humanity's stewardship awareness and a redeeming nature have helped restore the land and preserve Mesa Verde National Park. The future looks more promising than the past.

NOTES

1. Reginald Heber, "Missionary Hymn," cited in *Familiar Quotations* (Boston: Little, Brown and Company, 1992), p. 395. For background information on this chapter, see the Mesa Verde Superintendent reports, 1907–1928, Mesa Verde Research Center; Duane A.

Smith, *Mesa Verde National Park,* rev. ed. (Boulder: University Press of Colorado, 2002); Rosemary Nusbaum, *Tierra Dulce* (Santa Fe: Sun Stone Press, 1980); Correspondence File, Mesa Verde Research Center; and Duane A. Smith, "Cortez, Coal, and Cliff Dwellings," *Journal of the West* (October 1989).

2. Ted Warner (ed.), *The Domínguez-Escalante Journal* (Provo: Brigham Young University, 1976), pp. 13–14.

3. Frank Fossett, *Colorado . . . Tourist's Guide* (Denver: Tribune Steam Printing, 1878), pp. 428–429.

4. George Crofutt, *Crofutt's Grip-Sack Guide of Colorado* (Omaha: Overland Publishing, 1885), p. 163.

5. *Durango Morning Herald,* September 27 and November 12, 1887.

6. Mary Blake interview, Southwest Center, Fort Lewis College.

7. Fred Cowling letter, Wetherill Collection, Mesa Verde National Park; *Mancos Times,* June 2, 1893.

8. Gustaf Nordenskiold, *The Cliff Dwellers of the Mesa Verde* (Stockholm: P. A. Norstedt, 1893), pp. 8–9.

9. Hans Randolph, Superintendent's Report, September 4, 1908, p. 6.

10. Grazing Hearing Records, March 4, 1908, Mesa Verde National Park.

11. *Laws and Regulations Relating to the Mesa Verde National Park, Colorado* (Washington, D.C.: Government Printing Office, 1908), pp. 8–9.

12. Department of the Interior to Superintendent, December 11, 1913, Correspondence File, Mesa Verde National Park.

13. Thomas Rickner, "Report of the Superintendent," in *Reports of the Department of the Interior 1914* (Washington, D.C.: Government Printing Office, 1914), p. 796.

14. *Mancos Time-Tribune,* October 4, 1907.

15. Samuel Shoemaker, "Report of the Superintendent," *Report of the Department of the Interior 1913* (Washington, D.C.: Government Printing Office, 1913), p. 8.

16. Material on the grazing problem is found in Nusbaum, *Tierra Dulce,* pp. 77–78; Grazing Hearing Records, 1942, Mesa Verde National Park; Jesse Nusbaum Report, April 5, 1922, Mesa Verde National Park; and Nusbaum to Director, April 11, 1927, Mesa Verde National Park.

17. Jesse Nusbaum Report, April 5, 1922, Mesa Verde National Park.

18. Director to Nusbaum, April 20, 1927, Mesa Verde National Park.

PART IV

LOOKING INTO THE FUTURE: WHAT ARE THE HOPES AND CONCERNS FOR THE ECOLOGICAL INTEGRITY OF MESA VERDE COUNTRY?

The first three parts of this book provide a description, even a celebration, of the remarkable ecology and natural history of the country in and around Mesa Verde. We must end the book on a somewhat sobering note, as we look into the future and see possible signs of trouble. The ecological integrity of this region is still very great, and with some vision and care it will continue to be so into the indefinite future. Without vision and without care, however, we could lose many of the most important characteristics of this special place.

Chapter 22 reviews the key threats to long-term ecological integrity of old-growth piñon-juniper woodlands in Mesa Verde. Note that the ancient woodlands within the national park are protected from obvious forms of damage, such as woodcutting, mining, and harvest of wildlife. Yet even these "protected" areas may be vulnerable to the insidious effects of uncontrollable wildfires, invasion of non-native species, habitat fragmentation resulting from land-use changes outside the park, air pollution, and global climate change.

Chapter 23 is a thoughtful reflection on management challenges by the current director of resource management in Mesa Verde National Park. This chapter is a fitting epilogue to our book, for it provides a summary of all the issues raised in previous chapters as well as a vision for ensuring that future generations will be able to have the kind of Mesa Verde experience that our generation has been so privileged to enjoy.

THREATS TO THE PIÑON-JUNIPER WOODLANDS

William H. Romme, Sylvia Oliva, and M. Lisa Floyd

The ancient piñon-juniper woodlands of Mesa Verde and the surrounding region have endured centuries of climatic variability, fire, insect attacks, and human utilization. It is clear that these ecosystems are resilient to the kinds and magnitudes of disturbances and stresses that have occurred in the last several centuries. As we enter the twenty-first century, however, all wildland ecosystems on earth increasingly are being subjected to stresses and disturbances that exceed the magnitude or severity of previous centuries. They also are experiencing novel stresses, unprecedented in the evolutionary history of the native organisms. In this chapter we examine five kinds of disturbances and stresses that we believe pose the greatest threats to the long-term integrity and stability of piñon-juniper woodland ecosystems in southwestern Colorado: fire, invasion by non-native species, habitat fragmentation due to land-use change, air pollution, and global climate change.

FIRE

It is ironic that we include fire as a threat to the ecological integrity of piñon-juniper ecosystems in the Mesa Verde Country. Fire is a natural ecological process that has occurred repeatedly throughout the last several centuries, and all of the native organisms have adaptations that allow them to survive or even thrive in the face of fire. We regard fire as a potentially serious threat in the future, however, because the frequency, severity, and ecological effects of fire are likely to be different in the next century than they were in previous centuries. Moreover, the threat of fire is intertwined with nearly all of the other threats discussed in this part. For instance, invasion of burned areas by non-native plant species can result in a shorter interval before the next fire and can make fire behavior and fire effects more severe than they were in the past. Habitat fragmentation resulting from land-use change (e.g., building roads and homes in formerly uninterrupted tracts of native vegetation) enhances the spread of non-native species into burned areas and increases the risk of human-caused ignitions. Finally, current projections based on climate models all forecast rising temperatures, altered precipitation patterns, and increased frequency and severity of wildfires (see the following discussion for details on these other threats).

It is important to recognize that fire is *not* a threat to all types of vegetation in the Mesa Verde Country. As discussed in more detail in Chapter 16, fire actually stimulates many native plant species, including Gambel oak, serviceberry, aspen, and various native herbs. All of these fire-stimulated species share the characteristic of prolific root sprouting after damage to aboveground portions of the plant. Thus fire kills the stems and leaves but rarely damages the root system, because the roots are insulated by soil and can survive even without the aboveground portions of the plant. This undamaged root system responds by sending up abundant new, fast-growing stems (Brown 1958; Kunzler and Harper 1980; Tiedemann et al. 1987). Therefore, if the next century does indeed bring increasing frequency and severity of fires, then the abundance and areal extent of fire-stimulated shrubs and herbs likely will increase throughout southwestern Colorado and other portions of the West. Shrublands dominated by Gambel oak or serviceberry already are a major vegetation type in this region; they may become even more widespread in the wake of frequent and extensive wildfires during the next century.

If shrublands and other fire-stimulated vegetation types increase during the next century, they will do so at the expense of other kinds of vegetation, dominated by species that lack the ability to sprout from the roots and are completely killed by fire. A prime example of fire-sensitive vegetation in the Mesa Verde Country forms the centerpiece of this book: piñon-juniper woodlands.

Colorado piñon and Utah and Rocky Mountain juniper are easily killed by even relatively low-intensity fire, because their bark is thin and provides little insulation for the sensitive cambium located just below the bark. Their foliage also is very flammable and often low-hanging, so that flames from a fire burning on the ground can sweep into the crowns of the trees and consume all of the needles and small twigs (Barney and Frischknecht 1974; Koniak 1985; Leopold 1924). Once burned, piñon and the junipers are very slow to regrow in an area (Erdman 1970). Seeds from unburned trees must be transported by birds and mammals into the burned area and buried in suitable growing locations. The young seedlings are vulnerable to spring drought, winter freezing and thawing, and herbivory by birds and small mammals. A new piñon or juniper tree takes many decades to grow to maturity, and a stand of piñon-juniper woodland requires centuries to develop the old-growth structural characteristics described in Chapter 2.

At this time we have no evidence that fire frequency and severity actually have increased significantly in southwestern Colorado. There were more large, severe fires in piñon-juniper woodlands during the latter half of the twentieth century than during the first half, but similar large fires occurred during the second half of the nineteenth century (Floyd et al. 2000; Chapter 16). Thus the fire regime in piñon-juniper woodlands of southwestern Colorado does not yet appear to be outside the historic range of variability in fire frequency, extent, and severity that characterized previous centuries. The fire activity since about 1980, however, probably has been near the upper limit of that historic range of variability. In the last two decades we have seen four very large fires in Mesa Verde National Park as well as other large fires on private, Bureau of Land Management (BLM), and tribal lands near Durango, Red Mesa, and Naturita, Colorado.

A striking feature of these recent fires in piñon-juniper woodlands is an exceedingly slow regeneration of piñon and juniper trees. In the 1989 and 1996 Mesa Verde burns, intensive sampling has revealed almost no new tree establishment. Less rigorous observations in the other recent burns suggest a similarly slow establishment of new trees throughout the region. We do not yet know the reasons for this conspicuous paucity of tree regeneration in recently burned piñon-juniper woodlands in southwestern Colorado. It may simply reflect a normally slow successional process in this vegetation type, or it may be symptomatic of more profound habitat changes—related to climate changes, invasion of nonnative species, or other factors not yet identified. Many of the ancient stands in Mesa Verde National Park developed centuries ago, when the climate and overall environment were different than they are today (Chapter 2; Floyd et al. 2000). It is possible that when these old woodlands burn now, they cannot return to their previous state under current climatic and environmental conditions. We urgently need additional research to understand what controls postfire succession in piñon-juniper woodlands of southwestern Colorado. Nevertheless, given the extremely slow recovery of burned woodlands in this region, whatever the cause, the potential increase in fire frequency, extent, and severity that is forecast by climate models (see the following discussion) poses a serious threat to the long-term ecological integrity of this unique vegetation type.

There may be little that managers can do to reduce the threat of future wildfires in the Mesa Verde Country. Despite a policy of complete fire suppression in Mesa Verde National Park, the total area that burned within the park from 1951 to 2000 was equal to or greater than the area that burned from 1851 to 1900, when there was no attempt at fire control (Floyd et al. 2000). Piñon-juniper and mountain shrubland communities burn relatively infrequently, but they burn ferociously under certain weather conditions (Chapter 3). Even with modern fire-fighting technologies, it appears that in these vegetation types we mainly put out fires that would have been relatively small anyway. One thing that managers can do is to be very judicious in their use of prescribed fire. Prescribed fires, ignited by managers under low-severity weather conditions, have become an important tool for reducing fuel loads and restoring desirable ecological conditions in other vegetation types, notably ponderosa-pine forests and grasslands. Prescribed burning in piñon-juniper forests of the Colorado Plateau probably should be avoided, however, because of the slow regeneration of forests and also because of the risk of invasion by non-native species (as described in the next section).

INVASION BY NON-NATIVE SPECIES

One of the most important ecological events now occurring in the world is a sweeping change in the distributions of plant and animal species. Many species are dwindling in numbers, occupying a smaller range, or becoming extinct, while others are exploding in numbers and moving into new ranges that they never before occupied. The major driving force behind these changes is human activities. Humans are creating new kinds of habitats (e.g., large patches of bare ground), altering disturbance regimes (e.g., fire frequency and severity), and transporting species into places they never could reach before. A particular concern in piñon-juniper woodlands of the Colorado Plateau is colonization by plant and animal species that formerly did not exist in this region.

Colonization by new species is not necessarily a bad thing. In fact, it is a natural process. For example, elk evolved in Eurasia and did not exist in North America until they migrated across the Bering land bridge during the Pleistocene. Piñon and ponderosa pine were both absent from the Colorado Plateau region during the last ice age and only entered the Mesa Verde Country within the last 10,000 years (Betancourt 1990). Nobody would refer to elk, piñon, or ponderosa pine as "alien" or "exotic" or "bad" species in this region. The rate at which new species are arriving in the Colorado Plateau has accelerated dramatically in the last few hundred years, however, and some of the new arrivals have distinctly undesirable characteristics: they displace the native species and alter ecosystem processes in ways that reduce the ecological integrity of local ecosystems. It is with these latter species that this section is concerned.

Undesirable non-native species (especially plants) often are referred to as "weeds," but this term is vague and loaded with many different connotations. Non-native species also may be called "exotics" or "aliens"—terms that may be similarly loaded. In this chapter we refer to undesirable non-native species as "invaders" (Bazzaz 1986). These are species that can enter even relatively intact natural communities, where they displace the dominant native species or alter ecosystem processes.

Invader species can sometimes invade undisturbed ecological communities, but their invasion often is enhanced by disturbances to the native community (i.e., by natural or human-caused processes that kill some of the dominant native plants or even clear the vegetation from a piece of ground). One important natural process that can accelerate the spread of invader species is fire. Large fires occurred in piñon-juniper woodlands of Mesa Verde National Park in 1934, 1959, and 1972, each burning hundreds of hectares. Although detailed records were not kept, it appears that these large burns were not invaded extensively by non-native plants. Rather, native shrubs and perennial herbaceous species resprouted, and early successional species germinated from buried seed and other sources (Erdman 1970). A very different picture, however, emerged after the large fires in 1989, 1996, and 2000. Three years after the 1989 fire, burned piñon-juniper stands were dominated by non-native species: *Carduus nutans* (musk thistle), *Lactuca serriola* (prickly wild lettuce), *Alyssum minor* (yellow alyssum), and *Bromus tectorum* (cheatgrass). Musk thistle was particularly conspicuous, forming dense impenetrable patches of prickly stems up to 2 m tall. Less abundant but locally important non-native species included *Cirsium arvense* (Canada thistle), *Cirsium vulgare* (bull thistle), *Salsola iberica* (Russian thistle), *Centaurea debisa* and *C. repens* (knapweeds), *Ranunculus testiculatus* (bur buttercup), *Linaria vulgaris* (yellow toadflax), *Lepidium latifolium* (perennial pepperweed), *Agropyron intermedium* (intermediate wheatgrass), and *Bromus inermis* (smooth brome).

Why the dramatic change between 1972 (early postfire communities dominated by native species) and 1989 (early postfire communities dominated by non-native invaders)? The reason apparently is that the abundance and spatial distribution of invader species had been increasing rapidly throughout the park during the 1970s and 1980s. The increase in the national park paralleled an increase in these species throughout the entire region. When the fire occurred in 1972, seed sources of invader species were few and far away, but in 1989 the seed sources were numerous and close by. Three of these invader species are of particular concern in Mesa Verde National Park: musk thistle, Canada thistle, and cheatgrass.

Musk thistle's native range extends through North Africa, Europe, Asia Minor, and Siberia. It now has spread to New Zealand, Australia, and North America, where its range is still expanding (Shea and Kelly 1998). The plant was first detected in eastern Colorado in 1976 (Dunn 1976) and has since spread at an alarming rate throughout the state. Musk thistle is usually a biennial, reproducing exclusively from seed, with each plant producing up to 6,000 viable seeds. Seeds are dispersed by wind and less frequently by water, generally within 100 m of the parent plant (Beck 1991; Rutledge and McLendon 1998). Our observations in Mesa Verde, however, document large quantities of seeds dispersing by wind up to 300 m from the parent plants and at least a few seeds up to 3,000 m. Musk thistle thrives in places where the soil has been disturbed (e.g., agricultural lands, roadsides, trails, and burned areas). We have observed that it also can invade undisturbed piñon-juniper woodlands, where it persists in low numbers until a fire or other disturbance stimulates it to reproduce prolifically.

Canada thistle had a more limited distribution in the 1989 burn than musk thistle, but where it was found it was a cause for great concern. Canada thistle usually thrives in moist areas (e.g., along stream channels and around springs). Its horizontal adventitious roots may be 2 m deep and may extend for many meters outward around the plant. These roots send up new shoots and can form dense patches in which almost no other plant species can grow (Hodgson 1968; Rees 1990). These patches of nearly pure Canada thistle can expand outward at a rate of up to 2–4 m per year. The roots also resprout prolifically after fire. Canada thistle not only reproduces vegetatively but also is a prolific seed producer. Its small tufted seeds may be dispersed great distances by the wind. The most abundant populations of Canada thistle in Mesa Verde National Park are in the wetlands created by sewage treatment facilities near Morefield Campground and Far View Lodge. These artificial habitats, created to provide necessary visitor services, also create optimal conditions for Canada thistle, because they provide moisture, nutrients, and frequent disturbance. When some of these sewage wetlands burned in the fires of 1996, the thistle's roots survived the fire, resprouted prolifically, and produced a huge seed crop. Canada thistle is now abundant all along the stream channel that drains the sewage treatment facility.

Cheatgrass is of concern because of its potential for drastically altering the natural fire regime (Mack 1986). The plant is so named because it greens up early in the spring, giving an illusion of abundant future forage for livestock and wildlife, but soon it becomes dry and unpalatable. Cheatgrass invaded a wide variety of ecosystems over a vast geographic area during a relatively short period in the early twentieth century (Mack 1986). It owes its success in part to its extremely flexible life-history characteristics (Bazzaz 1986). It can complete its life cycle in a single summer, function as a winter annual, or even be biennial, depending on moisture conditions. Seed germination occurs over a period of several months, ensuring that an entire cohort will not be eliminated by a single period of drought (Mack and Pyke 1986).

Prior to the twentieth century, throughout much of the Great Basin of western North America, fires generally were patchy and infrequent because the native shrub-steppe vegetation provided only patchy, discontinuous fuel beds. Cheatgrass, however, creates a continuous fuel bed of light, easily ignited material that can carry a fire over extensive areas. Thus in some areas cheatgrass has changed the fire regime from a previous pattern of infrequent patchy fires to one of frequent extensive fires. Many of the native shrub species are poorly adapted to frequent fire and have been locally extirpated over large areas. In these areas where cheatgrass is now a dominant species, not only has community composition changed, but a basic ecosystem process—fire—also has been profoundly altered.

Although cheatgrass is widespread in Mesa Verde National Park, it does not yet appear to have caused any abnormal changes in fire behavior during recent fires. The rapid spread and severity of recent fires can be explained adequately by the severe weather conditions that existed at the time. Cheatgrass has the potential to change future fire behavior, however, in ways that would threaten the recovery of recently burned piñon-juniper woodlands. This recovery process takes place over hundreds of years (see Chapter 16 on fire). Frequent fires in burned areas now dominated by cheatgrass and other invader species could prevent normal recovery of the native piñon-juniper vegetation and lead to extensive stands of vegetation dominated by non-native plant species.

CONTROLLING THE INVASION OF NON-NATIVE PLANTS

When the 1996 Chapin 5 fire burned 1,972 ha of mesa tops, meadows, and deep canyons in Mesa Verde National Park, the postfire rehabilitation team recognized that invasion of non-native plants would be one of the most serious problems. Fortunately, research had been conducted on this problem after the 1989 Long Mesa fire, so in 1996 we knew where the invader problem was likely to be most severe and what kinds of steps could be taken to ameliorate the adverse impacts of non-native invaders. Two profoundly different successional trajectories were documented after the 1989 fire. In petran chaparral and other vegetation types with abundant shrubs and perennial herbs, resprouting of those shrubs and herbs returned total plant cover to prefire levels within two years. In these areas with adequate residual plants (i.e., plants that survived the fire and quickly resprouted) we saw relatively little invasion by non-native species, probably because the native residuals quickly exploited the bare ground and other resources made available by the fire such that non-native species had little opportunity to invade. Invaders quickly became abundant, however, in burned piñon-juniper woodlands and other vegetation types that lacked adequate residual plants. Therefore, after the 1996 fire, it was predicted that piñon-juniper woodlands would be most vulnerable to invasion by non-native plant species and that shrubland vegetation would be least vulnerable. Surveys conducted after 1996 verified this prediction.

Based on this prediction of relative vulnerability, efforts to reduce the invasion of non-native plant species were conducted most aggressively in the areas of burned piñon-juniper woodland. Three techniques were used: aerial seeding of native grass seed, mechanical removal of invading plants, and chemical removal. Seeds from local populations of *Poa fendleriana* (muttongrass), *Stipa comata* (needle-and-thread), *Oryzopsis hymenoides* (Indian ricegrass), *Agropyron trachycaulum* (slender wheatgrass), *Agropyron smithii* (western wheatgrass), *Koeleria cristata* (junegrass), and *Sitanion hystrix* (squirreltail grass) were mixed and applied by helicopter in early October 1996. Several dense musk thistle patches were removed by digging up the rosettes (hand-grubbing) from June to August 1997 and June 1998 in canyon bottomlands where native grasses were likely to resprout as well as on rocky canyon walls and mesa tops where residual vegetation was lacking. Twenty-three Canada thistle patches were sprayed in June and August 1998 with "Curtail" (3,6-dichloro–

2-pyridinecarboxylic acid, monthanolamine salt 7.5%, 2,4-dichlorophenoxy acetic acid, tris-propanolmaine salt 38.4%) (Gary Salamacha, personal communication, 1997). These were short-term treatments. In addition, a series of biological controls specific to Canada and musk thistles was released to reduce the populations over the next several decades (Deborah Kendall, personal communication, 1997).

Assessments conducted a year later indicated that all three treatment methods were at least somewhat effective in reducing the abundance of non-native invaders (Floyd et al. n.d.). One year after seeding treatment, grass density was significantly higher in the seeded areas than on nearby controls, and the difference was even greater in the following year (8,576 clumps/ha in seeded areas relative to 550 clumps/ha in unseeded areas). All species flowered and produced seeds in 1997 and in 1998. Slender wheatgrass, squirreltail grass, and Indian ricegrass were especially conspicuous. Two years after the seeding treatment, musk thistle density in seeded areas was about one-seventh its density in unseeded areas (13,850/ha in unseeded to 1,850/ha in seeded). The effectiveness of mechanical removal of musk thistle depended on whether or not native grasses became established in the same area. Where there were native grasses, average musk thistle density decreased from about 31,000 plants/ha to about 8,000 plants/ha. Where native grasses were lacking, the musk thistle simply resprouted after mechanical removal, and its density remained unchanged at about 35,000 plants/ha. The herbicide treatments killed 70–100% of Canada thistle stems within two months of spraying and maintained an 80% kill the next year, indicating at least a temporary reduction in Canada thistle.

The results of this experiment in controlling invasion of non-native plant species after fire indicated that we can achieve at least limited and temporary success from a combination of treatments, including the planting of native species as well as mechanical and chemical removal of invading plants. None of the three treatments prevented invasion, however: at best they simply reduced the density of invaders. The prospect for eliminating invader species and maintaining the natural ecological communities of the Colorado Plateau is not bright. The best approach would be to avoid creating suitable habitats for invaders and to avoid transporting invaders into suitable habitats, but neither of these actions can be achieved completely. Potential invaders like musk thistle and cheatgrass already have established high populations and a broad geographic distribution in the region, and disturbances that create bare ground are unavoidable. Ironically, fire—a natural ecological process that actually helped to maintain the ecological integrity of many ecosystems prior to the twentieth century—now facilitates invasion of non-native species that may impair ecological integrity (Hobbs and Huenneke 1992). One of the most pressing challenges for managers of natural areas is how to reintroduce the rejuvenating effects of a natural fire regime without accelerating the degrading effects of non-native invaders.

HABITAT FRAGMENTATION

The leading cause of species extinction in the world today is loss of habitat (Ehrlich and Ehrlich 1981). Therefore, habitat preservation is essential in any conservation program. Even a large total acreage of a particular kind of habitat, however, may not maintain all of the species that depend on that habitat if the individual patches of habitat are too small or too isolated (Forman 1995).

Many habitats are naturally fragmented, and most species are composed of more or less discrete subpopulations that interact only via occasional migrating individuals. The amphibians that live within the borders of Mesa Verde National Park provide an excellent example of naturally fragmented populations. Dr. Albert Spencer (professor emeritus at Fort Lewis College) has studied these populations for many years and provided the information presented here. Red-spotted and Woodhouse's toads live in isolated tinajas, small depressions in sandstone outcroppings. These depressions fill with water after even a small rain event because of the runoff from surrounding impermeable slickrock and may retain water for several weeks after a rain event. The toads lay their eggs in the pools; the young toads rapidly mature and in most years are able to emerge as adults before the pools dry completely. These habitats, however, are fraught with natural perils. During a prolonged drought the pool may dry before the tadpoles metamorphose into adults, in which case all perish. An intense local rainstorm may produce such a deluge that the pool is flushed out completely and all of its inhabitants are swept over the nearby cliffs. This local extinction event is only a temporary disaster, because the pool probably will be recolonized within a few to many years, as individuals dispersing from other pools wander into the one that had lost its former inhabitants. Thus if we were to monitor the toad populations in a number of these small, isolated tinajas over a long period, we would see local populations blinking in and out in response to periodic local extinction and recolonization.

Although this situation seems precarious, the toads apparently have persisted for hundreds or thousands of years in what is now Mesa Verde National Park, because after every local extinction event there has been the opportunity for recolonization from other surviving populations nearby. Nonetheless, conservation biologists now are concerned about the future prospects for amphibian populations within the park, because we do not know just where most of the recolonizing individuals are coming from. The major source may be the Mancos River, which lies just to the south of the park and is the only perennial water source of any size in the area. The Mancos River formerly was lined with marshes and oxbow ponds that provided excellent and stable habitat for amphibians. Because of upstream water diversions, however, the natural wetlands along Mancos Canyon near Mesa Verde are gradually drying up. As the wetlands decline, so will the core amphibian populations. A point may be reached at which too few toads remain along the Mancos River to disperse into the park and recolonize vacant tinajas. Thus, even though toads are protected from any kind of direct harm within the borders of the park, the park populations may be threatened by events taking place outside the park boundary. We do not know enough about the long-term population dynamics of red-spotted and Woodhouse's toads to predict their future survival prospects confidently. Unfortunately, we know even less about most of the other fauna and flora of the Mesa Verde Country, many of which exist within similarly fragmented patches of suitable habitat. More research is urgently needed to identify species at risk because of naturally fragmented habitats and human-caused changes in those habitats.

Even for species that are distributed more or less continuously over larger areas (e.g., the plants and animals of piñon-juniper woodlands), human-caused habitat fragmentation may be a threat to long-term persistence (Noss and Cooperrider 1994). Roads, gas wells, homes, agricultural fields, pasture improvement projects, and other developments break up patches of native vegetation into smaller patches (Knight 1997; Knight et al. 1995). Some species thrive in places where the vegetation has been disturbed or at the edge between disturbed and undisturbed vegetation. These "edge" species usually are widely distributed, do not have stringent habitat requirements, and are not of conservation concern. Many, in fact, could be considered "invaders" (see the preceding discussion). Examples of edge species include magpies and crows (native species) and cheatgrass and thistles (non-native invaders). As large patches of native vegetation are broken into smaller, isolated patches, edge species find opportunities to expand into places where they formerly could not thrive.

As the edge species expand into disturbed areas, other species decline. Species that require large tracts of native vegetation are referred to as "interior" species. These organisms often have demanding habitat requirements (e.g., shade, a deep layer of organic matter on the soil surface, cavities within old trees, or the dense crown of an old oak or serviceberry) and cannot survive in the open conditions created by clearing of the vegetation. Examples of interior species include the orange-crowned warbler, green-tailed towhee, and *Pedicularis centranthera* (a small lousewort of old-growth piñon-juniper woodlands). All of these sensitive interior species are native to the Four Corners region.

How serious a threat is human-caused fragmentation to the interior species of the Mesa Verde Country? Apparently no species in this region has yet become extinct because of an inadequate total amount of habitat or because the habitat has become too fragmented for individuals to find suitable places to live or to find one another. Because habitat fragmentation has been a major cause of local species extinction in other regions (e.g., some birds in deciduous forests of central and eastern North America: Terborgh 1989), however, we should be alert to early signs of fragmentation effects in our region. An analysis of the spatial distribution of critical wildlife habitat and rural subdivisions in La Plata County, Colorado, revealed that as of 1995, subdivisions occupied from 6 to 17% of several important kinds of habitats, including wetlands, critical wintering grounds for deer and elk, and black bear concentration areas (Romme 1997). If current trends in land development, oil-well drilling, and associated road building continue, we soon may reach a threshold after which increasing numbers of native species will become imperiled by human-caused habitat loss and fragmentation.

AIR POLLUTION

Mesa Verde National Park protects all native organisms from direct forms of exploitation, but threats related to changes in air quality are insidious. Even the most rigorous protection cannot keep out the sulfur oxides, nitrogen oxides, ozone, and particulate emissions produced by nearby power plants, gas wells, and automobiles.

Although it lies in a generally remote region, with sweeping scenic vistas, the Mesa Verde Country is unusually vulnerable to air pollution effects because it is situated immediately downwind from two large coal-powered, electric-generating plants in northwestern New Mexico (Figure 22.1), as well as other plants located elsewhere on the Colorado Plateau. The Arizona Public Service Four Corners Plant (FCP) and the Public Service Company of the New Mexico San Juan Generating Plant (SJGP) have been operating for thirty-seven and twenty-seven years, respectively. Both power plants are exempt from protective standards of the Clean Air Act (amended 1990), because legislators originally thought the plants would be phased out of operation as they aged. Emissions from these old power plants are up to ten times greater than those from newer or upgraded plants (Weinhold 2000).

Other sources of human-caused air pollution in the Mesa Verde Country are local automobile emissions, long-distance drifts from many sources within the airshed of the entire Southwest and Colorado Plateau, and waste products from regional gas and oil wells. Natural sources also cause episodes of air pollution and deposition. Smoke from wildfires throughout the entire West has contributed significantly to Four Corners air pollution on occasion, even to the point that light extinction from smoke particles apparently caused deciduous leaves to turn color earlier than the usual October 10 peak in 2000 (M. Colyer, personal communication, 2000; S. Oliva, personal observation). Unusual white streaks in rock formations and crumbling sandstone were noted after the eruptions of Mt. Pinatubo in June 1999 and Mt. St. Helens in 1980. Because volcanoes release vast quantities of carbon dioxide, sulfur dioxide, and hydrogen sulfide gases, M. Colyer and M. Griffitts (personal communication, 1999) believe that the resulting carbonic and sulfuric acids are responsible for the disintegration of local sandstones.

Two kinds of air pollutants pose the greatest threats to the biota of Mesa Verde Country: ozone and the acids produced from sulfur and nitrogen oxide emissions. The acids may arrive in the form of "wet deposition" (pollutants contained in rain and snow) or as "dry deposition" (settling, impaction, and absorption of pollutants in dust and aerosols). These are all products of power

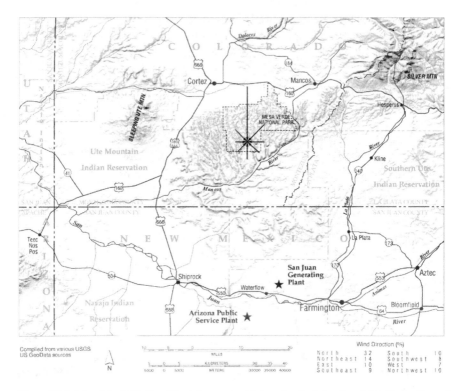

Figure 22.1. Mesa Verde National Park (center) with prominent wind vectors. Note the locations of two electrical power plants.

plants and automobiles. Other pollutants of concern—namely, particulate matter that reduces visibility and mercury that may contaminate water sources—are not discussed here (but see Oliva 2001 for details).

WET DEPOSITION OF ACIDS

Acidity (pH) is measured on a logarithmic scale from 1 to 14. A pH of 7 is neutral—neither acidic nor alkaline. Values less than 7 indicate acids; values above 7 indicate alkaline conditions. Because the scale is logarithmic, water at pH 5 is ten times more acidic than water at pH 6. Carbon dioxide naturally combines with water vapor to produce a mild acid with pH 5.6. Therefore, "acid rain" or "acid snow" has a pH value of less than 5.6. Such low pH values result mainly from sulfuric and nitric acids that form in the atmosphere from sulfates and nitrogen oxides produced by fossil fuel combustion. These acids can travel hundreds of

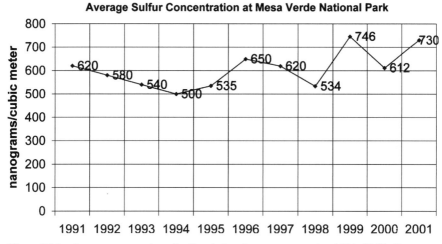

Figure 22.2. *Average concentration of sulfur during the summer months, 1991–2000. Data were collected twice each week with the National Park Service's IMPROVE Network.*

miles from the places where sulfur and nitrogen pollutants are produced (Birdsong 1999; Environmental Protection Agency 1980).

The park has monitored wet acid deposition since 1978. Average pH values for wet deposition in Mesa Verde National Park since 1981 have fluctuated between 4.7 and 5.1. For comparison, current pH values run from the low 4s in the northeastern U.S. to the low 5s throughout much of the West (National Atmospheric Deposition Network 1993–1999). From 1995 to 1998 the pH value averaged over each full year dropped from 5.0 to 4.8 in wet deposition collected in Mesa Verde National Park. Whether this represents the beginning of a long-term trend or merely annual fluctuation is not yet known. Annual hydrogen ion deposition (wet and dry combined) is currently 0.06 kg/ha.

DRY DEPOSITION OF ACIDS

Although monitoring of dry deposition began in 1992, the data have not yet been evaluated for possible trends.

SULFUR

Sulfur deposition at Mesa Verde National Park peaked in 1990 at 1,090 ng/m^3 (24-hour sampling for summer weeks). It then decreased to 500 ng/m^3 for the summer of 1993. Sulfur has been on the rise since then (Figure 22.2), but whether this will be a long-term trend is not yet known. Rivers, springs, and other water sources in and near the park have pH values from 7.1 to 8.4 (Mesa Verde National

Park Natural Resources, unpublished data). This indicates that the generally alkaline soils and rocks of the area are able to buffer or neutralize the acids being deposited on the soil surface before they reach the groundwater.

What are the biological effects of air pollutants in piñon-juniper woodlands of the Mesa Verde Country? A study of sulfur concentrations in soils and pine needles at Mesa Verde National Park in 1985 (Gladney et al. 1995) detected no measurable effects of current deposition rates on piñon or juniper trees, but these particular species may be relatively insensitive to air pollution effects. Fumigation studies elsewhere (summarized by Binkley et al. 1997) suggest that several other species of the Four Corners region—especially lichens—are very sensitive to SO_2 and SO_4. These acid-sensitive species include the grass *Agropyron smithii*, the trees *Pinus ponderosa* and *Pseudotsuga menziesii*, and a long list of lichens: *Buellia alboatra*, *Buellia punctata*, *Caloplaca cerina*, *Candelariella vitellina*, *Cladonia balfourii*, *Cladonia timbriata*, *Collema tenax*, *Lecanora saligna*, *Lecidea* sp., *Lepraria incana*, *Phaeophyscia* spp., *Physcia adscendens*, *Physcia dubia*, *Physcia stellaris*, *Physconia detersa*, *Rhizoplaca* spp., *Usnea hirta*, *Usnea subfloridana*, *Xanthoria fallax*, and *Xanthoria polycarpa*. Acid-sensitive plant species grow more slowly in the presence of acid deposition and may be killed outright if concentrations become too high. No direct effects of air pollution have yet been documented in Mesa Verde, but the species listed above should be monitored for early warning signs.

Certain animals also may be sensitive to acid deposition. Of special concern are amphibians and invertebrates that inhabit small pools or tinajas filled by rain falling on surrounding slickrock. Although the bedrock and soils have relatively high buffering capability for infiltrating precipitation, a brief episode of intense rainfall or snowmelt of low pH could completely change the water chemistry of these small pools. Larval stages of the chorus frog (*Pseudacris triseriata*) require water with pH greater than 6.0, and eggs of the tiger salamander (*Ambystoma tigrinum*) require a pH above 5.6 (Kiesecker 1996). These species may be at risk from current levels of acid deposition, although we need more research to determine just how serious this risk is. As described in the preceding section, the amphibians of Mesa Verde also are threatened by habitat fragmentation and environmental changes occurring outside the park. Thus even a small increase in stress caused by acidic deposition, combined with these other stresses, could imperil the species.

Ozone

This air pollutant ruptures plant cell chloroplasts and produces a dark mottling or stippling of the leaves (Heck and Cowling, in press). Although ozone high in the stratosphere is an important defense against the sun's ultraviolet radiation, ozone near the earth's surface (the troposphere) is a serious pollutant. Ozone exposures above 12,000 ppb-hrs (parts per billion per hour) represent a significant threat to sensitive plant species. Mesa Verde National Park measured ozone in excess of 30,000 ppb-hrs for the growing season in 1999 (Figure 22.3). Unfor-

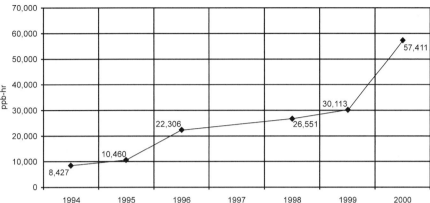

Figure 22.3. Accumulated ozone measurements of Mesa Verde National Park. Data were collected from April 15 to September 15, 1994, through 2000 by the National Park Service's Gaseous Air Network.

tunately, we have little specific information on effects of ozone on native species of the Mesa Verde Country. Several native species are known to be sensitive to high concentrations of ozone, however (Binkley et al. 1997), including the woody plants *Amelanchier alnifolia, Artemisia* spp., *Prunus virginiana, Rhus trilobata, Rubus parviflorus,* and the herbaceous *Mentzelia albicaulis.*

CLIMATE CHANGE

Climate is one of the master ecological factors controlling life on earth: it affects the distribution and abundance of species; the rates of productivity, decomposition, and other key ecosystem processes; and the frequency and severity of some natural disturbances. The fossil record reveals that dramatic changes in climate have taken place during the course of earth's history and that these climatic changes had powerful influences on the earth's biota. Whereas past climatic changes apparently occurred relatively slowly (over thousands of years), however, increasing evidence indicates that we are now living in a period of unusually rapid change in global climate (IPCC 1996). Evidence also increasingly points to human activities as the major cause of global climate change. What does this mean for the ecosystems of Mesa Verde Country? What tools are available to predict the effects of a rapidly changing climate?

Global circulation models (GCMs) are complex computer-simulation models that represent the earth's climate as a series of mathematical relationships (IPCC

1996). GCMs incorporate major climatic parameters such as the heat content of incoming solar radiation, escape of heat back into space, heat storage and release in the oceans and atmosphere, reflection and absorption of heat energy by landmasses and vegetation, and interactions among all the major components of the earth's climate system. The GCMs are imperfect, because the earth's climate is so complex and incompletely understood. Nevertheless, they are one of our best available tools for exploring the potential consequences of ongoing changes in the earth's atmosphere that result from industrial activities and land-use changes at a global scale.

One exercise that is commonly performed with GCMs is to simulate the earth's current climate (to make sure the model is operating correctly) and then to simulate the climate again with just one condition changed: the concentration of CO_2 in the atmosphere is doubled in the model to represent the situation likely to occur by the middle of the twenty-first century if the current rate of CO_2 production continues unabated. CO_2 is an especially important component of the climate models because it absorbs heat and warms the atmosphere. Plants and animals release CO_2 through their normal respiration processes, but all of this naturally produced CO_2 is removed from the atmosphere by plants (through the physiological process of photosynthesis), such that overall CO_2 levels in the atmosphere remain relatively constant. The burning of fossil fuels (petroleum, coal, natural gas) by industrial societies, however, has been releasing additional amounts of CO_2 over and above what can be readily taken out of the atmosphere by photosynthesis. The rise in atmospheric CO_2 probably began soon after the beginning of the industrial revolution, but it has been especially marked during the last half-century, increasing by about half a percent each year.

Because all of the global climate models are imperfect, the four major GCMs available today produce somewhat different results under the doubled-CO_2 experiment. Nevertheless, they all agree on the broad features of the climate likely to exist a half-century from now if no reductions are made in the rate of CO_2 production. Average temperatures worldwide increase by as much as 4° C, precipitation exhibits enormous local variability (less rain in some regions, more rain in others), the overall variability of climate becomes greater (e.g., greater temperature extremes in any given location), and the frequency of severe weather events (e.g., hurricanes) is increased (IPCC 1996). Because higher temperatures would cause more rapid rates of evaporation, plants could be subjected to greater moisture stress than they now experience even without an actual decrease in precipitation—unless the greater concentration of CO_2 in the air allows plants to conserve water more effectively than they do now. More research is needed to understand the details of how future climates may affect plant growth and survival.

Because the GCMs are imperfect, and because society may take effective steps to reduce CO_2 emissions before atmospheric concentrations reach such high levels, we cannot take the GCM results as certain predictions. The GCMs do represent our best current understanding of the earth's climate system and how that system is being changed, however, so it is prudent at least to consider the potential implications of these simulated climate changes for the ecosystems of the Mesa Verde Country. Because climate is one of the primary drivers of all ecosystem function and structure, even small changes in climate could have large ramifications.

The predicted average temperature increase of as much as 4° C does not sound very great. Yet from a meteorological or biological standpoint it is an enormous change. The average temperature of the earth's atmosphere was only about this much *cooler* during the last great ice age from about 65,000 to 15,000 years ago. During that time the San Juan Mountains of southwestern Colorado were covered by glaciers, and the ecosystems of the whole region were dramatically different from today. By looking at the kinds of ecological changes that took place during the transition from the glacial period to the modern period (roughly from 15,000 to 10,000 years ago), we can get some insights into the magnitude of ecological changes that may take place if the climate warms by several degrees during the next century.

Very little research has been conducted on ice age climates and vegetation in southwestern Colorado. Findings from other parts of the Colorado Plateau, however, especially from southeastern Utah, can be extrapolated cautiously to the Mesa Verde Country to suggest some of the major trends that may have occurred there (Betancourt 1990). Our understanding of ice age climate and vegetation in the Colorado Plateau region is based on fragments of plants collected and preserved in packrat middens. Over millennia, generations of packrats have collected bits of leaves, stems, fruits, and seeds from the area surrounding their nests in rocky cliffs and have created immense piles of material within caves and alcoves. Because these sites are protected from moisture or disturbance, they provide a glimpse of what was growing in the immediate vicinity of the nest (usually an area of about a hectare) at any time during the last several hundred to several thousand years. The oldest middens studied to date in the Colorado Plateau are nearly 20,000 years old (Betancourt 1990).

The midden data indicate that the vegetation in southwestern Colorado was dramatically different during the ice age of 65,000–15,000 years ago, when the climate was only a few degrees cooler than today (Betancourt 1990). Piñon was entirely absent from southwestern Colorado at that time. The species was restricted to more southerly latitudes (below 35° N) in what are now Mexico, New Mexico, and Arizona, perhaps because of a combination of spring frosts and summer droughts (Van Devender et al. 1984). At the elevations that now support piñon-juniper woodlands in southwestern Colorado, the major vegetation probably was forest or woodland composed of tree species that now grow 600–900 m higher on the flanks of the mountains (e.g., limber pine, Douglas fir, white fir, and blue spruce). Rocky Mountain juniper probably was present, and alkaline flats may have supported some of the same salt-tolerant shrubs that we now see in this region; but overall the plant communities of the late glacial period were almost entirely different in their composition of species and in their geographic distribution than today—all because of a global temperature difference of a few degrees plus the cascading effects of that small temperature change on seasonal patterns of precipitation, locations of major storm tracks, and other components of the earth's remarkably complex (and inadequately understood) climate system.

Even smaller fluctuations in climate can have powerful effects on ecosystems. Most of the Southwest experienced a severe drought from the early 1940s through late 1950s, during which time many piñon, juniper, and ponderosa pine trees died in Arizona and New Mexico (Swetnam and Betancourt 1998). This recent drought was less severe in southwestern Colorado than elsewhere and apparently did not result in much local tree death. An earlier drought in the late 1500s, however, probably did kill many trees in southwestern Colorado. In the nearby San Juan Mountains, and throughout the Southwest, one commonly finds old trees dating from the early 1600s but rarely finds a tree from before 1600 (Swetnam and Betancourt 1998; our own observations). The ancient woodlands on Mesa Verde (Chapter 2) are exceptional for this region, because they do contain many trees that survived the drought of the late 1500s. Most of the oldest trees on Mesa Verde became established soon after the Ancestral Puebloan people abandoned the region in the late 1200s (which was another drought period). The century that began in 1300 was relatively wet (Van West and Dean 2000) and therefore conducive to tree regeneration on the mesas that had been previously deforested by Ancestral Puebloan people. A more recent wet period began in 1976 with a shift in the El Niño–Southern Oscillation system and has been associated with increased growth and establishment of trees throughout the Southwest (Swetnam and Betancourt 1998).

So what will happen to the ecosystems of the Mesa Verde Country if average global temperatures do rise by a few degrees in the next century? With our present level of understanding, we cannot make any definite predictions, although the following speculations appear justified. First, the species composition and geographic distribution of plant communities may change as dramatically as they did at the end of the ice age period (15,000–10,000 years ago). We do not know which species will replace the ones now present, but a good guess is that piñon-juniper woodlands will be largely replaced by more drought-tolerant plant species that now grow at lower elevations in New Mexico, Arizona, and Utah (e.g., saltbush, blackbrush, and sagebrush). We could potentially lose nearly all of the ancient piñon-juniper woodlands that now exist on Mesa Verde and surrounding country, as the trees and associated plants succumb to increasingly severe moisture stress and fires. The piñon and juniper may migrate to higher elevations, filling the spaces vacated by ponderosa pine and other moisture-loving species of the foothills, as they too succumb to moisture stress and fire. We know enough to say that the vegetation of the Mesa Verde Country probably will look a great deal different a century or two from now, but we do not yet know enough to fill in any of the details.

Associated with the increase in average temperature and greater variability in precipitation is a likely increase in the frequency and severity of wildfires in many regions. It is predicted that a doubling of CO_2 would be associated with a near doubling of lightning fires and a 75% increase in annual area burned in the United States (Price and Rind 1994). Indeed, changes in the fire regime may cause more extensive and rapid changes in ecosystem structure than the direct physiological effects of elevated temperature and altered precipitation patterns (Franklin et al. 1992; Graham et al. 1990; Green 1989; Overpeck et al. 1990). Regional fire occurrence is strongly influenced by global-scale atmospheric processes such as the El Niño–Southern Oscillation system—processes that we are just beginning to understand. For example, El Niño years bring few fires to the southwestern United States because of moist conditions, whereas the dry La Niña years are associated with widespread fire activity (Swetnam and Betancourt 1998). A relatively small change in the frequency of La Niña years, or in the path of major storm tracks, could produce large changes in southwestern fire regimes. Unfortunately, our current level of understanding does not allow any specific predictions about future fire frequency or fire locations.

It may be that the new ecosystems of future centuries will be different but just as good as the ones we have today. There are reasons to be concerned, however, about the changes now taking place. The main reason is that the transition from glacial to modern climate and ecosystems occurred over a period of several thousand years—but this next major climate change is likely to occur over a period of only decades or centuries. A major concern is whether organisms will be able to migrate into new locations that are climatically suitable fast enough to avoid extinction. The kinds of species that will have the greatest difficulty are those with long generation times, slow and restricted dispersal processes, and demanding habitat requirements. This description characterizes piñon and many of the other distinctive species that make up the ancient woodlands described in Chapter 2. In contrast, the species that are likely to fare best under conditions of rapidly changing climate are those with short generation times, rapid and widespread dispersal processes, and the ability to thrive in disturbed habitats. This description characterizes many of the non-native invader species discussed above.

SUMMARY

The long-term ecological integrity of ancient piñon-juniper woodlands and other distinctive types of ecosystems in Mesa Verde Country may be threatened by fires of greater frequency and severity than occurred in recent centuries, by invasion of non-native species that displace the natives and alter basic ecological processes (e.g., disturbance regimes), by habitat fragmentation, by air pollution, and by global climate change. Of all these potential threats, climate change is the least well understood—but it is potentially the most serious threat of all. Climate change would directly impinge on nearly all organisms and ecological processes in our region, and it also would exacerbate the effects of the other threats described above. Climate change also is an especially difficult thing for land managers to deal with, because of the global scale of its causes and our current lack of definitive predictions. We need more research on habitat requirements, climatic tolerances, dispersal mechanisms, and ecological interactions of all of the native species in southwestern Colorado to gain a better understanding of what we may be facing. Managers and researchers also need to be alert to detect what may be the first signs of major ecological change resulting from global climate change. Even though no such ecological changes have yet been detected, some of the earliest warning signs may include widespread mortality of plants growing at the lower end of their elevational distribution (where they are already subjected to the greatest moisture stress) or failure of native plants to become reestablished after fires or other disturbances. The ecological integrity of Mesa Verde Country is generally very sound at present, but we need to be alert to what may be subtle signs of impending trouble.

REFERENCES CITED

Air Resources Division. 1999. Annual Data Summary: Mesa Verde National Park 1999. National Park Service. Gaseous Air Pollutant Monitoring Network, GAPMN.

Barney, M. A., and N. C. Frischkneckt. 1974. Vegetation changes following fire in the piñon-juniper type of west-central Utah. *Journal of Range Management* 27: 91–96.

Bazzaz, F. A. 1986. Life history of colonizing plants: Some demographic, genetic, and physiological factors. Chapter 6 in H. A. Mooney and J. A. Drake (eds.), *Ecology of biological invasions of North America and Hawaii*, pp. 96–110. Springer-Verlag, New York.

Beck, K. G. 1991. *Musk thistle: Biology and management*. Colorado State University Cooperative Extension No. 3.102, Fort Collins.

Betancourt, J. L. 1990. Late Quaternary biogeography of the Colorado Plateau. Chapter 12 in J. L. Betancourt, T. R. Van Devender, and P. S. Martin (eds.), *Packrat middens: The last 40,000 years of biotic change*, pp. 259–447. University of Arizona Press, Tucson.

Binkley, Dan, I. Dockersmith, and C. Giardina. 1997. *Status of air quality and related values in Class I national parks and monuments of the Colorado Plateau*. Department of Forest Sciences. Colorado State University, Fort Collins.

Birdsong, Jonathan. 1999. *Air pollution in our parks—1999: A record year for pollution in America's national parks*. Izaak Walton League of America, Washington, D.C.

Brown, H. F. 1958. Gambel oak in west-central Colorado. *Ecology* 39: 317–327.

Clary, W. P., and A. R. Tiedemann. 1986. Distribution of biomass within small tree and shrub form *Quercus gambelii* stands. *Forest Science* 32: 234–242.

Dunn, P. H. 1976. Distribution of *Carduus nutans, C. acanthoides, C. pycnocephalus* and *C. crispus* in the United States. *Weed Science* 24: 518–524.

Ehrlich, P., and A. Ehrlich. 1981. *Extinction: The causes and consequences of the disappearance of species*. Ballantine Books, New York.

Environmental Protection Agency. 1980. *Acid rain.* EPA-600/9-79-036. United States Department of the Interior, EPA, Washington, D.C.

Erdman, J. A. 1970. *Piñon-juniper succession after natural fires on residual soils of Mesa Verde, Colorado.* Biological Series Vol. 9 (2). Brigham Young University, Provo, Utah.

Flannigan, M. D., Y. Bergeron, O. Engmark, and B. M. Wotton. 1998. Future wildfire in circumboreal forests in relation to global warming. *Journal of Vegetation Science* 9: 469–476.

Floyd, M. L., D. D. Hanna, A. Loy, and W. H. Romme. n.d. A simple model of weed-risk following fire in Mesa Verde National Park, Colorado, USA. Submitted to *Journal of Weed Science.*

Floyd, M. L., W. H. Romme, and D. D. Hanna. 2000. Fire history and vegetation pattern in Mesa Verde National Park, Colorado, USA. *Ecological Applications* 10: 1666–1680.

Forman, R.T.T. 1995. *Land mosaics: The ecology of landscapes and regions.* Cambridge University Press, Cambridge.

Franklin, J. F., et al. 1992. Effects of global climatic change on forests in northwestern North America. In R. L. Peters and T. E. Lovejoy (eds.), *Global warming and biological diversity,* pp. 244–257. Yale University Press, New Haven and London.

Gladney, E. S., M. G. Bell, R. G. Bowker, R. W. Ferenbaugh, E. A. Jones, C. Llundstrom, J. D. Morgan, and L. A. Nelson. 1995. *An investigation of sulfur concentrations in soils and pine needles in Mesa Verde National Park, Colorado.* LA-12420-MS. Los Alamos National Laboratory, Los Alamos.

Graham, R. L., M. G. Turner, and V. H. Dale. 1990. How increasing CO_2 and climate change affect forests. *BioScience* 40: 567–575.

Green, D. G. 1989. Simulated effects of fire, dispersal and spatial pattern on competition within forest mosaics. *Vegetatio* 82: 139–153.

Heck, W. W., and E. Cowling. In press. The need for a long-term cumulative secondary ozone standard: An ecological perspective. *Journal of the Air and Waste Management Association.*

Hobbs, R. J., and L. F. Huenneke. 1992. Disturbance, diversity, and invasion: Implications for conservation. *Conservation Biology* 6: 324–337.

Hodgson, J. M. 1968. *The nature, ecology, and control of Canada thistle.* USDA Forest Service Technical Bulletin 1386.

IPCC (Intergovernmental Panel on Climate Change). 1996. *Climate change 1995: The science of climate change.* Cambridge University Press, Cambridge.

Kiesecker, J. 1996. pH-mediated predator-prey interactions between *Ambystoma tigrinum* and *Pseudacris triseriata. Ecological Applications* 6(4): 1325–1331.

Knight, R. L. 1997. Field report from the new American West. Chapter 12 in C. Meine (ed.), *Wallace Stegner and the continental vision: Essays on literature, history, and landscape,* pp. 181–200. Island Press, Washington, D.C.

Knight, R. L., G. N. Wallace, and W. E. Riebsame. 1995. Ranching the view: Subdivision vs. agriculture. *Conservation Biology* 9: 459–461.

Koniak, S. 1985. Succession in piñon-juniper woodlands following wildfire in the Great Basin. *Great Basin Naturalist* 45: 556–566.

Kunzler, L. M., and K. T. Harper. 1980. Recovery of Gambel oak after fire in central Utah. *Great Basin Naturalist* 40: 127–130.

Leopold, A. 1924. Grass, brush, timber, and fire in southern Arizona. *Journal of Forestry* 22(6): 1–10.

Mack, R. N. 1986. Alien plant invasion into the Intermountain West: A case history. Chapter 12 in H. A. Mooney and J. A. Drake (eds.), *Ecology of biological invasions of North America and Hawaii,* pp. 191–213. Springer-Verlag, New York.

Mack, R. N., and D. A. Pyke. 1986. The demography of *Bromus tectorum:* Variation in time and space. *Journal of Ecology* 71: 69–93.

National Atmospheric Deposition Network. 1993–1999. *Annual summaries.* National Atmospheric Deposition Program Office, Illinois State Water Survey, Champaign.

Noss, R., and A. Y. Cooperrider. 1994. *Saving nature's legacy: Protecting and restoring biodiversity.* Island Press, Washington, D.C.

Oliva, S. 2001. Air quality of the piñon-juniper woodlands. Unpublished report, on file, Mesa Verde National Park, Colo.

Overpeck, J. T., D. Rind, and R. Goldberg. 1990. Climate-induced changes in forest disturbance and regeneration. *Nature* 343: 51–53.

Price, C., and D. Rind. 1994. The impact of a $2 \times CO_2$ climate on lightning-causes fires. *Journal of Climate* 7: 1484–1494.

Rees, N. E. 1990. Establishment, dispersal, and influence of *Ceuthorhynchus litura* on Canada thistle in the Gallatin Valley of Montana. *Weed Science* 38: 198–200.

Romme, W. H. 1997. Creating pseudo-rural landscapes in the Mountain West. Chapter 8 in J. I. Nassauer (ed.), *Placing nature: Culture and landscape ecology,* pp. 139–161. Island Press, Washington, D.C.

Rutledge, C. R., and T. McLendon. 1998. *An assessment of exotic plant species of Rocky Mountain National Park.* Department of Rangeland Ecosystem Science, Colorado State University, Fort Collins. Northern Prairie Wildlife Research Center Home Page. http://www.npwrc.usgs.gov/resoruce/orthrdata/explant/explant.htm.

Shea, K., and D. Kelly. 1998. Estimating biocontrol agent impact with matrix models: *Carduus nutans* in New Zealand. *Ecological Applications* 8(3): 824–832.

Swetnam, T. W., and J. L. Betancourt. 1998. Mesoscale disturbance and ecological response to decadal climatic variability in the American Southwest. *Journal of Climate* 11: 3128–3147.

Terborgh, J. 1989. *Where have all the birds gone?* Princeton University Press, Princeton, N.J.

Tiedemann, A. R., W. P. Clary, and R. J. Barbour. 1987. Underground systems of Gambel oak (*Quercus gambelii*) in central Utah. *American Journal of Botany* 74: 1065–1071.

Van Devender, T. R., J. L. Betancourt, and M. Wimberly. 1984. Biogeographic implications of a packrat midden sequence from the Sacramento Mountains, south-central New Mexico. *Quaternary Research* 22: 344–360.

Van West, C. R., and J. S. Dean. 2000. Environmental characteristics of the A.D. 900–1300 period in the central Mesa Verde region. *Kiva* 66: 19–44.

Weinhold, Bob. 2000. Farmington power plants top polluters in country. *Durango Herald,* January 2, 2000.

23. EPILOGUE: MANAGEMENT CONSIDERATIONS FOR CONSERVING OLD-GROWTH PIÑON-JUNIPER WOODLANDS

GEORGE L. SAN MIGUEL

> Nothing short of defending this country during wartime compares in importance with the great central task of leaving this land even a better land for our descendants than it is for us.
>
> —THEODORE ROOSEVELT

THEODORE ROOSEVELT'S VISION LEFT A LEGACY IN OUR NATIONAL PARK SYSTEM, which, along with other federal reserves, preserves spectacular wildlife, grand scenery, unique geological formations, and the broad spectrum of our cultural heritage. Some of these reserves were designated specifically for a particular vegetative complex or individual plant species. Today America preserves land for giant sequoias, coast redwoods, Joshua trees of the Mojave Desert, the saguaro and organ-pipe cacti of the Sonoran Desert, and the tallgrass prairie. Not surprisingly, we do not celebrate piñon pines and juniper trees in this way, perhaps in part because about twenty million hectares of this seemingly common vegetative community occur in the United States. Only a small percentage of this area is made up of old-growth, however, and many of these woodlands already have been lost or badly degraded in historic times. This noncelebrity status of piñon-juniper poses important implications for land managers entrusted with overseeing the use of the remaining old-growth piñon-juniper woodlands.

This chapter is founded upon a few assumptions. The first is that there is a biological distinction between healthy old-growth piñon-juniper woodlands and the majority of the remnant stands of today. Chapter 2 discusses what actually constitutes old-growth piñon-juniper. Much of the scientific evidence and more detailed biotic accounts in the accompanying chapters show that this is the case. The presence of old-growth piñon-juniper stands within an ecosystem adds biological diversity to the region (Monsen and Stevens 1999). The second assumption is that there are compelling reasons and justifiable sentiments for conserving a significant portion of what is left of the old-growth piñon-juniper. The final assumption is that a new focus on piñon-juniper research, monitoring, and conservation management can succeed in maintaining this native southwestern landscape and its diverse biological components.

THE USES OF PIÑON-JUNIPER

For thousands of years piñon-juniper woodlands have helped sustain human populations. A wide variety of plants and animals of the piñon-juniper have offered nutritional and raw material resources for American Indians (Chapters 18–20). The vast majority of archaeological sites within Mesa Verde National Park and many parts of the Four Corners states are located within the piñon-juniper zone. Piñon trees long have held special meaning for indigenous peoples. Piñons formed the "nut orchards" for many tribes. Piñon seeds were a valuable component of the diet of many tribes of the Great Basin and Colorado Plateau and a critical staple for others. Spanish explorers and American settlers often fended off starvation by consuming the seeds. Piñon seeds generally are high in protein (particularly the amino acid tryptophan), carbohydrates, and especially oils. So it is not surprising that piñon pines appear in many oral traditions among the Indian nations (Chapter 18). The wood was used for construction as beams and poles and for tool handles. Piñon wood was favored for heating fuel, whereas juniper wood was preferred for cooking. Piñon pitch and other plant parts offered a variety of practical, medicinal, and ceremonial resources. Similarly, juniper wood was used in construction, the "berries" for food, and the bark as a tobacco substitute (Chapters 18–20).

It was not until the introduction of the steel axe to the West that piñon wood began to be harvested unsustainably (Lanner 1981). Like the slaughter of the bison on the Great Plains, piñon-juniper deforestation was a catastrophe for the lifeways of many tribes, which led to conflict and violence.

For most of America's brief history in the Southwest, homesteaders, ranchers, and other private and public land managers have had little positive use for piñon-juniper. Settlers often viewed the trees as an impediment to agriculture and livestock and as a wildfire hazard. Many of the West's natural piñon-juniper stands were deforested in historic times (Lanner 1981). The primary uses of the wood during this period were for heating fuel, cooking, railroad ties, fence posts, fuel for steam locomotives, and, in areas lacking tall tree forests, timbers for construction and mine shaft supports. Particularly in the Great Basin, vast piñon-juniper landscapes were completely denuded to produce charcoal for smelting the metal ores of the West's mining boomtowns. For the most part, however, piñon-juniper was cleared simply to get rid of it.

An enormous proportion of the original piñon-juniper stands has been cleared over the decades by cutting, chaining, burning, and spraying with herbicides (Lanner

1981). Beginning in the 1950s, government programs began the large-scale conversion of piñon-juniper to cattle pasture. Chaining is the dragging of ship anchor chains between two crawler tractors, which results in the rapid uprooting of vast piñon-juniper expanses. After chaining the land was sown with non-native forage grasses for livestock. Between 1950 and 1964 over 1.2 million ha of piñon-juniper were converted to pasture in the West. Between 1960 and 1972 over 140,000 ha of piñon-juniper were stripped on federal lands in Utah and Nevada alone. After clearing, piñon-juniper often is prevented from reestablishment by the use of roller chopping.

Even to this day managers of vast piñon-juniper landscapes view these trees as undesirable weeds that compete for space with forage species in a manicured environment. The definitions of some of the terms used in these management practices reveal a paradox in land-management goals. The clearing of piñon-juniper to improve livestock forage often is referred to as rehabilitation and restoration (Monsen and Stevens 1999). Although this may be accurate in shrublands and grasslands where piñon and juniper expanded during the twentieth century, in the case of old-growth stands these terms are inappropriate. Ironically, grazing productivity increased only slightly on lands cleared by chaining. Contrary to predictions by proponents of landscape-wide habitat conversions, large-scale chaining did not generally improve deer herds, reduce soil erosion, or improve surface water flows (Lanner 1981).

Much of the piñon-juniper that is left now is in a highly degraded state from the effects of long-term livestock grazing, deforestation, non-native weed invasions, erosion, habitat fragmentation, and the loss of biological diversity. In the early 1900s tens of thousands of sheep and cattle grazed seasonally on dense grasses from low elevations around Monticello, Utah, all the way to the La Plata Mountains of Colorado (McCabe 1975). Similarly enormous flocks and herds ranged over much of the region. During these seasonal livestock transmigrations, all unfenced land jurisdictions were affected by these vast droves, including the Indian reservations, national forests and BLM lands, and Mesa Verde National Park. For such livestock numbers to have flourished here, clearly the natural bounty in the Four Corners country must have been phenomenally high; but overuse led to decline. By the 1970s there were only half a dozen sheep ranchers with flocks of one to two thousand head still making annual treks from the dry valleys up to the subalpine meadows (Garlinghouse 1973), with considerable time spent within the piñon-juniper, but even this could not be sustained much longer.

In attempting to establish dryland farming, settlers cleared and plowed thousands of level hectares of piñon-juniper woodlands. Much of this was abandoned during the dry Dust Bowl years of the 1930s. In the early part of the twentieth century the area from between Cortez and Dove Creek, Colorado, to the Utah border was cleared of its piñon-juniper woodlands and successfully planted with pinto beans. These red loess soils around Mesa Verde continue under cultivation because of their greater natural fertility.

THE CONSEQUENCES OF OVERUSE

> Overgrazing and timber trespass had, in fact, combined to make the [piñon-juniper] woodland one of the worst-abused vegetation types in the West.
> —RICHARD M. LANNER, 1981

The piñon-juniper woodlands evolved in North America without a significant impact from large grazing mammals. These semiarid environments contain low levels of surface waters and support light grass densities. Additionally, the piñon-juniper's high elevations meant that winters generally are cold with little natural forage after a summer of grazing. This should have been a clue to homesteaders that the piñon-juniper could not support large numbers of domestic livestock over the long term. Ingenious and resilient ranchers developed dependable water sources that allowed stock to occupy new areas and more animals to stay on the land longer. After clearing the land of its natural piñon-juniper mantle, development of water sources from aquifers or mountain streams allowed for the irrigation of the new fields to produce hay as supplemental feed, without which livestock could not survive the winters above the desert valleys.

One account during the 1920s recalls a trainload of 1,200 cattle arriving in Mancos, Colorado (Goff 1985). Like several others, this private herd would spend its summers grazing on Mesa Verde and winter on the Ute Reservation to the south. When the national park was established in 1906, portions of Mesa Verde already were overgrazed. Besides large numbers of cattle, many hundreds of sheep grazed within the national park. Early park reports indicated that even more forage plants were lost to the sheep by trampling and erosion than were consumed by the animals (Torres-Reyes 1970).

Livestock may prevent native perennial species from developing seeds or replenishing root reserves, thereby providing weeds with a competitive edge. Small trees and shrubs remain, but forbs and native grasses disappear, and non-native weed species move in. This pattern occurs more definitely in the parts of the piñon-juniper's range that experience light summer rains (Monsen and Stevens 1999). As an example, in areas around Mesa Verde, which generally experiences moderate summer rains, the nutritious early-season muttongrass (*Poa fendleriana*) has become scarce due to livestock grazing. Inside the park, where grazing is prevented, the muttongrass remains abundant in the old-growth piñon-juniper. Similarly, bitterbrush (*Purshia tridentata*) is scarce outside the park, presumably as a result of heavy sheep browsing during the first half of the twentieth century. Salina wildrye (*Leymus salina*) has suffered a similar fate. In some areas, one of the least palatable native grasses, galleta (*Hilaria jamesii*), remains with little else (Colyer, personal communication, 2000). In the absence of a healthy native grass cover, decades of topsoil loss have contributed further to declining productivity on dry rangelands. Due to the fairly recent arrival of a wide variety of non-native invasive plants, consequences of clearing the piñon-juniper and increased grazing are becoming evident (Chapter 21).

Knapweeds, thistles, and other invasive species outcompete native perennial grasses and forbs and sustain themselves. Subsequent changes in vegetative cover and topsoil conditions lead to higher rates of erosion and invasion by more noxious weeds. In such a crisis, land managers have looked to chemical weed treatments and the sowing of non-native or hybrid forage grasses as a means of maintaining the land's commercial productivity. These short-term answers come at a price, in dollars and long-term environmental health.

Dependence on herbicides and other agricultural chemicals is an expensive option that can lead to soil and water pollution and harm nontarget plants. Alternative use of biological controls may be more environmentally friendly, although they are slow to establish, are not yet and may never be available for many weed species, and sometimes may attack nontarget plants. Non-native forage grasses may help to control erosion and provide livestock forage, but often these introduced species outcompete native grasses and forbs and do not provide much in the way of resources for native wildlife (Jones 2000). The biological impoverishment of rangelands is an unintended consequence of great concern.

Reseeding damaged lands is an important weed-control tool, but tactics should attempt to utilize native species whenever possible. It is important to recognize that west of the Rocky Mountains there are very few native rhizomatous (sod-forming) grasses. Accompanied by diverse assemblages of native forbs, various bunchgrasses evolved in this region in the absence of large herds of grazers such as bison. Reseeding with non-native rhizomatous grasses will discourage native bunchgrasses, forbs, and some woody species, thereby impoverishing biological diversity in recovering areas indefinitely into the future. Recent postfire rehabilitation in Mesa Verde has utilized only native bunchgrasses, and these have been successful in reducing non-native invasion into susceptible communities.

Heavy grazing also may prevent fires from maintaining some naturally open piñon-juniper stands. Many investigators have shown that in the presence of heavy livestock grazing since the late 1800s, which discouraged the spread of fire, some piñon pines and particularly junipers, alarmingly, have invaded millions of hectares of semiarid western grasslands and shrublands (Chambers 2001; Miller and Rose 1995; Miller and Wigand 1994). Some have gone so far as to express a belief that this "invasion" of piñon-juniper is responsible for high fawn mortality and declining mule deer populations in Colorado (Colorado Wildlife Federation 2000). Over time the canopies of these stands have closed in and degraded or eliminated the shrub and herbaceous layers (Miller et al. 1995). In addition, non-native invasive plants, a possible consequence of grazing, may change succession patterns and wildfire frequencies.

MANAGING FIRE IN THE PIÑON-JUNIPER

The piñon-juniper woodlands' high flammability under extreme burning conditions means that environmental values other than the piñon-juniper are at high wildfire risk. The quiet, wide-open spaces of the Colorado Plateau, the relative lushness of piñon-juniper compared with sagebrush and greasewood deserts, and the moderate climate have caused a recent housing boom in the region. High-intensity motorized recreation—activities that further fragment and degrade fragile habitats—increases along with the growing populations. The carving-out of residential areas in piñon-juniper woodlands means that more and more people, homes, and property are vulnerable to wildfire. Concurrently, there is greater pressure on federal and rural firefighting operations to protect life and property. Add to this the high density of prehistoric masonry structures and other Ancestral Puebloan (Anasazi) archaeological sites, plus many historic buildings in the Four Corners states, and it is not surprising that presuppression activities take a further toll on the region's piñon-juniper. The cutting of roads, utility rights-of-way, pipelines, fuel breaks, and the sprawling spiderwebs of oil and gas operations add to the fragmentation of remnant old-growth piñon-juniper stands. Habitats divided into smaller separated pieces cannot support as many species or maintain ecological health nearly as well as do extensive intact ones (Chapter 22).

Fire management in the West has changed since the 1960s. The benefits and necessity of fire in most western forest, shrubland, and grassland ecosystems are now widely accepted among professionals. Many contemporary wildland fire managers are anxious to apply prescribed burning in hazardous fuel reduction and habitat manipulation efforts. The role of fire in the ponderosa-pine forests of the Colorado Plateau has been thoroughly studied. Now there may be a tendency among some to apply the lessons learned in the ponderosa pines to the neighboring piñon-juniper woodlands. The wisdom of such an application has not yet been tested, and land managers should wait for scientific evidence before adopting such a strategy on a widespread basis.

Frequent low-intensity fires maintained many "pre-settlement" ponderosa-pine forests (Covington et al. 1997). Light ground fires thin out sapling pines, but the adult trees survive and shower the ground with seeds that readily sprout into a new generation of saplings. This kind of simple fire history does not occur in the piñon-juniper woodlands of Mesa Verde Country (Floyd et al. 2000). Here old-growth piñon-juniper develops in the absence of significant fires over perhaps several hundred years. The extensive old-growth piñon-juniper at Mesa Verde appears healthy and self-sustaining after many centuries (Chapter 16). When old-growth piñon-juniper ignites, it either fizzles out after smoldering over a hectare or less or the whole woodland goes up in a major conflagration that consumes most of the landscape's biomass. Fueled by large buildups of resinous deadwood and deep litter layers, the various species of piñons and junipers in the West tend to die easily in wildfires. Furthermore, for reasons not yet fully understood, regeneration of piñon-juniper after a stand-replacing fire may take decades. A more recent phenomenon is severe weed infestations in piñon-juniper occurring after high-intensity wildfires. The bare ground sterilized by the extreme heat leaves the land wide open to invasive weeds and soil erosion. Some of the piñon-juniper's major native plant species do not sprout or readily seed in after a hot fire.

Therefore, it is highly unlikely that in old-growth piñon-juniper the application of broad-scale prescribed burning to create open stands that are less likely to burn out of control will simultaneously allow for the conservation of that same old-growth. It seems far more likely that this kind of frequently repeated burning eventually would convert piñon-juniper to an environment dominated by non-native grasses, some native forbs and small shrubs, and invasive noxious weeds. If a land manager's only goal is to reduce an area's risk of severe wildfire or to increase forage for livestock, then this could succeed. But the costs and consequences of fighting weed infestations may offset much of the perceived early gain, and eventually the piñon-juniper will disappear.

For species as easily killed by fire as piñons and junipers, it makes sense that old-growth woodlands form only in the long-term absence of fire. For some of these trees to attain ages measured in the many hundreds of years, it seems clear that old-growth piñon-juniper in more or less its current state predates the recent age of fire suppression. This concept runs counter to conventional wisdom. How could a highly flammable fuel type growing in a semiarid climate, with frequent summer lightning and inhabited by peoples who freely manipulated vegetative cover with fire, remain unburned for centuries? (Certainly Indians would not have deliberately set fire to their piñon-nut orchards.) So it may take time for this idea to become widely accepted, assuming that scientific investigation continues to bear this out. Land managers find themselves in a very difficult situation. The piñon-juniper woodlands that they own or manage could continue standing untouched for centuries or could go up in smoke the very next summer. Complicating the manager's life further is the fear that a wildfire could escape one manager's jurisdiction with one set of land management goals and damage the resources and property within the jurisdiction of another manager with a different set of goals. In effectively managing a widespread ecosystem, good communication, understanding, and coordination are needed across barbed-wire fences.

PLANNING AND SETTING GOALS

Managers of public lands today are encouraged, if not required, to set management goals and objectives. A process must be open to participation from the community. A tiered system of planning documents is generated, derived from legislative requirements that are used to develop big-picture strategic plans that guide individual action plans directing work on the ground. Goals and objectives need considerable thought to help instill confidence, prevent conflicting directions, and provide avenues for success (Monsen and Stevens 1999). They must be definable, achievable, and measurable over time and space with desired future conditions spelled out for the whole woodland community. The management "problem" as stated must be understood even more clearly than the proposed management solution. This takes a great deal of foreknowledge about the ecosystem that is not often available.

During the planning process, goals are set that drive the priorities, funding allocations, and actions of public land managers. So it is here that the concerns of old-growth piñon-juniper management and conservation may be allowed a voice. Other contesting issues may include such compelling interests as health, life, safety, recreation, game production, infrastructure development and maintenance, tourist services, livestock production, endangered species, energy development and transmission, wildfire prevention and suppression, timber production, and others. It may end up taking more than just awareness for old-growth piñon-juniper issues to compete.

COMMUNICATION, EDUCATION, AND VALUES

Periodically meeting with neighboring landowners and managers provides opportunities to establish trust, exchange ideas and concerns, and introduce new information including research results. For public land managers this is especially important. If public lands are going to be managed in an innovative way, the public land manager will need to "sell" these changes and allow for open discussion of the merits. The sales pitch is made easier when the changes are backed by the results of good science, but acceptance may take many years of careful effort. Despite some notable recent setbacks, one of the most successful education efforts in recent years has been the explanation that, under certain circumstances, forest and range fires can be desirable and necessary for the health of the land. For piñon-juniper, that message would need considerable modification.

It seems likely that the present-day degraded condition or senescent appearance of much of these dwarf evergreen forests contributes to a lack of appreciation for old-growth piñon-juniper. But piñon-juniper stands are extremely variable in structure and composition over great expanses of seemingly uniform woodlands. A "one-size-fits-all" understanding and management scheme would be inappropriate. Perhaps what the piñon-juniper needs is a good public relations campaign to improve its image. It is most likely that any major effort to promote the values of old-growth piñon-juniper would start with public land management agencies.

Certainly public opinion and values will continue to influence public land management policy and priorities. If people are made aware of the treasure that old-growth piñon-juniper really is, there will be greater support for the conservation of what remains and for true restoration of some of what has been mismanaged or neglected. Few land management units have specific mandates to conserve old-growth piñon-juniper woodlands. Mesa Verde National Park is one unit that does have such a legal responsibility. The U.S. Congress requires the National Park Service at Mesa Verde by law to preserve "all timber, natural curiosities," and wildlife in the park unimpaired for the benefit and enjoyment of present and future generations. But without such specific congressional orders, some public land managers may need encouragement from public opinion to make old-growth piñon-juniper protection an issue and a priority. Only then can piñon-juniper conservation finally begin to compete in the struggle for attention and scarce fiscal resources.

What a fascinating story there is to tell about the modest piñon-juniper woodland. When one contemplates the long odds for reaching its old-growth stage and the array of interconnected organisms in the ecosystem, we should not be disappointed by what the piñon-juniper is not. We should be astonished by its richness and be grieved by how much has been lost; to appreciate it fully, however, we may need to experience some of the real old-growth firsthand at a deliberate pace.

Interpretive centers in piñon-juniper country need to allow ample space to showcase this biotic community. Indoor and outdoor exhibits, trail guides, audiovisual programs, and published literature would provide important learning opportunities. Interpersonal educational programs in the field and in classrooms would add that distinctive vital connection so important for making the piñon-juniper story come to life. The themes of interpretive and educational exhibits, classroom curricula, literature, and presentations must extend beyond the basic facts about how humans have lived in the piñon-juniper and exploited its resources. For the piñon-juniper to stand out in the public consciousness, a deeper message about this community's intrinsic ecological values must come through as well. We already get this kind of discourse when visiting other more famous regions and environments; we should expect no less for the piñon-juniper woodlands.

MANAGING FOR CONSERVATION

There are definite limits to how much sustainable productivity can be extracted from piñon-juniper woodlands. Exceeding those limits has resulted in erosion and weed invasions followed by chemical pollution, loss of biological diversity, and further declines in productivity. Private, public, and tribal land managers need to find low-impact land-use practices that do not exceed the environment's sustainable carrying capacity. For lands already damaged, a healing process must be established.

Another important resource to maintain for restoration efforts is biomass capital, the frozen sunshine and mineral reserves in the wood. Once wood is removed, the site's resilience is greatly diminished. Woody debris on the ground can provide favorable microsites to promote the growth of herbaceous plants, support invertebrates, and retard soil erosion.

The Pandora's Box of alien weeds is open and they can never be eradicated. The most realistic goal for land managers is to carefully reduce the presence of invasive species to an acceptable level that does not significantly diminish the land's health and productivity. This requires the use of integrated weed-management techniques employing moderated use levels of the most effective and environmentally sensitive chemicals, mechanical controls, and biological controls with the initiation of low-impact land-use practices that resist new invasions. Reseeding with native species is an important tool in achieving recovery, but even this is an imperfect solution. Seeds of most native species are not available commercially, and many that are obtainable may not represent locally adapted genotypes (Monsen and Stevens 1999). Also, regenerative source materials generally are not going to be available for less conspicuous yet important ecosystem components such as mycorrhiza, lichens, mosses, tiny flowering plants, and soil invertebrates.

The piñon-juniper woodland's high wildfire danger enormously complicates the land manager's work. Total fire exclusion would safeguard the human and property values at risk as well as the old-growth woodlands themselves and avoid postfire weed invasions, but this is not realistic. There is no option in the piñon-juniper except an active fire management program. Managers will continue to target the protection of life, property, cultural resources, critical habitat for threatened and endangered species, and escape across jurisdictional borders, but hazardous fuel removal must be kept limited if conserving the mature stands and old-growth is a goal. The use of prescribed fire in the piñon-juniper needs to be strictly limited to prevent wholesale habitat conversion. Fuel management also could be designed to inhibit fires from sweeping through remnant stands of old-growth piñon-juniper.

The age of fire suppression has lasted no more than 125 years with the onset of heavy livestock grazing followed later by technological fuel and fire management. In the absence of fire, once-isolated old-growth islands have coalesced as more open interspaces have closed in. Now crown fires can speed across much larger interconnected areas, including old-growth stands that previously and naturally were protected from large fires. It seems highly unlikely that anyone could completely exclude wildfire from the piñon-juniper. Sometimes the fires simply cannot be controlled for days and sometimes the danger level to the firefighters is unacceptable. Yet a critically important part of an old-growth piñon-juniper conservation strategy may be suppression, because recent and current loss rates from habitat conversion and degradation together with wildfires probably exceed the rate of old-growth regeneration. Currently this may be the most conflicting dilemma in old-growth piñon-juniper management. About 6,624 ha of old-growth piñon-juniper woodlands have burned on the Mesa Verde cuesta during the period for which data exist, from the early 1930s through 2001 (Floyd et al. 2000). This amounts to about 15% of the cuesta's historic piñon-juniper woodland but close to one-third of that within Mesa Verde National Park. This is an alarming loss rate for old-growth in one of its few areas protected from development and extractive resource uses. Even more old-growth was lost in 2002 and 2003.

Large stand-replacing wildfires may be the inevitable conclusion for these old dense stands. Like any climax vegetative community, old-growth piñon-juniper woodlands cannot last forever, no matter how desirable it is to retain them. Each site is in a constant state of flux over decades and centuries. So managing the old-growth piñon-juniper will take a landscape-level approach, attempting to maintain certain proportions in various seral stages instead of managing variably on a site-by-site basis or uniformly for a particular static outcome. Perhaps the best we can plan for is to preserve as much as possible of what is left through time so that areas undergoing natural recovery now may have time enough to regenerate into old-growth. In this way there always could be enough old-growth in the ecosystem to sustain biological diversity and serve as refugia. Some recovering areas could receive limited and infrequent mechanical and prescribed burning treatments to maintain a successional mosaic of different-aged and functional piñon-juniper patches in strategic locations among the old-growth stands with the purpose of trying to keep future piñon-juniper wildfires more containable. These management strategies represent a commitment for the centuries, a time frame to which most land managers are not accustomed. It would mean an infusion of natural resource conservation awareness and values into fuel and wildfire management programs that traditionally and exclusively have focused on fire suppression.

Making matters more difficult to interpret and control are indications that the world's atmospheric carbon dioxide levels are rising along with temperatures and the persistence of droughts (Chapter 22). In combination with more flammable growth stages in the piñon-juniper stands adjacent to remnant old-growth, sustaining the woodlands for the long term will not be easy. A worst-case scenario could be large-scale deforestation in the Intermountain West driven by climate change and fierce wildfires, with natural regeneration shut down by drought and invasive weeds. A warming climate forces piñon-juniper northward and to higher elevations. At Mesa Verde the upper end of the piñon-juniper runs into a fire-adapted mountain shrub community with a much shorter fire return interval that cannot sustain piñon-juniper woodlands.

A fuel management strategy may need to begin by protecting special values at risk from current conditions, such as areas prone to erosion and invasive species, structures and other property, aboveground archaeological sites, jurisdictional boundaries, rare and sensitive species habitats, and old-growth stands. With proper treatment, seral stages then may be reset in limited areas adjoining the sensitive sites. Once these areas are protected from the full force of wildfires, natural fires may be allowed a more active role in determining what will grow on the land, when, and for how long within set parameters. This may be the best and most practical way for ensuring a full range of natural variability in the piñon-juniper ecosystem.

Land managers must set goals for many competing interests and track progress toward meeting those goals. A dynamic interaction develops in the goal-setting process. Two of the factors that influence goal setting are public opinion and the results of scientific investigations. Education and interpretive efforts based on research results can introduce the old-growth piñon-juniper woodlands as a favorable asset. At the same time, land management goals and priorities gain credibility when backed by science. So land managers and their clientele are best served by a good science program and professional resource management and public education programs. Even this late in the game these objectives could be achieved for piñon-juniper, but only with enough vision and courage. If the scientific evidence and public opinion both indicate that old-growth piñon-juniper conservation is a valid goal, land managers need to listen and respond. Then desired future conditions can be defined for the piñon-juniper, and management plans can be developed for maintaining or restoring old-growth.

There is much that we still do not know about the piñon-juniper woodlands. To maintain ecological health, the woodlands that are left need to be retained in scales that are as large as possible, kept as close to each other as possible, and linked by suitable corridors whenever possible (Chapter 22). One important management tool useful in the research and conservation of old-growth piñon-juniper is the locating, inventorying, and mapping of these woodlands, with the remaining high-integrity stands a priority. The logical place to start would be the pooling of existing knowledge and databases from the Bureau of Land Management, U.S. Geological Survey, National Park Service, U.S. Forest Service, Bureau of Indian Affairs, natural resource departments among the American Indian nations, state natural areas programs, and conservation partnerships such as The Nature Conservancy's state natural heritage programs. Also, results from past, present, and future scientific investigations into the identified subjects would help determine

- the relative health and constitution of remnant old-growth stands;
- presettlement and postsettlement changes in community structure and composition;
- fire histories in the piñon-juniper in various geographic areas;
- current rates of old-growth loss from habitat degradation, wildfires, or deliberate conversion and how those rates compare with regeneration rates back into old-growth;
- what controls postfire succession in the piñon-juniper in southwestern Colorado;
- how to reintroduce the rejuvenating effects of a natural fire without accelerating the degrading effects of non-native invasive plants;
- how invasive non-native plants influence long-term successional patterns;
- which remnant stands have the highest potential for long-term preservation;
- where threatened, rare, endemic, and other unique species exist in the piñon-juniper;
- how much piñon-juniper woodland will need conservation to ensure sustainability of large healthy stands and the rare and unique species and plant communities they support;

- the effect of changing climates on piñon-juniper woodlands; and
- what compatible and sustainable uses and tolerable use levels, including recreation, are appropriate for piñon-juniper woodlands and which practices are not consistent with piñon-juniper habitat conservation.

In addition to this research, managers, landowners, and investigators need to monitor current conditions for alarming signs, such as whether

- wildfires in the piñon-juniper are getting more frequent, larger, and more severe;
- prescribed herbicide use applications, reseeding, and the release of biological control insects are having the desired effect on invasive plants without serious side effects;
- interior old-growth species are being affected by habitat fragmentation;
- piñon-juniper species are beginning to show sensitivity to acid deposition ozone, or other windborne pollution; and
- possible ecological alteration resulting from global climate change is occurring.

The information obtained from these research and monitoring investigations would be crucial in developing long-term land management plans with old-growth piñon-juniper conservation as a stated goal.

SUMMARY

The old-growth piñon-juniper woodlands and their associated species are not a famous or even familiar environment to most Americans. Their biological significance is only now becoming recognized. We have grown too accustomed to viewing the current depauperate state of much of these lands as all these poor semideserts can offer without major development efforts. Today even some of the most experienced land managers in the Four Corners area are unaware of the amazing richness, diversity, and productivity that once was the aboriginal ecosystem here. Many species have disappeared or retreated to remote areas, further misleading us as to the land's natural abundance.

With education, people can learn that these woodlands really are aesthetically pleasing and provide their own flavor of recreational, cultural, and spiritual opportunities. The "desert scrub pines," the short bushy juniper trees, and the other members of the old-growth piñon-juniper community live modest lives between dry deserts and tall forests. Like the indigenous peoples that lived here before us, perhaps the thrifty piñon-juniper community would make a good role model for modern peoples of the Southwest, if only we would make the effort to learn its lessons.

Many of the original woodland stands have been lost or changed already, and only some of those that remain are likely to be saved in the long run. With the cultivation of interest and understanding, the piñon-juniper woodlands' conservation needs finally would be widely recognized. Perhaps the greatest threats today to the piñon-juniper really are not habitat degradation, climate change, and wildfire. Maybe the greatest long-term threats are ignorance and neglect. Land managers should consider the benefits of promoting new values of old-growth piñon-juniper conservation by celebrating one of the Southwest's true natives. The piñon-juniper woodlands are an authentic part of our natural heritage that deserve their places in the sun.

REFERENCES CITED

Chambers, J. C. 2001. *Pinus monophylla* establishment in an expanding pinyon-juniper woodland: Environmental conditions, facilitation, and interacting factors. *Journal of Vegetation Science* 12: 27–40.

Colorado Wildlife Federation. 2000. Fauna survival study yields new clues to mule deer decline. *Colorado Wildlife* (February–June).

Covington, W. W., P. Z. Fule, M. M. Moore, S. C. Hart, T. E. Kolb, J. N. Mast, S. S. Sackett, and M. R. Wagner. 1997. Restoring ecosystem health in ponderosa pine forests of the Southwest. *Journal of Forestry* 95: 23–29.

Floyd, M. L., W. H. Romme, and D. D. Hanna. 2000. Fire history and vegetation pattern in Mesa Verde National Park, Colorado USA. *Ecological Applications* 10(6): 1666–1680.

Garlinghouse, Ethel. 1973. Oral history interview with representatives of Mesa Verde National Park, Colo.

Goff, Walter. 1985. Personal telephone communication with representatives of Mesa Verde National Park, Mancos, Colo.

Jones, Lisa. 2000. He's worried about weeds. *High Country News* 32(10) (May 22).

Lanner, R. M. 1981. *The Piñon Pine: A Natural and Cultural History.* University of Nevada Press, Reno.

McCabe, Henry. 1975. *Cowboys, Indians and Homesteaders.* Desert Press, Tucson.

Miller, R., J. Rose, T. Svejcar, J. Bates, and K. Paintner. 1995. Western juniper woodlands: 100 years of plant succession. In D. W. Shaw, E. F. Aldon, and C. LoSapio (technical coords.), *Desired future conditions for piñon-juniper ecosystems,* pp. 5–8. USDA Forest Service General Technical Report RM-258.

Miller, R. F., and J. A. Rose. 1995. Historic expansion of *Juniperus occidentalis* (western juniper) in southeastern Oregon. *Great Basin Naturalist* 55: 37–45.

Miller, R. F., and P. E. Wigand. 1994. Holocene changes in semiarid piñon-juniper woodlands. *BioScience* 44: 465–474.

Monsen, S. B., and R. Stevens. 1999. *Proceedings: Ecology and management of pinyon-juniper communities within the intermountain West.* Rocky Mountain Research Station. U.S. Forest Service, Fort Collins, Colo.

Torres-Reyes, Ricardo. 1970. Mesa Verde National Park: An administrative history 1906–1970. National Park Service. On file, Mesa Verde National Park.

CONTRIBUTORS

EDITORS

M. Lisa Floyd (also known as M. Lisa Floyd-Hanna) has been active in ecological research in southwestern Colorado and the surrounding area for the past two decades. She received her Ph.D. from the University of Colorado in 1981. Floyd's recent research has involved fire and disturbances in piñon-juniper woodlands, fire history and fire-risk modeling in the San Juan Mountains, and numerous projects in national parks of the region. She is professor and chair of Environmental Studies at Prescott College.

David D. Hanna received his M.S. from Antioch University in 1982. For the past two decades he has been active in fire-related research for the National Park Service and the U.S. Forest Service in the San Juan Mountains, principally lending his hand in Geographic Information Sciences. Hanna is currently an instructor of GIS at Prescott College.

William H. Romme has led many research projects concerned with fire history, fire effects, and twentieth-century changes in forest composition in the Four Corners region as well as in the northern Rocky Mountains. He received his Ph.D. from the University of Wyoming in 1975 and taught biology and ecology at Fort Lewis College. He is currently a professor at Colorado State University in the College of Natural Resources.

CONTRIBUTORS

Marilyn Colyer has extensive experience in Mesa Verde Country, both as an archaeologist with the Wetherill Project and as a resource management specialist for the past thirty-five years at Mesa Verde National Park. She has accumulated a wealth of information on the flora, fauna, geology, and climate of Mesa Verde. Colyer obtained her B.S. from Colorado State University and has continued her training through the National Park Service and on her 250-acre ranch in southwestern Colorado. She is currently a resource management specialist at Mesa Verde National Park. Colyer is committed to preservation of southwestern ecosystems.

CONTRIBUTING AUTHORS

Jayne Belnap is a research scientist for the U.S. Geological Survey–Biological Resources Division in Moab, Utah. She received her Ph.D. from Brigham Young University in 1991 and was a resource manager with the National Park Service, Canyonlands National Park, during which time she was involved in studies of invasive plants and nitrogen fixation by soil crusts. Belnap's area of specialty is microbiotic crusts throughout the world. She has published extensively on the effects of disturbance on nitrogen fixation in soils.

Michael A. Bogan is a wildlife research biologist with the U.S. Geological Survey's Midcontinent Ecological Science Center (MESC) in Albuquerque, New Mexico. He is the project leader of the Arid Lands Project and a research professor of biology at the University of New Mexico. His M.S. degree from Fort Hays (Kansas) State University focused on the ecology of small mammals, and his Ph.D. at the University of New Mexico concerned the taxonomy of bats. Since completing his degrees he has been a curator of mammals at the National Museum of Natural History and a research supervisor of vertebrate ecology with the MESC and its predecessors. Throughout his professional career Bogan has had an abiding interest in bats and has published many scientific articles on bats and other mammals. He currently serves on several committees and advisory boards in the general area of mammalian conservation and continues to have an active field program on mammals.

Alice L. Chung-MacCoubrey is a research wildlife biologist with the USDA Forest Service Rocky Mountain Research Station in Albuquerque, New Mexico. Her research currently focuses on the ecology of bats and herps in southwestern ecosystems. She received her B.S. in biochemistry with highest honors from Rutgers University. Her M.S. in fisheries and wildlife sciences at Virginia Tech (VPI & SU) focused on wildlife nutrition and toxicology. Chung-MacCoubrey earned her Ph.D. at the University of New Mexico. Her dissertation focuses on the maternity roost requirements of bats in piñon-juniper woodlands and ponderosa-pine forests of New Mexico.

Nolan J. Doesken is the assistant state climatologist for Colorado and has worked at the Colorado Climate Center, Department of Atmospheric Science, Colorado State University, since 1977, studying all aspects of the recent (since the late 1800s) climate of Colorado. Before coming to Colorado, he earned his B.S. in meteorology and oceanography from the University of Michigan in 1974 and his M.S. in atmospheric science from the University of Illinois in 1976.

Patricia Robins Flint-Lacey received her Ph.D. in anthropology from the University of Oregon in 1982. She worked with the Dolores Archeological Program in 1982–1983 and conducted excavations under the North Rim of Mesa Verde on Pueblo II sites from 1985 to 1986. Currently she is the technical editor for Mesa Verde National Park. Flint-Lacey was recently appointed to the Colorado Historic Preservation Board. Her newest projects include working with a geologist on identifying the lithic sources and quarries in the Canyon of the Ancients National Monument and developing a brochure on the seven cliff dwellings on the Sand Canyon Trail in southwestern Colorado.

Mary L. Gillam is an independent geologist who specializes in Quaternary stratigraphy, geomorphology, soils, geoarchaeology, and geologic hazards. She completed her Ph.D. at the University of Colorado in 1998 and earlier took her M.S. and B.S. at Stanford University in 1975 and 1974. In addition to research in the Four Corners area, she has worked on applied projects in most western states for consulting firms, government agencies, and private clients.

Timothy B. Graham specializes in reptile and amphibian research in many locations on the Colorado Plateau. He earned his Ph.D. at Utah State University, studying plant herbivore interactions of grasshoppers and grasses in the Grand Canyon. Since 1984 he has been a researcher with the Park Service and is currently a biologist with the U.S. Geological Survey–Biological Resources Division in Moab, Utah. Graham's current projects include invertebrate and amphibian responses to disturbances (grazing and roads).

Mary Griffitts has been active in geologic research at Mesa Verde National Park for the past two decades and is the author of *Mesa Verde Geology*. She received her Ph.D. from the University of Colorado in 1944. Griffitts taught general physical and historical geology at Bryn Mawr College, Pennsylvania; historical geology, mineralogy, optical crystallography, and petrology at the University of Illinois; and historical geology, invertebrate paleontology, and stratigraphy at the University of Colorado. She was a ranger naturalist at Yellowstone National Park. Griffitts helped establish and teach geology at Boulder Junior Natural Science School in Boulder (now expanded at several places along the Front Range). For the last twenty years she has worked intermittently at Mesa Verde National Park in whatever capacity needed in relation to geology.

CONTRIBUTORS

Deborah M. Kendall earned her Ph.D. in entomology in 1987 from Colorado State University, her M.S. in entomology in 1981 from the University of Colorado, and her B.S. in zoology and entomology in 1976 from Colorado State University. Her research interests include insect ecology and biological control of weeds. She has published in *Science, Journal of Chemical Ecology, Environmental Entomology,* and *Southwestern Entomologist,* among others. Kendall is an associate professor of biology at Fort Lewis College.

Boris C. Kondratieff is a professor of entomology and curator of the C. P. Gillette Museum of Arthropod Diversity at Colorado State University, Fort Collins. His primary research interests focus on the systematics and ecology of insects. He has published over 120 research publications and is the author of two books.

David A. Leatherman was born and raised in central Ohio and received his Master of Forestry degree in forest protection (emphasis on entomology) from Duke University in 1974. He has been forest entomologist for the Colorado State Forest Service (CSFS) from 1974 to the present. During his career with CSFS he has made over fifty visits on entomological business to southwestern Colorado, including several to Mesa Verde National Park. Leatherman is an adjunct faculty member in the Department of Bioagricultural Science and Pest Management at Colorado State University, where he teaches an "Integrated Forest Pest Management" class. He is the former editor of *Colorado Field Ornithologists' Journal* and is an avid birder.

J. Page Lindsey received her B.A. in biology at Hendrix College in 1970, M.S. in botany from the University of Arizona in 1972, and Ph.D. in plant pathology from the University of Arizona in 1975. Lindsey has identified and catalogued the fungi of Mesa Verde. Her areas of specialty are the ecology and taxonomy of basidiomycete wood decay fungi. She is a professor of biology at Fort Lewis College, where she has taught since 1978.

William J. Litzinger is a botanist and plant ecologist who received his Ph.D. from the University of Colorado in 1983. He specializes in plant-human interactions. Since 1974, when he began work at Hovenweep National Monument, he has participated in numerous archaeological projects. Litzinger currently studies the flora of the Southwest, focusing on plant families that have a long history of relationship to humans (Solanaceae). He is a professor of environmental studies at Prescott College.

Thomas B. McKee was the state climatologist for Colorado, working in the Department of Atmospheric Science at Colorado State University from 1974 until his retirement in 1999. Prior to coming to CSU he received his B.S. in physics from the University of North Carolina in 1958 and his M.S. in 1963 from William and Mary, followed by a Ph.D. in atmospheric science in 1972 from Colorado State

University. He worked for the National Aeronautics and Space Administration (NASA) from 1958 to 1971, specializing in atmospheric radiation.

Sylvia Oliva is a former physicist and computer consultant. She is currently the air quality contractor for Mesa Verde National Park. Oliva has resided in piñon-juniper country for twenty-five years, during which time she has been active in exploring the backcountry on horseback and in xeric gardening.

Doug Ramsey has been a soil scientist and project leader for the U.S. Department of Agriculture–National Resources Conservation Service in Cortez, Colorado, since 1978. He attended Fort Lewis College and then Colorado State University, where he received a B.S. in soil science. Since 1980 he has been conducting soil investigations throughout the San Juan Basin and has spent four years working within Mesa Verde National Park, mapping the soils and assisting with soil-related investigations. Ramsey is currently in the process of conducting a soil survey of the Ute Mountain Indian Reservation in Colorado and New Mexico.

George L. San Miguel earned his bachelor's degree in analysis and conservation of ecosystems at the University of California–Los Angeles. He also graduated from the National Park Service's Natural Resources Specialist Training Program. San Miguel has worked for the National Park Service since 1980 as a research assistant, ranger-naturalist, biological technician, and natural resource manager. He specializes in avian studies, ecology, and restoration and has written several natural resource–related management plans and articles. San Miguel has been the branch chief of Natural Resource Management at Mesa Verde National Park since 1998. One of his professional goals is to inspire an appreciation of the old-growth piñon-juniper woodlands among Mesa Verde's managers, staff, and visiting public so that this unique habitat may endure unimpaired within the park for each future generation.

Duane A. Smith has been a professor of history and Southwest studies at Fort Lewis College since 1964. He earned his Ph.D. at the University of Colorado. Smith has published numerous books on Colorado mining, urban history, and western history, including *Mesa Verde National Park: Shadows of the Centuries*, Revised Edition (University Press of Colorado, 2002).

Preston Somers earned his Ph.D. in zoology at the University of Colorado. He has been active in inventorying small mammals and birds in southwestern Colorado for three decades, often including students in his work. He has published numerous articles and reports for the Bureau of Reclamation. Somers is a professor and chair of biology at Fort Lewis College.

CONTRIBUTORS

CONTRIBUTING ARTISTS

Walt Anderson is an artist and ecologist in Prescott, Arizona. He is professor of environmental studies at Prescott College.

Noni Floyd is an artist and art teacher in Kailua, Hawaii.

Elizabeth Griffitts is an artist and botanist who lives in Cortez, Colorado.

Susie Harvin is an artist who has spent many years drawing the natural landscape of southwestern Colorado.

Sarah Luecke is a graduate of Prescott College, Prescott, Arizona.

Pat Oppelt is a natural historian who volunteers much of her time at Mesa Verde National Park.

Agnes Suazo is a graduate of Fort Lewis College, Durango, Colorado.

Mary Vozar is an artist, retired from Mesa Verde National Park.

Amy Wendland is an assistant professor of art at Fort Lewis College, Durango, Colorado.

INDEX

Abandonment, of Mesa Verde, 1, 38, 288, 310, 316
Acid rain, 350
Acid snow, 349, 350, 352
Actinomycetes, 79
Adaptations to arid conditions, 115
Age-diameter models, 14
Agricultural chemicals, 365
Agricultural depression, 327, 334
Air pollution, 337, 339, 349, 352, 357
Albedo, 82, 85
Aleurodiscus mesaverdensis, 66
Alluvial, 33, 194, 199, 201, 203, 205, 206, 208, 209, 210, 236, 238, 240, 241
Alluvial aquifers, 225, 226
Alluvium, 304, 305, 316
Altering fire regimes, 342, 344, 346, 357
Amaranthus species, 304, 315
Ambystoma tigrinum, 352
Amelanchier utahensis, 7, 32, 264, 284, 310, 313, 353. *See also* Utah serviceberry
Amphibians, 1, 7, 10, 12, 111, 151, 152, 153, 156, 157, 159, 163, 193, 223, 228, 229, 244, 347, 352; decline of, 158, 159, 163
Anasazi, 368
Ancestral Puebloans, 1, 4, 11, 12, 25, 30, 102, 152, 192, 218, 223, 226, 235, 237, 249, 259, 268, 288, 295, 303, 309, 313, 315, 316, 317, 321, 356, 366
Ancient piñon-juniper woodlands, 1, 3, 4, 13, 131, 234, 271, 273, 339, 356, 357
Ancient woodland, 8, 61, 151, 337, 356, 357
Animas Formation, 194, 220, 221
Animas glacier, 202, 203
Antarctica, 201
Antrozous pallidus, 132, 138, 141
Ants, 161, 167, 168, 201, 280, 281
Anuran amphibian populations, 158
Aquifers, clay slope, 226
Arceuthobium divaricatum, 168
Archeology, 319
Arid-land mammals, 118
Arizona, 132, 136, 170, 180, 189, 190, 205, 229, 349, 355, 356
Arrows, 310, 311, 313
Artemisia nova, 32, 114, 264. *See also* Black sagebrush
Artemisia tridentata, 32, 35, 114, 316. *See also* Big sagebrush
Astragalus deterior, 35, 36
Astragalus schmolliae, 33, 37
Atriplex canescens, 310, 311
Avifauna, 89, 93, 102, 103

381

INDEX

Bacteria, 125, 126
Badgers, 125
Bark gleaners, birds, 97
Basketmakers, 288
Bats, 21, 27, 94, 95, 111, 112, 131–141, 145, 146, 147, 167, 223, 226; Allen's big-eared, 140; big brown, 137, 138; big free-tailed, 140; Brazilian free-tailed, 140; California myotis, 132, 136; distribution of, 131; foraging and roosting habitats, 10, 21, 83, 93, 98, 99, 122, 131, 132, 135, 139, 141, 146, 157, 168; fringed myotis, 134, 136, 138; habitat of, 139, 141, 147; hoary, 133, 138; little brown, 138; long-legged myotis, 132, 135, 138; pallid bat, 132, 137, 139
Battleship Rock, 237, 274
Bead potato, 299
Bears, 18, 126, 232, 236
Bedrock, 242, 243, 244, 352
Beetles, 63, 95, 156, 159, 168, 170, 172, 173, 228, 238, 283
Bentonite, 34, 193, 223, 225, 226, 238, 239
Berry pickers, birds, 100
Big sagebrush, 114, 264. *See also Artemisia tridentata*
Bioessential nutrients, 80
Biological complexity, 112
Biological diversity, 361, 363, 365, 369, 371
Biological soil crusts, 63, 75, 76, 78, 80–83. *See also* Biotic crust; Cryptobiotic; Cryptogamic
Biotic crust, 89. *See also* Biological soil crusts
Bircher fire, 120, 263, 270, 274
Biscuit-roots, 298, 299, 301
Bitterbrush, 1, 20, 32, 34, 91, 95, 101, 118, 119, 120, 151, 217, 306, 364. *See also Purshia tridentata*
Black sagebrush, 114, 264. *See also Artemisia nova*
Black stain root rot, 19, 21, 66, 95, 168. *See also Leptographium wageneri*
Blue-gray gnatcatcher, 98
Bluestain fungi, 170, 172
Bluff Formation, 189
Bromus tectorum, 343. *See also* Cheatgrass
Brown Springs, 225
Brush Basin Member, 190
Brush mouse, 122
Bryophytes, 22, 61, 63, 64, 65
Bufo punctatus, 153
Bufo woodhousii, 153
Bullfrog, 158, 159, 163
Bullsnakes, 162
Bull thistle, 343

Burro Canyon Formation, 204, 206

C^{14} dating, 218
Caching behavior, of birds, 120
Cairns, 161
Canada thistle, 236, 242, 343, 344, 346. *See also Cirsium arvense*
Canopy: disturbances, 31; gaps, 20, 32
Canyon mouse, 122
Canyon wren, 99, 241, 244
Carbon, 341
Carbon dioxide, 350, 371
Carduus nutans, 343
Carmel Formation, 189
Carnivores, 112, 113, 124
Cattle grazing, 230, 234, 243, 291, 327, 331, 332, 334, 363
Caves, 134, 136, 137, 138, 139, 140, 146, 355
Cavities, in trees, 13, 21, 22, 24, 89, 90, 91, 94, 95, 139, 141, 146, 160, 228, 229, 348
Cecidomyiid gall midges, 174
Cedar Mountain Formation, 190
Cenozoic uplift, 198
Centaurea dehisa, 343
Centaurea repens, 343
Chaparral, 345
Cheatgrass, 343–346, 348
Cheno-ams, 304
Chenopodium fremontii, 283
Chickadees, 91, 95, 96, 97, 99, 173
Chihuahua desert, 84
Chinle Formation, Triassic, 188, 190
Chokecherries, 298
Chorus frogs, 152, 157, 158, 163, 226, 352
Christianity, 322
Chrysothamnus nauseosus, 7, 115. *See also* Rabbitbrush
Cirsium arvense, 343. *See also* Canada thistle
Cirsium vulgare, 343. *See also* Bull thistle
Clark's Nutcrackers, 91, 92, 93, 120
Clean Air Act, 349
Clerid beetles, 172
Cliffhouse Sandstone, 24, 35, 37, 153, 193, 204, 206–209, 219, 223, 226, 228, 236, 237, 239
Cnemidophorus velox, 161
Coal, 190, 191, 192, 193, 198, 328, 349, 354; mining of, 332, 333, 334, 335
Collomia grandiflora, 38
Colluvial, 36, 206, 208, 216, 219, 220, 221; soils, 34, 219, 220
Colorado Climatic Center, 248
Colorado Plateau, 3, 4, 12, 32, 62, 66, 78, 79, 84,

382

91, 315, 342, 346, 349, 355, 362, 366; bats of, 137; fire history on, 260, 269, 272, 273; geology of, 184, 199, 202; insects of, 167; reptiles of, 163
Colorado State Species of Concern, 38
Commandra, 31, 33
Conks, 61, 62, 66
Conophthorus edulis, 173
Conservation, 103, 135, 217, 219, 330, 334, 347, 348, 362, 367, 368, 369, 370, 371, 373, 374; value, 12
Corn layer, 24
Cottontails, 22, 112, 113, 115, 116
Coyote, 18, 113, 115, 116, 124, 126, 305; water attracts, 225, 226, 227, 228, 236, 238, 239, 240, 241, 242, 243, 244
Cradleboards, 310, 311
Creede, Colorado, 200
Cretaceous, 183, 190, 191, 193, 194, 198, 199, 206, 207, 210
Cross-bedded sandstone, 188
Crotalus viridis, 162
Crotapytus collaris, 161
Cryptobiotic, soils, 75, 85
Cryptogamic, soils, 75
Cryptogams, 63
Cucurbita pepo, 309
Culturally scarred trees, 274
Cultural resources, at risk by fire, 261, 264, 273, 274
Curecanti National Recreation Area, 158
Curtail (herbicide), 345
Curtis Formation, 189
Cyanobacteria, 31, 62, 63, 75–78, 80, 81, 83, 84, 85; cyanobacterial sheath, 80, 81
Cyanolichen, 62, 63

Dakota Belt, 205
Dakota Sandstone, 153, 190, 191, 204–206, 225, 226, 228, 241–244
Datil, 300
Dead wood, 366
Debris flow deposits, 202, 204
Deer, mule, 18, 22, 91, 103, 111, 112, 113, 114, 115, 118, 120, 121, 122, 125, 309, 316, 317, 349, 363, 365; water attracts, 225, 227, 236, 237, 241
Deer mouse, 18, 72, 111, 112, 115, 118, 121, 122, 125, 126. *See also Peromyscus maniculatus*
Deforestation, 310, 315, 356, 362, 363, 371
Dendroctonus species, 170, 265
Density, of plants, 7, 13, 14, 17, 18, 19, 20, 25, 27, 79, 112, 121, 122, 160, 322, 346, 366; ethnobotanical relationship, 296; relationship to soils, 215, 221
Desertification, 81, 83
Desert shrub vegetation, 146
Desired future condition, 368, 372
Devonian Elbert Formation, 187
Die-off, of trees, 19
Differential mortality hypothesis, 17–18
Digging sticks, 296, 310, 313, 316
Dinosaur National Monument, 132
Dioryctria albovittella, 173
Dioryctria ponderosae, 174
Disturbance, of piñon-juniper woodlands, 6, 9, 10, 11, 13, 17, 19, 24, 27, 31, 32, 79, 81, 82, 83, 84, 85, 172, 191, 222, 261, 264, 339, 342, 343, 344, 353, 355, 357, 358
Diurnal, 103, 116, 124, 140, 151, 152, 160, 249, 253; birds of prey, 90
Dobbins Stockade, 314
Dolores Formation, Late Triassic, 188
Dolores River, 138, 159, 206, 209, 229, 241, 244, 322, 323, 324, 325, 326
Douglas fir, 6, 9, 32, 34, 91, 135, 141, 145, 310, 355. *See also Pseudotsuga menziesii*
Downed wood, 20, 21, 161, 269
Down woody debris, 20, 24, 26, 369
Downy woodpecker, 95, 97, 172
Drought, 2, 10, 13, 19, 85, 95, 155, 168, 172, 173, 235, 238, 247; effect on fire cycles, 255–257, 263; effect on woodland health, 309, 316, 317, 340, 344, 347, 356, 391
Dry deposition, 349, 351
Dryland farming, 363
Durango, Colorado, 194, 202, 203, 206, 210, 220, 253, 254, 255, 324, 325, 326, 328, 341
Dust trapping, in soils, 79

Early successional communities, 280
Ecological dysfunction, 163
Ecological integrity, 8, 337, 341, 342, 346, 357, 358
Ecosystem-based management, 23, 151, 153, 163, 337, 341, 362
Ecotone, 31, 32, 299
Ecotherms, 156, 159
El Niño, 18, 250, 268, 356, 357
Elymus cinerus, 35
Endemic, 10, 37, 38, 111, 372
Endotherms, 160
Engraver beetle, 170
Environmental crisis, 322
Eolian deposition, 189, 205; soils, 217–219

383

INDEX

Eptesicus fuscus, 133
Eriogonum corymbosa, 7, 35
Euderma maculatum, 139, 141
Evapotranspiration, 214, 215
Extinction, local, 162, 348, 349

Facultative exploitation, 113
Federal Candidate Species, 146. *See also* Threatened and Endangered species
Fendlera rupicola, 264, 311
Fewkes Canyon, 153; Plunge Pool, 237
Fire: effects, 261, 266, 275, 279, 285, 340; fire-sensitive vegetation, 264, 340; fire-stimulated species, 304, 340; frequency, 7, 83, 267, 268, 273, 275, 341, 342, 357; history, 17, 20, 264–275, 366, 372; management plan, Mesa Verde National Park, 274, 366, 370; natural barriers to spread, 290; regime, 12, 20, 265, 267, 268, 269, 270, 272, 273, 275, 341, 344, 346, 357; scars, 20, 264, 265, 269, 274, 286; severity, 272, 339, 340, 341, 344, 357; suppression, complete, 261, 267, 341; threat of, 340; written fire records, 265
Flies, 95, 98, 115, 159, 168, 172, 174, 223, 226, 235, 238, 280
Fluvial deposition, 188, 189, 190
Flycatchers, 226
Folding and faulting, 184, 198
Foliage gleaners, birds, 97–98
Food webs, 93, 99, 167
Formations, geologic, 23, 183, 185, 188, 199, 200, 201, 205, 206, 208, 209, 210, 220, 225, 226, 349, 361
Fort Lewis, 114, 248, 347
Four Corners, 4, 6, 157, 172, 182, 186, 187, 191, 198, 202, 206, 209, 211, 215, 217, 218, 249, 256, 268, 273, 305, 321, 348, 349, 352, 362, 363, 366, 373
Frogs, leopard, 158, 163, 226, 241; northern leopard frog, 158
Fruitland Formation, 193
Fuel management, 370, 371
Fuel reduction, 290, 366
Fungi, 10, 27, 61–67, 79, 80, 112, 125, 170, 172; wood-rotting, 62, 66

Gambel oak, 6, 7, 32, 98, 122, 226, 236, 242, 264, 265, 274, 284, 296, 310, 340. *See also Quercus gambelii*
Gametophyte, 63
GCM (global climate model), 354
Generalist, birds, 90, 91, 97, 101, 112

Geologic map, 183, 206
Geomorphology, 197
Glacial deposits, 198, 201, 202
Glacial ice, 198
Glades, The (Mesa Verde National Park), 268
Glen Canyon Group, 188
Global climate change, 7, 31
Globe mallows, 242
Goodman Point Springs, 242
Gray fox, 22, 112, 225, 227, 238, 241, 244
Grazing, cattle and sheep, 267, 290, 327, 331, 332, 333, 334, 335, 363, 364, 365, 370
Grazing permits, 333, 334
Great Basin, 10, 19, 27, 78, 84, 159, 167, 273, 289, 344, 362
Green algae, 62, 63, 75, 77, 80
Greenland, 201
Green-tailed towhee, 348
Ground foragers, 100
Gunnison County, Colorado, 158
Gymnorhinus cyanocephalus, 109. *See also* Pinyon jay
Gypsum, 187, 189, 198, 220, 305

Habitat fragmentation, 337, 339, 340, 346, 348, 352, 357, 363, 378
Hackelia gracilenta, 38
Hammer and ax handles, 312
Hantavirus, Sin Nombre strain, 121, 126
Hawks, 10, 90, 94, 244
Hemotoxins, 162
Herbicides, 229, 291, 346, 362, 365, 373
Herbivores, 92, 113, 114, 115, 122, 280, 281, 283, 284, 285
Hermosa Formation, 189
Hesperus, Colorado, 139, 140, 141, 202, 328
Heterotrophs, 61
Historic range of natural variability, 341
Hogback, 204, 209, 210
Homesteading, 327, 335
Hopi, 289, 292, 316
Horse Springs (Horse Canyon Springs), 239, 240
Hovenweep National Monument, 25, 153, 213, 214, 215, 244
Human activity, 1, 85, 342, 353
Human coprolites, 295, 304, 305, 316
Human disturbances, 11
Human impact, 221
Hummingbirds, 99, 100
Hyphae, 62

Idionycteris phyllotis, 140, 145
Ignacio Quartzite, 186, 187

Igneous rock, 185, 198, 201, 203, 208, 209
Iliamna rivularis, 38
Incision, 203, 209, 211
Increasing density hypothesis, 16, 20
Indigenous foragers, 287, 288
Indigenous plant foods, 288
Inoceramus grandis, 192
Insect eaters, 22
Insectivores, 91, 95, 112
Insect outbreaks, 13
Insect recolonization, 279
Instrumental climatic record, 247, 248
Integrated weed management, 365, 369
Interdecadal variations, climatic, 29, 246, 253
Invasive species, plants, 66, 156, 236, 272, 280, 339, 340, 341, 342, 343, 345, 346, 348, 357, 365, 367, 369, 370–373
Ips confusus, 20, 168
Ips latidens, 170
Ips pini, 170
Irrigation, 153, 154, 158, 229, 241, 242, 326, 327, 364

Janitiella coloradensis, 174
Juniper bark (bast), 111, 310, 311
Juniper titmouse, 89, 91, 97
Juniperus osteosperma, 6, 215, 284
Juniperus scopulorum, 6, 35, 217

Karst surface solution depressions, 187
Kayenta Formation, 188
Kelly, Charles, 328, 331
Kirtland Shale, 199
Kit foxes, 115, 116, 118, 124, 241

Lactating, 113, 134, 135, 137, 139
Lactuca serriola, 343
Lagomorphs, 112, 162
Landforms, 4, 181, 197, 198, 200, 201, 202, 203, 205–211, 218, 219, 271
Landslide deposits, 200, 204, 206, 208
La Niña, 237
La Plata, 210, 229
Laramide orogeny, 198, 199, 209
Larval feeding galleries, 170
Lasionycateris noctivagans, 133, 137
Lasiurus cinereus, 133, 138
Lava flows, 194, 198
Leadville Limestones, 187
Leaked nitrogen, 79
Leptographium wageneri, 19, 66, 168. *See also* Black stain root rot

Lewis Shale, 193
Lichens, 10, 22, 31, 61–67, 75–78, 81–84, 100, 352, 370; crustose, 63; foliose, 63; fruticose, 63
Lightning ignitions, 259, 261, 267, 269, 270, 273, 360, 367
Lithologic units, 183, 188, 189, 190, 192
Little Long House Pothole Series, 237
Liverworts, 63, 64
Livestock, 2, 39, 81, 84, 362, 363, 364, 365, 367, 368, 370
Lizards, 90, 123, 124, 159, 160, 161, 162, 163; northern plateau, 161; sagebrush, 151, 160, 161
Long Mesa fire, 263, 270, 279, 281, 285, 345
Low-severity disturbance, 9, 342
Lumbering, 2
Lupines, 32

Magnesium, 79, 80, 225, 226
Maize, 288
Mancos–Mesa Verde belt, 24, 206
Mancos Shale, 25, 35, 37, 38, 191, 192, 204, 205, 206, 209, 220, 221, 226, 227, 236, 240, 241, 301, 302
Mancos Valley, 12, 58
Manipulated piñon forests, 291
Masting, 18, 25, 92, 93, 120, 121, 290, 292
Maternity colonies, 134, 136, 137, 138, 139
Matsucoccus acalyptus, 168, 173
McElmo Creek, 132, 207, 227
Medicinal plants, 295, 305
Menefee Shale, 25, 34, 35, 154, 204, 206, 225, 226, 227, 235, 236, 238, 239, 240
Mesa Verde cuesta, 7, 10, 13, 25, 30, 32, 33, 37, 38, 111, 191, 215, 218, 219, 221, 223, 225, 226, 235, 239, 259, 261, 265, 267, 272, 274, 275, 296, 310, 370
Mesa Verde escarpment, 181, 208
Mesa Verde Group, 192, 193, 206, 207, 209
Mesozoic Era, 188, 194
Mesozoic sediments, 186
Metamorphic rock, 186, 198, 201, 203, 208, 209
Metamorphs, 156, 158
Microcoleus vaginatus, 77
Microphytic soil crusts, 75
Mines, 135, 136, 137, 140, 147, 328, 332, 333, 334
Mining, 12, 147, 325, 332, 333, 334, 336, 337, 362
Miocene, 199, 210
Mistletoe, 23, 100, 168
Moab, Utah, 154, 157
Moccasin Mesa fire, 236

Mojave, 84, 361
Monitoring, 163, 248, 255, 256, 264, 351, 362, 373
Montezuma County, Colorado, 132, 133, 138, 140, 145
Montezuma Creek, 4, 206, 209
Morefield, Mabyn, 331
Morefield Spring, 235
Morrison Formation, 25, 189, 190, 204, 206
Mosses, 22, 30, 63, 64, 75, 76, 77, 78, 80, 83, 85, 370
Moths, 90, 94, 95, 168, 173, 174
Mountain lion, 91, 124, 225, 227, 228, 234, 236, 237, 238
Mountain sheep, 115
Mountain shrubland, 24, 94, 281, 296, 299, 341
Mummy lake, 334
Muskthistle, 236, 343, 344, 345, 346. *See also Carduus nutans*
Mutton grass, 1, 33, 34, 113, 215, 217, 302, 345, 364. *See also Poa fendleriana*
Mutualistic, 61
Mycelium, 62
Mycobionts, 62
Myotis californicus, 135, 136, 137
Myotis lucifugus, 138
Myotis thysanodes, 136
Myotis yumanensis, 140, 145
Myotoxins, 162

National Park Bill, 328
Native understory grasses, 33
Natural Bridges National Monument, 140, 162
Natural disturbances, 6, 13, 261, 353
Navajos, 289, 295, 298, 306
Navajo Sandstone, 188, 189
Nectar sippers (birds), 99
Needle Mountains, 199, 206
Neodiprion edulicolus, 174
Neotropical migrants, 95, 98, 102, 103
Neurotoxins, 162
Nevada, 188, 362
Newhall Simulation Model, 213, 214
New Mexico, 4, 6, 19, 28, 33, 131, 132, 133, 134, 136, 137, 139, 140, 141, 145, 146, 157, 158, 189, 190, 194, 202, 205, 304, 322, 355, 356
Nicotiana attenuata, 37, 304
Nitric acid, 350
Nitrogen, 10, 20, 62, 75, 78, 79, 80, 82, 84, 85, 159, 280, 283, 284, 285, 304, 349, 350, 351; isotopic ratios, 34; fixation, 61, 78, 83; input, 79, 82, 83
Nitrogenase, 82, 84
Nocturnal, 90, 91, 94, 118, 124, 160, 162

Nonfatal exposure to wildfire, 172
Non-native forage grasses, 362
Nonvascular biota, 61
Nonvascular plants, 10, 39, 61, 67
Nordenskoild, Gustaf, 328
NRCS (Natural Resources Conservation Service), 217, 219, 220, 221
Nuptial chamber, 170, 172
Nusbaum, Jesse, 331
Nut birds, 89, 91
Nutrient cycling, 61
Nyctinomops macrotis, 139, 140, 145
Nymphs, 168, 173

Off-road vehicles, 81
Old-growth, piñon-juniper, 7, 8, 9, 10, 11, 12, 14, 17, 20, 24, 26, 27, 35, 37, 38, 39, 65, 66, 89, 90, 94, 102, 110, 135, 145, 152, 167, 172, 195, 215, 222, 274, 279, 280, 348, 361, 367, 368, 370, 371, 372, 373; remnant stands, 372
Old Ugly, 225
Omnivores, 89, 91
Orographic lifting, 250
Oryzopsis hymenoides, 33, 345
Outwash, 201, 210
Overexploitation, of piñon-juniper woodlands, 291
Owls, 90, 91, 94, 95, 226, 236, 244
Oyster beds, 25
Ozone, 249, 352, 353, 373
Packrat middens, 118, 249, 355
Paiute Indians, 289, 291
Paleozoic-Ancestral Rocky Mountains, 189
Paleozoic sediments, 189
Parasitic, 24, 61, 63, 66, 125, 126, 306
Park Sensitive Plants, 38
Parrot City, 324
Pathogens, 9, 13, 32, 123, 170, 306
Pedicularis centranthera, 33, 37, 348
Pennsylvanian Hermosa Group, 189
Penstemon linearoides, 33, 34, 99, 283
Perennial streams, 153
Permanent water sources, 138
Peromyscus boyleii, 122
Peromyscus crinitus, 122
Peromyscus maniculatus, 18, 111. *See also* Deer mouse
Peromyscus truei, 18, 122. *See also* Piñon mouse
Petran chaparral, 345
Petroglyphs, 162
pH, 80, 219, 220, 225, 226, 241, 244, 350, 351, 352
Phloeosinus furnissi, 172
Phloeosinus hoferi, 172

Phloeosinus scopulorum neomexicanus, 172
Phosphorus, 79, 304
Phrynosoma douglasii, 161
Phycolichen, 62
Physical environments, 11, 122, 163
Pictured Cliffs Sandstone, 193
Pine River, 4, 210, 229
Pinnacled (microbial crusts), 79
Piñon cone beetle, 172
Piñon dwarf mistletoe, 168
Piñon ips, 168, 170, 172–173
Piñon jay, 96, 121
Piñon mouse, 18, 122, 125
Piñon needle scale, 168, 173
Piñon nuts, 93, 121, 289, 316, 369
Piñon pine dispersal, 121
Piñon pitch, 174, 300, 302
Piñon pitch mass borer, 174
Piñon snags, 66, 134, 135, 138, 141, 146
Piñon spindle gall midge, 172
Piñon tip moth, 173
Pinus cembroides, 288
Pinus edulis, 6, 133, 268, 284, 288, 289, 310
Pinus monophylla, 120, 288
Pinus ponderosa, 6, 311, 352
Pinyonia edulicola, 172
Pinyon jay, 89, 91, 93
Pioneers, 324, 325, 326, 329
Pipistrellus Hesperus, 133, 139, 141
Pituophis melanoleucus, 162
Pityophthorus, 168
Plunge pool, 153, 226, 227, 234, 237, 238, 239
Poa fendleriana, 33, 113, 215, 245, 264. *See also*
 Mutton grass
Point Lookout Sandstone, 25, 37, 193, 204, 206, 207, 220, 225, 228, 240
Pollen analysis, 315, 317
Polygamous, 170, 172, 283
Polygonum sawatchensis, 34
Pony fire, 263, 270, 274, 303
Populus fremontii, 70
Porcupines, 112, 123
Potassium, 79, 80, 304, 309
Potholes, 10, 153, 154, 223, 225, 228, 229, 236, 237, 238, 241, 243, 244
Prater, Albert, 331
Prayer sticks, 310, 312, 313
Precambrian rocks, 186, 199, 201
Precipitation/evapotranspiration (PPT/ET), 214, 215
Predator-prey, 124
Prehistoric fire history, 265

Prescribed fire, 342
Presettlement structure, 12
Prokaryotes, 75
Pseudacris triseriata, 157, 352
Pseudotsuga menziesii, 6, 31, 310, 352. *See also*
 Douglas fir
Pteromalid wasp, 173
Public lands, management, 273, 362, 368, 369
Pueblo ruins, 145
Purshia tridentata, 20, 32, 118, 217, 364. *See also*
 Bitterbrush
Pyronema domesticum, 66

Quartzite, 186, 187
Quercus gambelii, 6, 32, 264, 274, 284, 310. *See also*
 Gambel oak

Rabbitbrush, 7, 115. *See also Chrysothamnus nauseosus*
Raccoons, 124, 226
Railroad, 324, 325, 326, 328, 362
Raindrop erosion, 80, 85
Rana catesbeiana, 158
Rana pipiens, 158
Randolph, Hans, 329, 332
Range of natural variation, 24, 222, 272
Recovery time, 84
Red-spotted toads, 152, 153, 154–157, 228, 237, 239, 243
Reference period, model of, 12, 268
Reproductive investments, 155
Reptiles, 1, 7, 27, 111, 123, 151, 152, 156, 159, 160, 162, 163, 193, 223, 229
Resident status, 112
Resprouting (shrubs), 264, 282, 343
Restoration, 85, 363, 368, 369
Rhizoids, 64
Rickner, Thomas, 332
Rico Formation, 187
Ringtails, 124
Riparian habitats, 112, 152, 298
Rock art, 199
Rock Canyon, 235, 238, 239
Rock units, 199
Rodent, 79, 90, 95, 112, 113, 115, 118, 120, 121, 122, 125, 162, 163, 309, 316, 317
Roosevelt, Theodore, 328, 361
Roosts, 131, 134, 135, 136, 138, 139, 140, 141, 146
Rugose (microbial crusts), 77, 78
Russian thistle, 345

Safe sites, 79

Sagebrush, 114, 122, 221, 264, 281, 306, 315, 316, 327, 356, 366
Salix, 7, 310
Salsola iberica, 343
Salt Wash Member, 190
Sandstone aquifers, 225
San Juan Basin, 5, 186, 199, 205, 210
San Juan Generating Plant, New Mexico, 349
San Juan Mountains, 5, 20, 184, 186, 194, 198, 200, 201, 202, 205, 206, 208, 210, 224, 249, 250, 273, 322, 323, 324, 355, 356
San Luis Valley, 199
San Rafael Group, 188, 189, 190
Saprotrophs, 61
Scale insects, 168
Sclerocactus mesae-verdae, 37
Scolytid bark beetle, 168
Scrub-jay, 18, 91, 92, 93, 121, 244
Sedimentary rocks, 5, 183, 184, 185, 198, 199, 201, 203
Seed eaters (birds), 100
Seeps, 37, 152, 153, 197, 208, 223, 225, 237, 244
Sego lilies, 302
Septa, 62
Shale aquifers, 225
Sheep grazing, 33, 331
Short-horned lizard, 161
Shoshone, 289, 291
Shulman piñon pine, 14
Side-blotched lizard, 161
Silver-haired bats, 137
Single-leaf piñon, 120, 288. *See also Pinus monophylla*
Sitanion hystrix, 33, 345
Slender wheatgrass, 345, 346
Slump pool, 227
Snakes, 123, 124, 125, 159, 160, 162, 163, 226, 228, 238, 306; Mesa Verde Night snake, 160; midget-faded rattlesnake, 162
Snakeweed, 7, 114
Soda Canyon, 234, 235, 237, 238; Spring, 237
Soil: crust, 7, 10, 76–85, 264; deterioration, 309; hydrology, 80; moisture balance, 214; soil surface stability, 82
Sonoran (desert), 361
Southern Oscillation Index (SOI), 18
Southern Utes, 329
Spanish, 5, 295, 322, 323, 362
Spea bombifrons, 157, 159
Spea intermontanus, 159
Spea multiplicatus, 156
Species of concern, plants, 38, 62, 63, 147. *See also* Threatened and Endangered species
Spiritual uses (of plants), 206
Spotted bats, 139
Spotted skunk, 22, 125
Spruce-fir forests, 14
Stand-replacing fires, 234, 269, 226, 367, 371
Stand structure, 14, 17, 24, 28, 292
Stanleya pinnata, 335
Steel axe, 362
Steller's jay, 91, 92, 93, 225, 236
Stephen's woodrat, 113
Striped plateau whiptail, 161
Subalpine fir, 14
Successional trajectories, 345
Sulfur oxides, 349
Summer thunderstorms, 156, 163, 213, 234, 234, 239, 240
Summerville Formation, 189
Surficial deposits, 197, 200, 201, 202, 204, 207, 208, 210, 211
Surficial geologic features, 184
Swallows, 90, 94, 230, 238, 240, 244

Tadarida brasiliensis, 133, 140, 141
Tadpoles, 152, 154, 155, 156, 157, 158, 159, 234, 239, 240, 347
Talus flatirons, 204, 208
Tectonic forces, 198, 200
Telluride, Colorado, 199
Termites, 84, 157, 167
Terpenes, 113, 284
Terraces, 201, 203, 205, 208, 210, 316, 317
Thamnophis elegans, 162
Threatened and Endangered (T and E) species, 37, 370
Tiger salamanders, 152, 241
Time-since-fire cumulative curve, 265
Tinajas, 154, 345, 347, 352
Toad, New Mexico spadefoot, 157
Tree bats, 137
Tree lizard, 160
Tree mortality, 19, 20, 172
Triassic, 188
Tufa, 208
Turkey Springs, 236
Turnover time, 265, 266, 268, 269, 271, 274
Twig beetles, 172
Two Raven House, 314
Typha latifolia, 7

Uncompahgre Plateau, 5
Unconformity, geologic, 187

Understory diversity, 61
Urosaurus ornatus, 161
Utah, 4, 6, 14, 22, 32, 79, 82, 84, 85, 133, 134, 139, 140, 151, 154, 156, 157, 159, 162, 163, 167, 172, 189, 190, 202, 215, 217, 250, 269, 340, 356, 363
Utah serviceberry, 32, 35, 98, 284, 298. *See also Amelanchier utahensis*
Uta stans buriana, 161
Ute Mountains, 7, 194, 198, 199, 205, 207, 208, 226
Ute Mountain Ute, 239, 327; Reservation, 5, 226, 241, 334; Tribal Park, 270, 271, 272
Ute Peak, 226
Utes, 1, 12, 240, 267, 291, 295, 296, 298, 301, 303, 306, 322, 324, 325, 327, 331, 334, 364
UV-protective pigments, 78

Vascular plants, 9, 10, 60, 62, 63, 64, 65, 78, 79, 80, 82, 83, 85
Vegetation pattern, 265
Vertebrates, 111, 113, 115, 117, 160
Vireo, 98, 100, 103
Virus, 112, 121, 125, 126

Waterpockets, 152
Waters, Ellen, 331
Weasels, 112, 125

Weeds, 342, 363. *See also* Invasive species
Western bluebirds, 24, 94
Western pipistrelle, 132, 137, 139, 145
Western rattlesnakes, 162
Western terrestrial garter snakes, 162
Wet deposition of acids, 350
Wetherill Mesa Archeological Project, 302
White-breasted nuthatch, 91, 97, 99
White settlement, 329
Wild tobacco, 37, 304
Wild turkey, 102
Wind erosion, 81, 82
Wingate Sandstone, 188
Winter habits, 137, 139, 140
Wood borers, 95, 172
Woodland structure, 11, 24, 89

Yarrow, 305
Yellow jacket Creek, 299
Yucca, 31, 33, 296, 300, 310, 314
Yucca House, 226, 229, 241
Yuma myotis, 140

Zadiprion rohweri, 174
Zea mays, 288, 309
Zion National Park, 159
Zuñi, 289, 292, 316